Agricultural and Synthetic Polymers

ACS SYMPOSIUM SERIES 433

Agricultural and Synthetic Polymers

Biodegradability and Utilization

J. Edward Glass, EDITOR
North Dakota State University

Graham Swift, EDITOR
Rohm and Haas Company

Developed from a symposium sponsored
by the Divisions of Cellulose, Paper and Textile Chemistry;
and Polymeric Materials: Science and Engineering
at the 197th National Meeting
of the American Chemical Society,
Dallas, Texas,
April 9-14, 1989

American Chemical Society, Washington, DC 1990

Library of Congress Cataloging-in-Publication Data

Agricultural and synthetic polymers: biodegradability and utilization / J. Edward Glass, editor, Graham Swift, editor.

p. cm.—(ACS Symposium Series, 0097–6156; 433).

"Developed from a symposium sponsored by the Division of Cellulose, Paper, and Textile Chemistry; and Polymeric Materials: Science and Engineering at the 197th Meeting of the American Chemical Society, Dallas, Texas, April 9–14, 1989."

Includes bibliographical references and index.

ISBN 0–8412–1816–1

1. Polymers—Deterioration—Congresses. 2. Polymers—Biodegradation—Congresses. 3. Plastics—Deterioration—Congresses. 4.Plastics—Biodegradation—Congresses. I. Glass, J. E. (J. Edward), 1937– . II. Swift, Graham 1939– . III. American Chemical Society. Meeting (197th: 1989: Dallas, Tex.) IV. Series
QD381.9.D47A35 1990
620.1'9204223—dc20 90–39043
CIP

The paper used in this publication meets the minimum requirements of American National Standard for Information Sciences—Permanence of Paper for Printed Library Materials, ANSI Z39.48–1984. ♾

PRINTED IN THE UNITED STATES OF AMERICA

ACS Symposium Series

M. Joan Comstock, *Series Editor*

1990 ACS Books Advisory Board

Foreword

The ACS SYMPOSIUM SERIES was founded in 1974 to provide a medium for publishing symposia quickly in book form. The format of the Series parallels that of the continuing ADVANCES IN CHEMISTRY SERIES except that, in order to save time, the papers are not typeset but are reproduced as they are submitted by the authors in camera-ready form. Papers are reviewed under the supervision of the Editors with the assistance of the Series Advisory Board and are selected to maintain the integrity of the symposia; however, verbatim reproductions of previously published papers are not accepted. Both reviews and reports of research are acceptable, because symposia may embrace both types of presentation.

Contents

Preface xi

OVERVIEWS

1. **Degradability of Commodity Plastics and Specialty Polymers: An Overview** 2
   Graham Swift

2. **Biodegradative Processes and Biological Waste Treatment: Analysis and Control** 13
   Gary S. Sayler and James W. Blackburn

3. **Insect Symbionts: A Promising Source of Detoxifying Enzymes** 33
   Samuel K. Shen and Patrick F. Dowd

4. **Plastics Recycling Efforts Spurred by Concerns About Solid Waste** 38
   Ann M. Thayer

5. **Plastic Degradability and Agricultural Product Utilization** 52
   J. Edward Glass

DEGRADABILITY OF COMMODITY PLASTICS AND SPECIALTY POLYMERS

6. **Polyethylene Degradation and Degradation Products** 60
   Ann-Christine Albertsson and Sigbritt Karlsson

7. **Biodegradation of Starch-Containing Plastics** 65
   J. Michael Gould, S. H. Gordon, L. B. Dexter, and C. L. Swanson

8. **Constraints on Decay of Polysaccharide–Plastic Blends** 76
   Michael A. Cole

9. **Biodegradation Pathways of Nonionic Ethoxylates: Influence of the Hydrophobe Structure** 96
L. Kravetz

10. **Biodegradation of Polyethers** 110
Fusako Kawai

11. **Biodegradable Poly(carboxylic acid) Design** 124
Shuichi Matsumura and Sadao Yoshikawa

12. **Biodegradation of Synthetic Polymers Containing Ester Bonds** 136
Yutaka Tokiwa, Tadanao Ando, Tomoo Suzuki, and Kiyoshi Takeda

13. **Biodegradable Polymers Produced by Free-Radical Ring-Opening Polymerization** 149
William J. Bailey, Vijaya K. Kuruganti, and Jay S. Angle

14. **In Vitro and In Vivo Degradation of Poly(L-lactide) Braided Multifilament Yarns** 161
Brian C. Benicewicz, S. W. Shalaby, Alastair J. T. Clemow, and Zale Oser

15. **Bioabsorbable Fibers of *p*-Dioxanone Copolymers** 167
R. S. Bezwada, S. W. Shalaby, and H. D. Newman, Jr.

AGRICULTURAL POLYMER UTILIZATION: MONOMER SOURCE

16. **Monomers and Polymers Based on Mono- and Disaccharides** 176
Stoil K. Dirlikov

17. **Polymers and Oligomers Containing Furan Rings** 195
Alessandro Gandini

AGRICULTURAL POLYMER UTILIZATION: ALTERNATE CROP STRATEGIES

18. **Microbial Fructan: Production and Characterization** 210
Y. W. Han and M. A. Clarke

19. **Coatings Based on Brassylic Acid (An Erucic Acid Derivative)** 220
David E. Chubin, James P. Kaczmarski, Zeying Ma, Daozhang Wang, and Frank N. Jones

20. **Synthesis and Characterization of Chlorinated Rubber from Low-Molecular-Weight Guayule Rubber** 230
Shelby F. Thames and Kareem Kaleem

21. **Chemical Modification of Lignocellulosic Fibers To Produce High-Performance Composites** 242
Roger M. Rowell

22. **Cellulose and Cellulose Derivatives as Liquid Crystals** 259
Richard D. Gilbert

AGRICULTURAL POLYMER UTILIZATION: CORN-BASED FEED STOCKS

23. **Specialty Starches: Use in the Paper Industry** 274
Kenneth W. Kirby

24. **Saponified Starch-*g*-poly(acrylonitrile-*co*-2-acrylamido-2-methylpropanesulfonic acid): Influence of Reaction Variables on Absorbency and Wicking** 288
George F. Fanta and William M. Doane

25. **Use of Hemicelluloses and Cellulose and Degradation of Lignin by *Pleurotus sajor–caju* Grown on Corn Stalks** 304
D. S. Chahal and J. M. Hachey

INDEXES

Author Index 313

Affiliation Index 313

Subject Index 314

Preface

Two themes are prevalent in three of the five sections of this book: the waste problems associated with the convenience use of plastics in packaging, and the more universal acceptance of agricultural products. Overviews of the current understanding of the degradability of synthetic polymers, the complexity of biodegradation, and the promise of insect symbiont processes in achieving acceptable waste control are discussed in the first three chapters of the opening section. The remaining chapters focus on the necessity of recycling commodity plastics as part of the solution to the waste problem and discuss the commercial factors involved in the success of commercial processes and products.

The second section deals with the degradability of commodity plastics and specialty polymers. Emphasis is on the biodegradation of polyethylene, its blends with starch, and constraints in the decay of such composites. Additionally, the biodegradability of different functional groups (polyethers, carboxylic acids, esters, and dioxanones) is examined with respect to composition and microstructure.

The concluding section suggests alternative uses of agricultural products in nonnutritional areas: first, as a source of monomers; then as an alternate crop strategy, particularly under less-than-optimum farm conditions. Their suitability for both low-cost and specialty applications are considered, as well.

We express our appreciation to those who participated in the symposium at Dallas and to Fred Hileman of the Monsanto Corporation for his assistance in organizing the symposium.

J. Edward Glass
North Dakota State University
Fargo, ND 58105

Graham Swift
Rohm and Haas Company
Spring House, PA 19477

April 25, 1990

OVERVIEWS

Chapter 1

Degradability of Commodity Plastics and Specialty Polymers

An Overview

Graham Swift

Rohm and Haas Company, 727 Norristown Road, Spring House, PA 19477

This overview is an attempt to briefly cover the history and recent developments in environmentally degradable commodity and specialty polymers and plastics. Degradation pathways are mentioned, polymer types, including blends, are reported and the limitations of current testing protocols raised. The chapter concludes with generalizations on structural requirements for degradable polymers.

The review is not meant to promote degradable plastics, but rather to highlight them as one possible route to a cleaner environment. In many applications, recycling of polymers or incineration of waste products may offer more protection to the environment. The final decision for any application requires careful deliberation and not emotional judgements.

For the purpose of this overview, commodity plastics are packaging and agricultural products based largely on polyolefins, poly(vinyl chloride), and poly(ethylene terephthalate). Specialty polymers represent a wider range of applications for generally lower volume products; they find use, for instance, in the home in detergents and superabsorbent polymers, e.g. poly(alkylene oxides) and poly(acrylic acid), in industry, e.g. poly(vinyl alcohol), and in medicine for sutures and drug delivery, e.g. polyesters. Degradability refers to their stability in a given environment, including in vivo for mammalian implants used in medicine. Any or all of the following degradation mechanisms may be operative in the degradation process; with biodegradation being the ultimate desirable goal with complete removal from the environment.

Biodegradation is promoted by enzymes and may be either aerobic or anaerobic. Operative in all environments, burial, surface exposure, waterways, in vivo, etc., and may lead to complete removal from the environment.

0097–6156/90/0433–0002$06.00/0

- Photodegradation is promoted by irradiation, e.g. sunlight, and is restricted to surface environment exposure. It rarely leads to complete removal, though fragments produced may biodegrade.

- Environmental Erosion is promoted by the elements - wind, rain, temperature - and by larger animals. Usually, a surface environment phenomenon, and fragments produced may biodegrade.

- Chemical Degradation is promoted by chemical reaction through additives in the plastic, e.g. metals, functional groups and is operative under all conditions. Fragments produced may biodegrade.

Commodity plastics and specialty polymers should either have predictable life times and then degrade on exposure to the chosen release environment, or be non-degradable and recycled or incinerated.

Research on (bio)degradable synthetic polymers is not new. Several reviews have appeared over the last twenty years(1-7). However, beginning in the early 1980's, the recognized lack of biodegradability of many large volume commercial polymers, particularly commodity plastics, used in packaging, e.g. fast food industry, and in agricultural films, focussed public attention on a potentially huge environmental accumulation and pollution problem that could persist for centuries. Plastics have reached about 7 weight percent and 30 volume percent of solid waste in the U.S.A(8)and Japan(9), a large proportion of which are commodity plastics referred to above. Their lack of degradability is impacting significantly on the rate of depletion of landfill sites and their persistence also adds to the growing water and land surface litter problem. This has prompted intense research activity world-wide to either modify current products to promote degradability(10,11) or to develop new, suitable, alternatives that are degradable by the mechanisms mentioned.

Specialty polymers represent a broad compositional and use range. Some enter the environment as landfill, along with commodities; e.g. poly(acrylic acid) as super absorbents in diapers, poly(acrylic acid) and poly(alkylene oxides) enter streams in detergent effluent and others from industrial effluents, e.g. poly(vinyl alcohol) from paper and textile mills. Some of these polymers are biodegradable and of little concern; others are not, and will receive attention over the course of time as priorities change. Medical polymers are a unique case of specialty polymers. These have been designed to degrade in vivo where they perform such functions as controlled drug release and temporary supports as sutures in post-operation recovery. They are widely accepted and used.

Degradable plastics do not represent the only route to an improved environment. Plastics can be collected after use for recycle and/or incineration.(8) Collection looms as a problem for both approaches, and separation of plastic types is needed for recycling to be successful. Without debating the 'pros' and 'cons' of recycling, incineration, and degradable plastics, it does seem likely that all three will play an important part in a cleaner environment. Each will offer particular advantages for certain applications of plastics.

This overview, for simplicity of presentation, is divided into addition and step-growth polymers rather than commodity plastics and specialty polymers. It is not all encompassing, rather a skim across the field. A brief review of testing protocols is included, for appreciation of the difficulties involved in establishing (bio)degradability.

Fate analysis deserves special comments since it does bear strongly on the environmental issues being discussed with regard to degradable plastics. Unless total fate analysis is done on the newly developed degradable plastics, it will be extremely difficult to pronounce them environmentally friendly. In the course of degradation, small molecules are produced (from polymers) which are considered to be biodegradable. Without identification and testing, they may equally be considered non-biodegradable. Since they are produced in oxidation processes, they may be expected to be polar and water soluble (to some extent) and could enter water tables and rapidly become part of our ecosystems. Questions that should and must be raised, therefore, relate to toxicity, both as the small molecules produced and with any heavy metal affinity they may possess. All forms of life should not be jeopardized by a precipitous but well meaning switch from an inert to a degradable polymer.

Testing Protocols

A testing protocol should include an environment that fairly represents that to which the substrate polymer will be exposed. This has not always been the case; many polymers that are disposed of in landfills have been tested for (bio)degradation under aerobic conditions when anaerobical is probably closer to the condition that pertains. Another shortcoming in tests that have been reported on commercial polymers is that no indication as to the degree to which degradation has occurred is reported. Results are based merely on whether fungal or bacterial life is supported, or whether physical change occurs on exposure to the environment, e.g. U.V., from sunlight. Many 'biodegradable' plastics, based on polyolefin blends will disintegrate in the environment, but the recalcitrant polyolefin remains. Can this be called biodegradation, and is it acceptable? In spite of these objections, much of the work that has been done is valuable and sets the tone for future developments, as we recognize these deficiencies.

Biodegradation testing with plastics and polymers falls into two categories: water soluble and water insoluble. Testing of water soluble polymers is simpler and benefits from the testing that has occurred with water soluble detergent additives over many years. These include biological oxygen demand (BOD), carbon dioxide evolution, and semi-continuous activated sludge tests (SCAS) (12), which are reviewed by R.D. Swisher in his monumental work on surfactant degradation.(13) Many workers have extended these tests to analyze metabolites and identify the mechanism of degradation. Labelling with ^{14}C has played an important role.

Water insoluble polymers are more difficult to evaluate. Early tests were subjective and based on a relative growth rating of bacterial and fungal colonies.(14,15) They ignored low molecular weight species and additives present, both of which could make positive contributions to growth. Extension to physical testing after the bacterial and fungal growth gave some indication of property loss,(16,17,18), but again, no distinction was made for the contribution of impurities, making it difficult to compare different samples of the 'same' polymer or plastic. Loss of physical form(19) and weight(20) have also been used to measure the biodegradation of plastics buried in soil; some measure of loss of polymer can be obtained in these tests. Finally, ^{14}C labelled materials have been used to follow $^{14}CO_2$ evolution as a measure of biodegradation and its rate.

As of this time, there is no real satisfactory test for degradation in any polymeric system. Complete fate analysis is needed. Definitions have to be established and degree of analysis accepted for many different polymers, something that will take time. Tests will be required for each type of degradation mentioned earlier.

Addition Polymers

Early work by Potts and coworkers(21) indicated that high molecular weight polymers did not support fungal growth. He tested polymers such as poly(methyl methacrylate), polyethylene, polystyrene, poly(vinyl chloride), poly(vinylidene chloride) and poly(vinyl acetate). In the few instances where some growth occurred, he attributed this to low molecular weight fragments or the presence of plasticizers. Potts(22) also demonstrated that low molecular weight linear polyethylene was biodegradable by fungal growth support, whereas branching stopped biodegradation as shown in Table I.

Table I

Microbial Growth in Hydrocarbons

Compound	Substitution (Methyl)	M.W.	Growth
n-Dodecane	-	170	4
Dodecane	2,6,11	212	0
n-Hexadecane	-	226	4
Hexadecane	2,6,11,15	283	0
n-Dotriacontane	-	451	4
n-Hexatriacontane	-	507	0

(0 = no growth; 4 = maximum growth)

More recent work on hydrocarbon oligomers by T. Suzuki established the biodegradation of oligomers of cis-1,4-polyisoprene,(23) polybutadiene,(24) polyethylene,(25) and polystyrene(26).

The most fundamental work on polyethylene degradation is being done by A. C. Albertsson in Sweden(27,28,29,30,31) using elegant science with ^{14}C labelled polymers to measure the rate and quantity of carbon dioxide evolution from buried high and low density polyethylene films. Degradation occurs at about

0.5% per year; powders which have larger surface area than film degrade slightly faster; irradiation prior to burial also slightly enhances degradation rate. Albertsson observed slightly enhanced degradation for HDPE versus LDPE, which she attributes to less branching and probably slightly lower molecular weight.

Polyethylene degradation rate enhancement has been claimed for blends with mercaptans,(32) ketones,(32) ethers,(32) and specifically by photodegradation in the presence of benzophenone,(33) and metal promoted oxidations.(34) Copolymers with carbon monoxide are also subject to photodegradation.

Major contributors to this area of 'doped' polyolefins are G. J. C. Griffin, J. E. Guillet, and G. Scott, and many of the blossoming degradable packing polymers are based on their pioneering work.

Polyvinyl alcohol is unique in being the only biodegradable carbon-carbon backbone polymer.(35,36) The degradation is oxidative, followed by hydrolytic cleavage, Suzuki claims acids and ketones as metabolites, and Watanabe acids and alcohols. This difference is shown below, schematically.

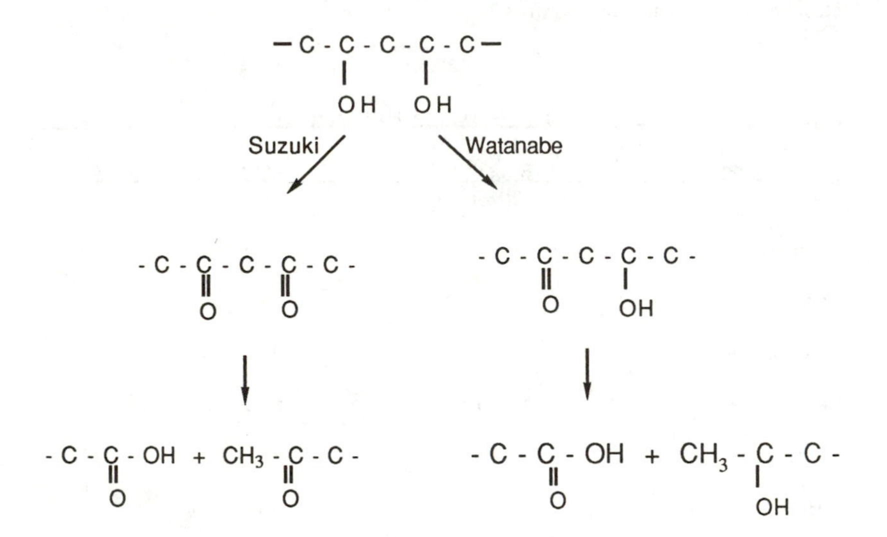

The ease of degradation of the polyketone intermediate of Suzuki, obtained by oxidation of poly(vinyl alcohol), has been demonstrated by Wang and Huang(37) at the University of Connecticut.

Poly(carboxylic acids), based on acrylic and maleic acids, are used widely in the detergent industry as builders (low M.W.) and in superabsorbents diapers (high M.W.). Neither the low molecular weight nor the high molecular weight polymers are biodegradable in sewer systems or landfill. S. Matsumura has published many significant papers in the detergent area, and concludes that oligomers of acrylic acid of degree of polymerization (D.P.) <7 biodegrade. These have no detergent value. Copolymers have no advantage over the homopolymer in biodegradation;(38-45) higher D.P.'s are recalcitrant, showing ~20% biodegradation in BOD tests and by gel permeation chromatographic product analysis. Two patents claim high biodegradation for acrylic acid/acrolein copolymers, but the experimental conditions would favor low molecular weight(46,47). Grafts of acrylic acid onto biodegradable substrates, such as starch,(48) have been claimed as biodegradable; however, results indicate that only starch is biodegrading.

Polyacids, based on anionically polymerized glyoxylic esters were cleverly developed by Monsanto and the work published in a series of patents(49) and papers(50,51). The polymers are polyacetals, stable in alkali, but not in acid; hence, in use as detergents, they are stable, but hydrolyze as the pH falls to ~7 in the sewer systems, after use, to biodegradable monomer.

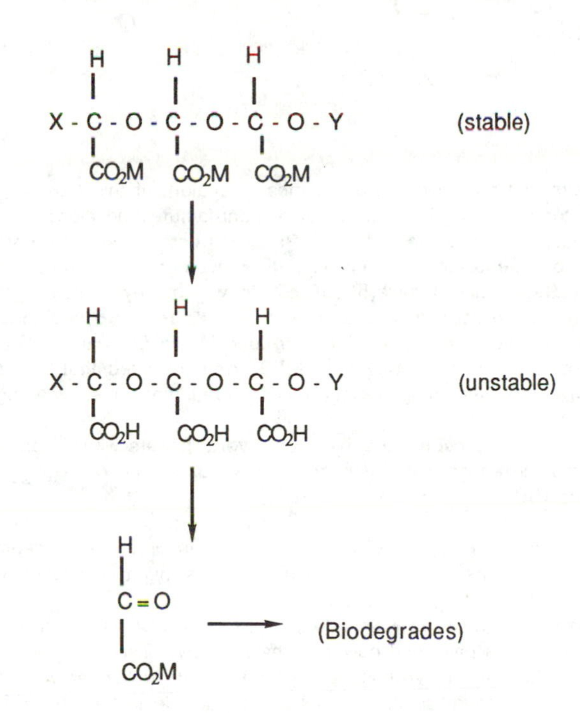

They have not found commercial acceptance due to cost.

W. J. Bailey's(52,53) work with ketene acetals deserves mention as potentially a route to biodegradable addition polymers. Its novelty resides in the instability of the vinyl radical and rearrangement to introduce a polyester linkage into a radically produced polymer. As we shall see in the next section, polyesters are biodegradable; hence, their introduction into a polymer with a C-C backbone produces weak links which fracture the polymer into oligomers which we have seen are biodegradable. This chemistry is exemplified schematically, below.

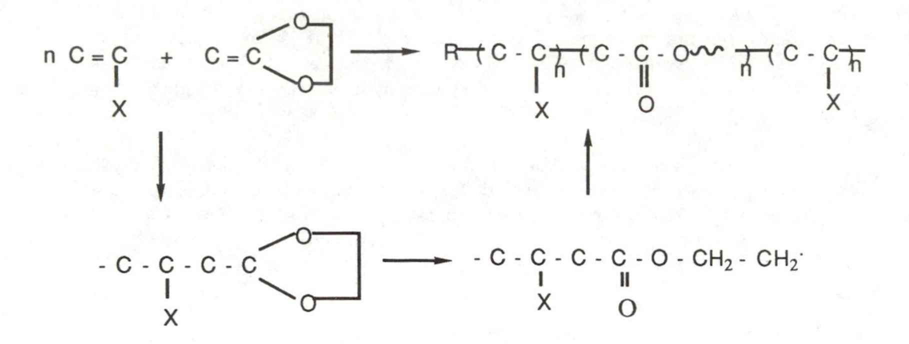

Step-Growth Polymers

Polyesters and polyamides are the most prevalent of this type of polymer. Poly(ethylene terephthalate) is used in bottle manufacture and along with other packaging plastics is not biodegradable. Potts(54) established very early that only low melting and low molecular weight aliphatic polyesters were biodegradable. Later work by Suzuki and Tokiwa(55,56,57) in which they evaluated the stability of polyesters to lipases confirmed the work of Potts. Potts(58) demonstrated the biodegradation of polycaprolactone which he used in the fabrication of agricultural articles, such as plant pots. J. P. Kendrick(59) demonstrated that amorphous regions of polyesters were more readily biodegraded than crystalline regions.

Poly(amides) have been the subject of several papers which indicate that structure around the amide link and hydrophilicity control the ease of biodegradation.(60,61)

Poly(urethanes) based on polyethers are more resistant to biodegradation than poly(urethanes) based on polyesters,(62,63) as might be expected.

Polyethers have received wide attention because of their use in detergent manufacture and in the paint and polyurethane industry. They have water solubility, particularly the poly(ethylene glycols). Low molecular weight poly(ethylene glycols) were established as biodegradable in the 60's.(64) The whole field is well reviewed by F. Kawai(65) who established the symbiotic nature of the biodegradation of poly(ethylene glycols).(65) Poly(propylene glycol) and poly(tetramethylene glycol) are reviewed in the same paper.(65) Substituted alkylpolyethers are widely used surfactants and their environmental fate has been reviewed by several authors(66,67,68) and in Shell literature on Neodol Chemicals.(69) Conclusions are that the low molecular weight hydrocarbon tail degrades in the order non-branched > aliphatic branched >> aliphatic phenyl, and the polyether degrades, as described earlier.

Polymer Blends

Polymer blends, particularly olefins with biodegradable polymers, are gaining popularity as an approach to degradable packaging plastics. The materials are at best only partially biodegraded, but will lose form and bulk as the plastic disintegrates. This may be sufficient in landfill as volume diminishes, leaving room for more solid waste. It should be understood, though, to date, only the biodegradable part of the blend has been shown to be removed from the environment.

This work began at the U.S.D.A. with Otey,(70) where he developed blends, based on starch and ethylene/acrylic acid copolymers. Potts also has patents in this area with olefin/polycaprolactone blends(71). Starch is recognized as a cheap filler for polyolefins which enhances the biodegradability, or apparent biodegradability; depending on the interpretation of results,e.g. starch/poly(vinyl alcohol),(72) starch/polyurethane,(73) and starch/poly(methyl methacrylate) (74) also biodegrade. Polyethylene has also been blended with metals(75) and metals plus autoxidizable substrates76 in attempts to enhance degradation by U.V. irradiation and/or oxygen.

Other uses of blends include controlled rate of fertilizer release(77) based on ethylene/vinyl acetate/carbon monoxide polymers which is U.V. sensitive, polyolefin blends with any biodegradable polymers,(78) and polyolefins blended with metals and autoxidizable substrates.(79) Doane and co-workers(80) at the U.S.D.A. have used grafted starches in many applications, including soil stabilization.

Starch, being cheap, will continue to be an attractive substrate(81) unless the blends and grafts are determined to be unacceptable substitutes for packaging materials.

Medical Polymers

This field has been well reviewed by B. J. Tighe.(82) The polymers, for the most part, are polyesters. Poly(glycolic acid) (83) is widely used in sutures under the trade name of DEXON. Poly(lactic acid) is also used.(84) A copolymer of 92/8 mole percent poly(glycolic acid)/poly(lactic acid) (85,86) is another alternative.

Recent advances have seen both Tighe(87) and Lenz(90)developing procedures for carboxylated polyesters. Tighe has evaluated biologically produced polymers.(88) American Cyanamid Company continues to modify poly(glycolic acid) by copolymerizing ethylene oxide.(89)

Generalizations

From this brief overview, several generalizations can be established concerning (bio)degradable polymers.

1. Addition polymers with C-C backbones do not biodegrade to any significant extent.

2. Non-branched oligomers of C-C polymers (M.W. <1000) tend to be biodegradable.

3. Addition polymers with hetero-atoms in the backbone biodegrade.

4. Step-growth polymers are biodegradable to a greater or lesser extent, depending on

   - chemical coupling, esters > ethers > amides > urethanes
   - molecular weight, lower is faster
   - morphology, amorphous degrades faster than crystalline
   - hardness, (Tg), softer degrades faster
   - hydrophilicity, more hydrophilic degrades faster

5. Recalcitrant polymers may be rendered susceptible to degradation by

   - incorporation of weak linkages which are hydrolytically unstable (esters), photo reactive (ketones), etc.
   - blending with biodegradable additives (starch)
   - addition of activators to promote oxidative degradation.

References

1. W. M. Heap and S. H. Morell, J. Appl. Chem., 1968, 18, 189.
2. F. Rodriguez, Chemtech., 1970, 1, 409.
3. W. Coscarelli, Polymer Stabilization, W. L. Hawkins Edn. Wiley, N. Y. 1972, 377.
4. J. E. Potts, Aspects of Degradation and Stabilization of Polymers, H. H. J. Jellinek Edn. Elsevier Amsterdam, 1978, 617.
5. E. Kuster, J. Appl. Polymer Sci., Appl. Polym. Symp., 1979, 35, 395.
6. G. S. Kumar et. al., J. Macromol. Sci. Rev., Macromol Chem. and Phys., 1981-83, C22(2), 225.
7. G. S. Kumar, Biodegradable Polymers, Marcel Dekker, Inc., New York and Basel, 1987.
8. Ann M. Thayer, C & E.N. News, 1989, Jan. 30, 7.
9. Plastic Waste Resource Recovery & Recycle in Japan, Plastic Waste Institute, 1987.
10. S. L. Wilkins, Chem. Week, 1989, April 5, 19.
11. NTIS Report #PB 89-859599, 1989, April.
12. A.S.T.M. Standard, D2667.
13. R. D. Swisher, Surfactant Biodegradation, Marcel Dekker, 2nd. Edn., 1987.
14. A.S.T.M. Standard, # G21-70 (1980).
15. A.S.T.M. Standard, D1924 (1963).
16. A.S.T.M. Standard, D26276T.
17. A.S.T.M. Standard, D882-83.
18. A.S.T.M. Standard, D638-84.
19. J. E. Potts, et. al., Great Plains Agric. Council Publ., 1974, 68, 244.
20. R. E. Burgess and A. E. Darby, Brit. Plast., 1964, 37, 32.
21. J. E. Potts, et. al., Polymers and Ecological Problems, Plenum Press, 1973.

22. J. E. Potts, et. al., E.P.A. Contract, 1972, CPE-70-124.
23. T. Suzuki, et. al., Agric. Biol. Chem., 1979, 43(12), 2441.
24. T. Suzuki, et. al., Agric. Biol. Chem., 1978, 42(6), 1217.
25. T. Suzuki, et. al., Report of the Fermentation Institute, Japan, 1980.
26. T. Suzuki, et. al., Agric. Biol. Chem., 1977, 41(12), 2417.
27. A. C. Albertsson, Proc. Int. Biodeg. Symp., 1976, 743.
28. A. C. Albertsson, J. Appl. Poly. Sci., Apl. Poly. Sci. Symp., 1979, 35, 423.
29. A. C. Albertsson, Europ. Poly. J., 1979, 16, 123.
30. A. C. Albertsson, J. Appl. Poly. Sci., 1978, 22(11), 3419, 3435.
31. A. C. Albertsson, J. Appl. Poly. Sci., 1980, 25(12), 1655.
32. Owens- Illinois, U.S. Patent 4,056,499.
33. Biodeg. Plastics, U.S. Patent 3,888,804.
34. G. J. C. Griffin, J. Am. Chem. Soc., Div. Org. Coat. Plast. Chem., 1973, 33(2), 88.; J. Poly. Sci., 1976, 57, 281, Proceedings: Autumn Meeting of the Biodeterioration Society, 1985, 12-13.
35. T. Suzuki, et. al., J. Appl. Poly. Sci., Appl. Poly. Symp., 1973, 35, 431.
36. Y. Watanabe, et. al., Arch. Biochem. Biophys., 1976, 174, 535.
37. I. Wang, U. Conn., D155. Abst., 8,429,769.
38. S. Matsumura, et. al., Yukagaku, 1984, 33, 211.
39. S. Matsumura, et. al., Yukagaku, 1984, 33, 228.
40. S. Matsumura, et. al., Yukagaku, 1985, 34, 202.
41. S. Matsumura, et. al., Yukagaku,1985, 34, 456.
42. S. Matsumura, et. al., Yukagaku, 1985, 35, 167.
43. S. Matsumura, et. al., Yukagaku, 1986, 35, 937.
44. S. Matsumura, et. al., Yukagaku, 1980, 30, 31.
45. S. Matsumura, et. al., Yukagaku, 1981, 30, 757.
46. Deutche Golde und Silber, U.S. Patent 3,923,742.
47. Deutche Golde und Silber, U.S. Patent 3,896,086.
48. Sanyo, J. Kokai, 61-31498
49. Monsanto, U.S. Patents, 4,144,226, 4,146,495, 4,204,052, 4,233,422, 4,233,423.
50. W. E. Gledhill and V. W. Saeger, J. Ind. Microbiol, 1987, 2 (2), 97.
51. W. E. Gledhill, Appl. Environ. Microbiol., 1978, 12, 591.
52. W. J. Bailey, et. al., Contemp. Topics Polym. Sci., 1979, 3, 29.
53. W. J. Bailey, et. al., J. Poly. Sci., Poly. Lett. Edn., 1975, 13, 193.
54. J. E. Potts, et. al., Aspects of Degradation and Stabilization of Polymers, Elsevier, 1978, p. 617.
55. T. Suzuki, et. al., Agric. Biol. Chem., 1986, 50 (5), 1323.
56. T. Suzuki, et. al., Agric. Biol. Chem., 1978, 42 (5), 1071.
57. T. Suzuki, et. al., Nature, 1977, 270 (5632), 76.
58. J. E. Potts, et. al., Plastics Engr. & Tex. Pap., 1975, 217, 567.
59. J. P. Kendrick, Diss. Abst., 82329391.
60. W. J. Bailey, Proc. of the 3rd. Int. Biodeg. Symp., 1975, Appl. Sci. Publ., 1976, p. 765.
61. S. J. Huang, Proc. of the 3rd. Int. Biodeg. Symp., 1975, Appl. Sci. Publ., 1976, p. 731.
62. R. T. Darby & A. M. Kaplan, Appl. Microbiol, 1968, 16, 900.
63. W. A. Shuttleworth, Appl. Microbiol, 1986, 23, 407.
64. E. L. Fincher and W. J. Payne, App. Microbiol, 1962, 10, 562.
65. F. Kawai, C.R.C. Critical Reviews in Biotechnology, C.R.C. Press, 1987, 6, 273.
66. P. Ruter, Cell Czeckoslov Chem. Comm., 1968, 33, 4083.

67. J. Steber, Tenside Detergents, 1983, 20, 983.
68. R. J. Larsen, Env. Sci. and Tech., 1981, 15 (2), 1488.
69. Shell Publication.
70. F. H. Otey, Org. Coat. Plastic Chem., 1977, 37 (2), 297.
71. J. E. Potts, U.S. Patents 3,901,838, 3,921,333.
72. U.S.D.A. U.S. Patent 3,949,145.
73. Coloroll, B.P. 824821.
74. R. J. Dennenburg, et. al., J. Appl. Poly. Sci., Chem. A., 1978, 22, 459.
75. Coloroll, U.S. Patents 4,016,117 and 4,021,388.
76. Rotterdam Manage SE, FR 2,617,857A.
77. Chisso Corp., EP 252555-A.
78. Techocolor Celebran, FR 2611732-A.
79. G. J. L. Griffin, W.O. 8,809,354-A.
80. G. J. L. Doane and G. F. Fanta, Modified Starches: Properties and Uses, 'ed O. B. Wartzburg, 1986, C.R.C. Press, Inc.
81. NTIS, PB 89-859599, PB 89-857114, 88-863295.
82. B. J. Tighe, et. al., J. Controlled Release, 1988, 4,155.
83. E. J. Frazza, et. al., J. Biomed. Mat. Res. Symp., 1971, 1, 43.
84. A. K. Schneider, U.S. Patent 3,636,956.
85. D. Wasserman, B.P. Patent 1,375,000.
86. D. Wasserman, B.P. Patent 3,839,297.
87. B. J. Tighe, et. al., Biomaterials, 1987, 8 (4), 353.
88. B. J. Tighe, et. al., Biomaterials, 1987, 8(4), 289.
89. American Cyanamid, U.S. Patent 4,716,203, 4,438,253.
90. R. W. Lenz, et. al., Am. Chem. Soc. Div. Polym. Chem. Prepr., 1979, 20, 608.

RECEIVED March 19, 1990

Chapter 2

Biodegradative Processes and Biological Waste Treatment

Analysis and Control

Gary S. Sayler[1] and James W. Blackburn[2]

[1]Center for Environmental Biotechnology and [2]Department of Chemical Engineering, University of Tennessee, Knoxville, TN 37932

Opportunities exist to dramatically enhance the extent and kinetics of biodegradation of agricultural, industrial and domestic wastes through an integrated systems approach. Such a systems approach combines the strengths of modern molecular biology, and ecological and engineering science to achieve new levels of process understanding and control that can be applied to the development and optimization of efficient biodegradation systems for a variety of wastes. A fundamental research agenda consisting of microbial strain development, bioanalytical monitoring methods and environmental and reactor systems analysis has been identified as a critical framework for the integration of science and engineering disciplines to achieve process optimization.

Among the available methods to provide needed quantitative monitoring of critical biodegradative populations, gene probe technology has been shown to be particularly useful. This technology has been successfully integrated with frequency response analytical techniques in developing a new system analysis protocol for biodegradation process control. A new bio-analytical method that utilizes engineered bioluminescent bacteria has also been developed as a remote, on-line sensor of biodegradative activity in waste treatment processes. The coupling of new measurement technology with system analytical methods provides new insight for process design and operation and prediction of optimal regimes of biodegradation performance.

0097–6156/90/0433–0013$06.00/0

The abilities of microorganisms to degrade the vast majority of natural and manmade organic substances is well known and is the basis of domestic and industrial waste treatment practice, composting technology, biomass conversion, major carbon cycling components, pesticide degradation and problem soils, and environmental remediation and restoration of contaminated areas (1). Often major problems arise when rates of degradation or conversion occur much slower than accumulation of the organic substrate, such as the case of solid waste accumulation in cattle feedlot operations, or occur too slowly to make commercial fermentations profitable such as biomass to methane. Some of these rate limitations may be due to an inherent biological resistance to degradation, such as hydrolytic depolymerization of polymers, like cellulose, lignin, nylon, (2,3) etc. or the fact that sufficient numbers of active microbes or nutrients cannot be brought into contact rapidly enough to process the organic material (4).

Recent advances in genetic engineering may contribute to partial solutions of these problems in terms of developing more versatile and active organisms for degradation processes. However, by itself genetic engineering cannot solve ecological and engineering problems that may ultimately control rate limiting processes.

There are major opportunities for integrated biological, engineering and ecological research in order to develop truly effective, predictable and safe applications of biotechnology for chemical control and waste treatment. Collectively, the development, application and control of biological processes for waste treatment is an integral component of environmental biotechnology (5). Operationally, environmental biotechnology for hazardous wastes must proceed as an integrated science and engineering effort in order to achieve successful lab to field scale-up operation. However, such an integrated research strategy is seldom available on the national scene. The resulting limitations for hazardous waste control have been documented by the Office of Technology Assessment (6) and by a recent NSF Environmental Biotechnology Research Planning Workshop (5). From this workshop a consensus research agenda has been formulated that encompasses the major research needs in environmental biotechnology. This agenda is described in Table 1.

While research accomplished in any one of the areas in the agenda will contribute to advancing technology for hazardous waste control; major advances are anticipated by interfacing both science and engineering research across the research agenda. This is particularly important because of the rapid advances made in the molecular understanding of the genetic elements and pathways involved in biodegradation, and an unprecedented ability to utilize genetic engineering technology to develop new and improved biodegradative microorganisms (7), which in itself may be an object of federal regulatory oversight (8,9). The objective of this report is to summarize some of the modern molecular approaches for the analysis of microbial populations involved in biodegradative processes and to examine the applications for systems analysis protocols that contribute to better processes control and optimization.

Table 1. Fundamental components for a research agenda leading to the development of effective Environmental Biotechnology for hazardous wastes

---

Agent (Strain) Development
- Source and Selection
- Characterization
- Modification and Improvement
- Model Systems (mixed and pure)
- Applications
- Evolutionary Relationships/Diversity
- Stress Responses
- Collections and Libraries

Process and System Analytical Tools
- Quantitative Analytical Techniques (chemical/physical measurements)
- Bio-analytical Methods
- Molecular Analysis Methods
- Monitoring Applications
- Remote Sensing
- Biomonitors
- Reporter-Signal Analysis (structure and function)

Environmental System Analyses
- Ecological Interactions
- Environmental Fate and Abiotic Processes
- Population Dynamics (organisms and genes)
- Environmental Stability
- Determination of Kinetic Parameters
- Micro-habitats (niche invasions)
- Organismal or Genetic Mobility
- Controllability and Environmental Modification
- Stress-induced Effects

Reactor System Analysis
- Reactor Design
- Transient Outcomes and Perturbations
- Dynamic Analysis
- Ecological Interactions
- System Stability and Component Stability
- Online Analysis and Control
- Kinetic Parameters Analysis

Science/Economic Policy Analysis
- Regulatory and Other Constraints to Application
- Costs and Benefits of Competitive Cleanup Technologies
- Potential Risks and Liabilities

Molecular Tools. The application of biodegradative processes in control of environmental contaminants is ultimately determined by understanding and controlling microbial community structure and activity. Since microbial communities are heterogenous and composed of both biodegradative and non-degradative populations, specific and quantitative methods are needed to predict the genetic potential of a community to degrade specific contaminants, and to evaluate controlled interventions designed to establish or enhance biodegradative activity.

Successful application of biological processes for bioremediation and control of hazardous environmental contaminants is dependent on the understanding and control of the structure and activity of biodegradative microbial communities. Because these communities are complex mixtures of microbes involved in degradation of specific toxicants (7), as well as microbes involved in biogeochemical transformations of non- anthropogenic materials, there are major needs to develop and apply specific and quantitative monitoring technology to discriminate those organisms whose specific role is the removal of contaminants of environmental concern (7,8). This apparent need is further reinforced by the development of improved biodegradative microorganisms (10), by classical methods or genetic engineering techniques, that are intended for bioaugmentation purposes and process development. In addition, the whole area of biostimulation presupposes that naturally occurring biodegradative organisms and processes can be selectively enhanced to achieve more efficient rate of biological transformation under environmental contaminants.

Occurring simultaneously with increases in the molecular understanding of biodegradative processes has been the development of new molecular analytical monitoring techniques that permit both sensitive and specific quantitation of microbial populations involved in biodegradation (Table 2). Many of these techniques offer advantages over more conventional measures of microbial population abundance and activity. In general, they are best applied in combination with conventional enumeration methods to develop new information on the fate and dynamics of individual populations. The concern over recombinant DNA (rDNA)-containing or genetically engineered microbes (GEM) in the environment has driven the search for gene-or DNA-specific molecular monitoring methods to provide information on the fate, persistence, amplification, and transfer of genes or rDNA sequences in the environment(8). This is precisely the same information needed to monitor and control genes involved in biodegradation processes relative to long term system optimization. Over the past decade, a variety of new techniques have emerged in the field of molecular biology that have demonstrated great utility in developing more sophisticated and specific approaches for the analysis of microbial communities(Table 2). These approaches include new immunological and protein analysis techniques, as well as methods to directly analyze the genetic structure and information of microbial systems. Plasmid and nucleic acid recovery and analysis methods when integrated with genotype specific, nucleic acid hybridization techniques have shown great utility in enhancing our understanding of the biodegradative potential of microbial communities and the response of these communities to engineering practice designed to promote in situ or reactor

Table 2. Some example molecular methods offering improved specificity and sensitivity for quantifying biodegradative organisms or genes in environmental samples

Currently Available Technology
Immunological Techniques
Fluorescent Antibodies
Enzyme-linked immunosorbant assays
Nucleic Acid Analysis Techniques
rRNA sequence analysis
DNA/RNA probe technology
Plasmid analysis
Restriction digestion RFLP analysis
Genetic Reporter Strain Techniques
Bioluminescent strain analysis
Selectable phenotypic markers
Analytical Chemical Techniques
Signature fatty acid analysis
Pyrolysis GC/MS analysis

(See Ref. 9 for specific examples.)

level biodegradation. Nucleic acid hybridization (gene probe) detection technology can also be used successfully to rapidly screen microbial populations for organisms with degradative genes and the eventual recovery of organisms with new or enhanced biodegradative activity. Such organisms represent new source material for bioaugmentation and strain improvement using classical selection or genetic engineering technology.

Technologies that focus on the specific nucleic acids responsible for the degradative capacity of a biological process hold promise for predicting the genetic potential of a biodegradative system. One such example may be plasmid fingerprinting. Plasmids, which are extra chromosomal genetic elements, are often associated with specific biodegradative pathways (7,11). Many of these plasmid can be readily isolated from bacteria involved in biodegradation and similar plasmids can be diagnosed by digesting the plasmid DNA with specific restriction endonucleases (DNA cutting enzymes that recognize particular nucleic acid base pair sequences) and examining the molecular weight distribution of the resulting DNA fragments using electrophoresis techniques. Such an example is presented in Figure 1 for a plasmid associated with 4-chlorobiphenyl degradation. Such information can be used to determine if a unique plasmid is present and is stably maintained in microorganisms promoting a particular degradative process. While such techniques are good for molecular characterization and ecological monitoring for genetic elements in biodegradative processes they can be difficult to apply to real time, quantitative process monitoring.

As a group, nucleic acid hybridization techniques and sequence analysis provide molecular monitoring for dynamic populations and/or genes involved in biodegradation (12,13,14,15). Potentially hybridization technology can be applied to real time, quantitative process monitoring. The strategy of nucleic acid hybridization as a monitoring technology is described in Figure 2 (9). Fundamentally, target DNA (unknown) from bacterial colonies or DNA extracted from environmental samples (16,17) is denatured from a double stranded to a single stranded form and conveniently bound to a solid support such as a nylon membrane. At an appropriate buffer and temperature condition, a known probe DNA (also in single stranded form) is added to the reaction mixture. Under the proper conditions, the single stranded probe DNA forms base pair hydrogen bonds with homologous complementary base sequences of the target DNA; if these sequences are present in the unknown target DNA. This process of hybridization or reassociation of the single stranded probe DNA and single stranded complementary target DNA results in the restoration of a double stranded hybrid DNA molecule.

In order to detect and discriminate these positive hybrid molecules, it is necessary to have a detectable tag or label on the probe DNA. This is accomplished by cross-linking various fluorescent, chromogenic, enzymatic or immunological reagents to the probe DNA or by biochemical synthesis of radio-isotopically labeled DNA. While the non-isotopic detection methods offer many advantages and are rapidly progressing, isotopically labeled DNA probes (most often labeled with ^{32}P), are the most sensitive and generally used. Detection of the positive hybrids, indicating the presence specific DNA

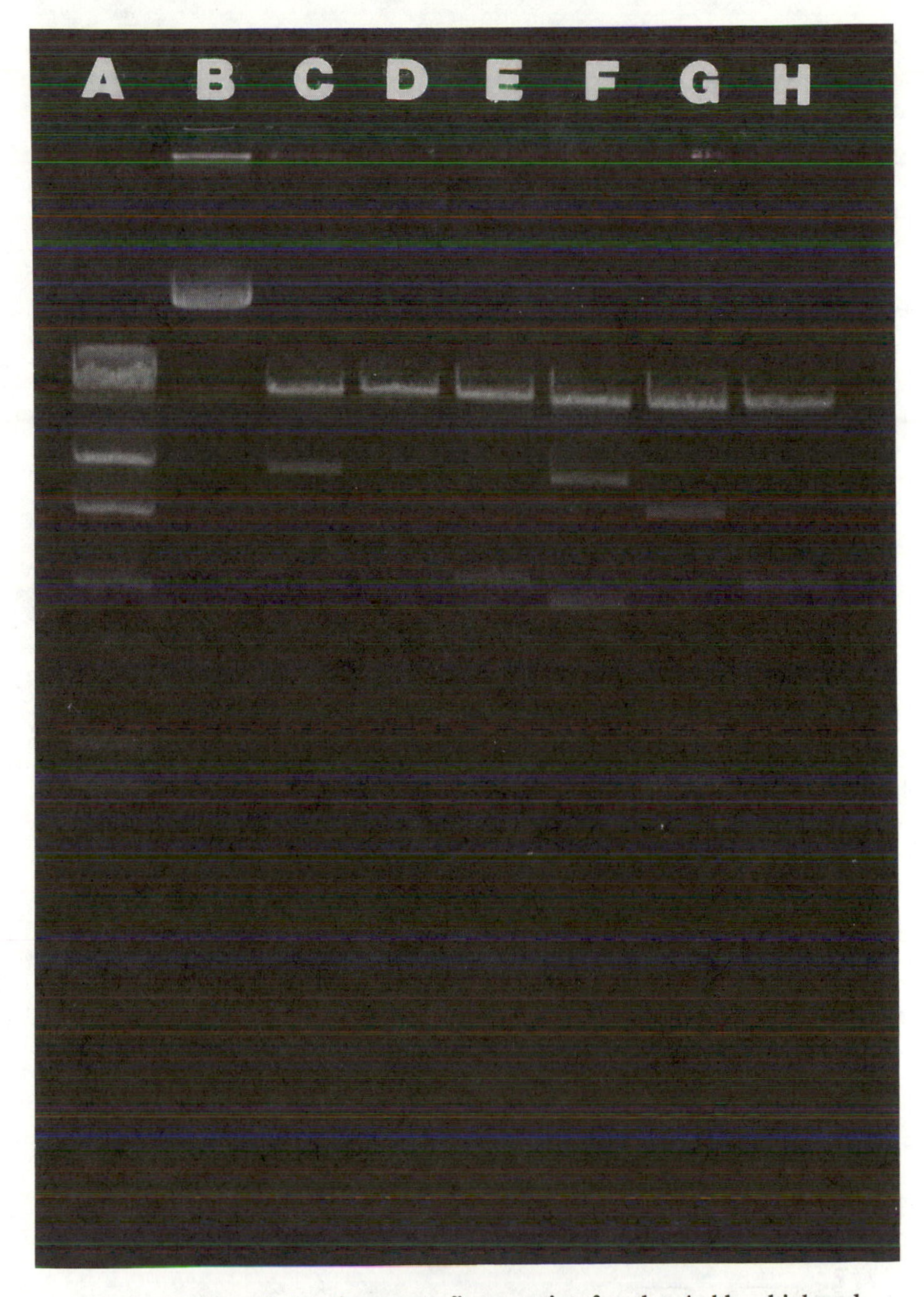

Figure 1. Restriction fragment finger print for the 4-chlorobiphenyl catabolic plasmid pSS50 determined by agarose gel electrophoresis. (Lanes:A, λ Hind III standard; B, undigested; C-H enzymatic digests of pSS50, C, Eco RI; D, Bam HI; E, Hind III; F, Eco RI and Bam HI; G, Eco RI and Hind III; H, Hind III and Bam HI.)

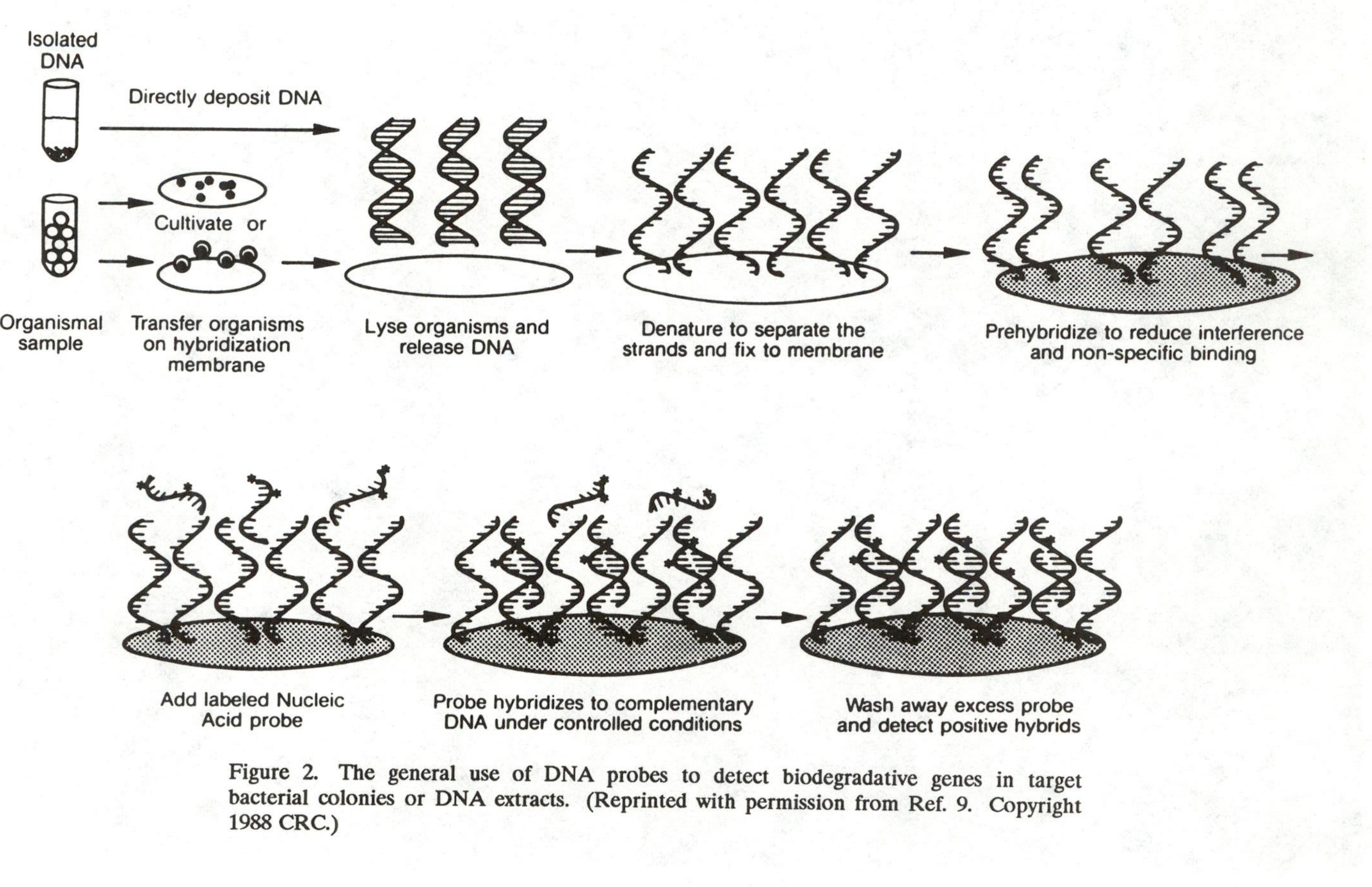

Figure 2. The general use of DNA probes to detect biodegradative genes in target bacterial colonies or DNA extracts. (Reprinted with permission from Ref. 9. Copyright 1988 CRC.)

sequence in the unknown target DNA mixture, is easily quantitated by liquid scintillation spectrometry or by radioautography when employing ^{32}P labeled DNA probes (9).

This technology has found wide applications for detection of bacterial populations and individual species in the environment or in waste treatment systems that contain specific genes for catabolism of environmental contaminants (12,14,15). The technology is easily integrated with conventional microbiological cultivation, plating and enumeration procedures to both enumerate bacterial colonies containing degradative genes of interest (12,14,16) and recovery of pure cultures of degradative organisms(18). A typical example of detecting the genes for naphthalene degradation in bacterial colonies from a mixed culture is given in Figure 3. In this example ^{32}P-labeled probe DNA from a *Pseudomonas putida* naphthalene catabolic plasmid was used to demonstrate that Bacteria capable of naphthalene degradation predominated in the inoculum of a continuous stirred soil slurry biotreatment reactor.

Another example application of this technology is to estimating the genetic potential of bacterial populations in subsurface soils at a Manufactured Gas Plant (MGP) site to degrade aromatic hydrocarbons. Again DNA specific for naphthalene degradation was used as a ^{32}P-labeled probe to determine which bacterial colonies contained gene sequences for naphthalene degradation and to determine relative concentration of these bacteria at different depths in contaminated MGP soils. Table 3 indicates the occurrence and differences naphthalene degradative cell densities in subsurface-contaminated MGP soils as determined by plate count enumeration and DNA colony hybridization.

Thus far, these examples have demonstrated the use of DNA probe technology to detect catabolic genes in bacterial colonies that have been cultured by conventional microbiological methods. Since the technology is a molecular technology, there is no requirement that colonies of organisms must be used as the source of target DNA. In fact, it is now possible to directly extract total DNA directly from an environmental sample or waste treatment population and to use this total DNA as a target to directly determine the frequency or abundance of catabolic genes in a given environment (14,16,17). In this case aliquots of DNA from an environmental sample are bound in single-stranded form to the hybridization membrane as a blot of DNA. This DNA target is probed with an appropriately labeled catabolic DNA probe. Figure 4 is an example of this approach where DNA has been extracted directly from chemically contaminated reservoir sediments and probed directly for the qualitative abundance of genes associated with 4-chlorobiphenyl degradation (plasmid pSS50). Recently, this technology has been advanced in detection sensitivity using polymerase chain reaction (PCR) amplification of specific DNA sequences recovered in environmental extracts (19). PCR is an enzymatic in vitro method for exponentially copying specific sequences in a complex DNA sample. Theoretically 1 gene or DNA fragment represented as single

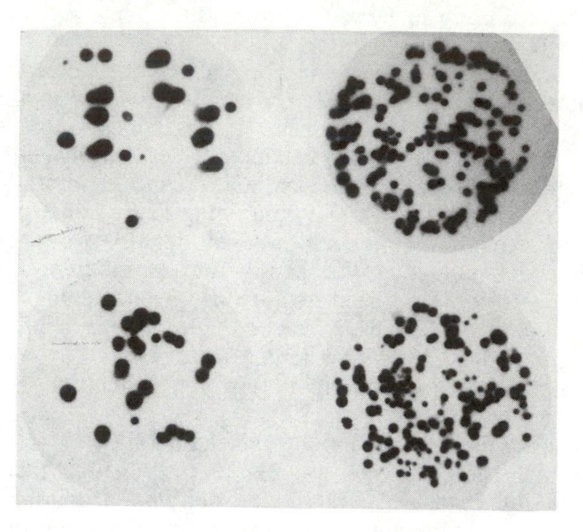

Figure 3. Autoradiographic detection of naphthalene degradative bacterial colonies from MGP soil enrichments used as inoculum for continuous stirred soil slurry bioreactors.

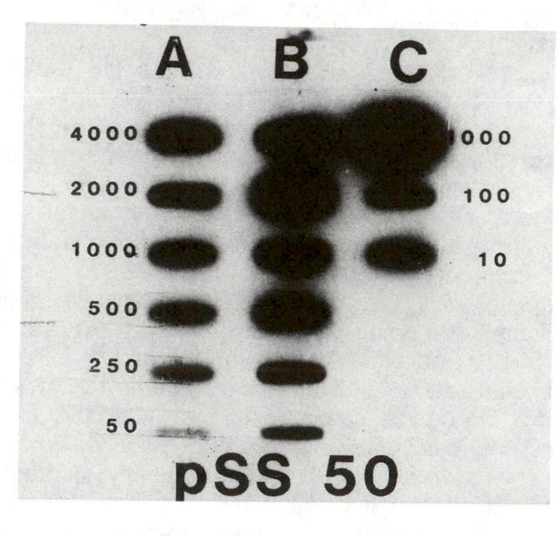

Figure 4. Comparative detection of 4-chlorobiphenyl catabolic gene abundance by blot hybridization of pSS50 plasmid DNA probe to target DNA extracted from sediments. A & B spatially separated sediments; C control self hybridizations. (Units A & B, ug total DNA; C ng of probe DNA.)

Table 3. DNA probe detection and distribution of naphthalene biodegradative bacteria in Manufactured Gas Plant soils

	Culturable Microorganisms			
Sample Depth (ft)	YEPGA		Naph. vapor plates	
	Total[1]	DNA+ (2)	Total[1]	DNA+
Surface	$4.0x10^7$	<DL	$1.6x10^6$	<DL
4-4.5	$1.8x10^6$	<DL	$2.6x10^4$	<DL
Water table	$4.9x10^5$	$3.0x10^5$	$1.0x10^5$	$1.0x10^5$
10-11	$5.2x10^5$	$4.0x10^5$	$1.4x10^5$	$3.1x10^4$
15.5-16.5	$5.1x10^6$	$1.7x10^6$	$3.1x10^6$	$2.3x10^5$
18'11"19.5	$9.0x10^7$	$1.0x10^7$	$4.3x10^7$	<DL

(1) = Colony forming units g^{-1} soil.

(2) = DNA+ indicates number of organisms containing naphthalene degradative genes detected by colony hybridization with naphthalene specific catabolic DNA probe. (This work performed in collaboration with Dr. W. Ghiorose, Cornell University.)

<DL = Below detection limit.

YEPGA = Yeast extract, peptone and glucose agar.

copy in a DNA extract can be amplified a million fold overnight. This technology, given sufficient sequence information, could be used for detection of very rare (low abundance) DNA fragment in the environment. Consequently, the technology is receiving considerable attention for environmental monitoring.

These preceding examples are clear evidence of the new sophistication available in environmental microbiology that contribute more specific and quantitative information on the structure of biodegradative microbial communities. Such information will lead to additional experimental study on the degree to which specific chemical biodegradation can be controlled and enhanced (in terms of total population densities) as a means for increasing the relative rate of removal of specific hazardous chemicals from the environment.

The second area of molecular contributions to analyzing and enhancing biodegradation relates is bioanalytical methods to assess the specific activity of microorganisms; activities that are related directly to the processes of biodegradation and, hence, ultimately tied to the genes involved in degradation. One specific example in developing new significant molecular methods for assessing biodegradative activity is the development of bioluminescent reporter strains. The rationale for such strains is to develop on-line biosensor capability to determine if biodegradative bacterial strains are active in the environment and to determine the environmental regimes that result in optimized or maintainable levels of biodegradative activity. Currently this is both difficult and expensive to do.

To obtain biodegradative bioluminescent reporter bacteria it is necessary to develop bioluminescent strains that only give off visible light when the genes for degradation of a specific chemical are active (being transcribed to messenger RNA, precedent to protein and enzyme synthesis (20)). This can be accomplished by introducing the bioluminescence (lux) genes from marine bacteria *Vibrio fischeri* (21) directly into the genetic operon for a catabolic pathway. This strategy was used to place the *lux* bioluminescence genes, carried by a vector transposon Tn4431, directly into the naphthalene catabolic genes harbored on a naphthalene plasmid in an environmentally isolated *Pseudomonas* spp. This molecular construction is made in such a manner that in the presence of naphthalene or a degradative intermediate, salicylate, light emission is induced in the bacterial cells. A typical example of the result of such molecular engineering is given in Figure 5. This figure is an example of bioluminescent light given off by bacteria that are engineered to produce light when biodegradative activity is induced by the presence of naphthalene.

Preliminary testing of such bioluminescent strains has demonstrated that light production can be enhanced 2-3 orders of magnitude during naphthalene exposure. In addition, the strains can be maintained in continuous flow waste treatment reactor systems and, importantly, light emission can be quantitatively measured by remote fiber optic technology in complex environmental matrices such as sand (Figure 6). The development of such reporter strains for this and other toxic chemicals will provide unique monitoring capabilities for evaluating and optimizing biodegradative microbial community processes.

Figure 5. Bioluminescent light emissions from colonies of engineered naphthalene degrading bacteria which respond as bioluminescent reporters of degradative activity.

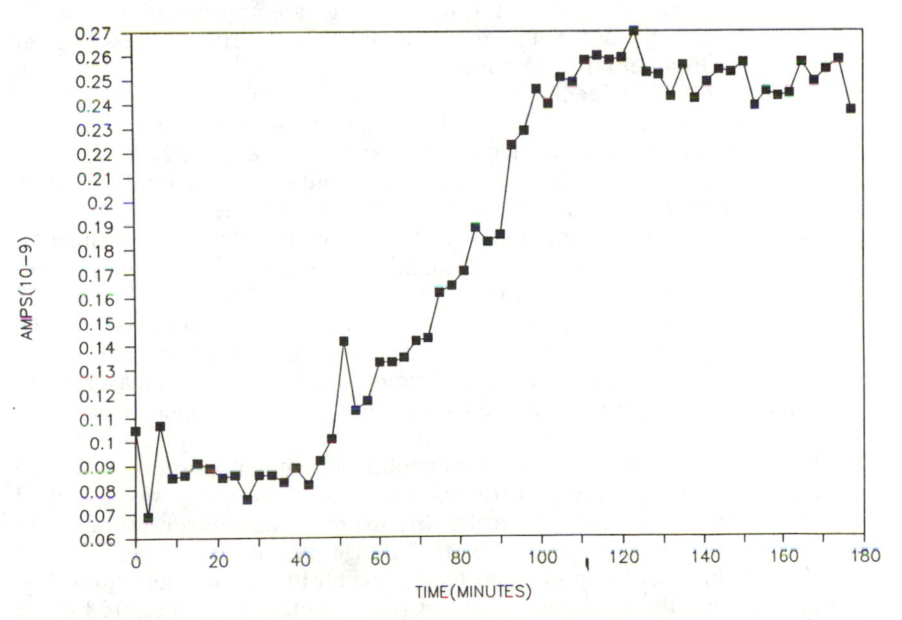

Figure 6. Remote sensing of light emission from naphthalene degradative bioluminescent reporter bacteria in sandy aquifer lab simulation. (Y axis, relative light output 10^{-7} amps.)

These have been somewhat limited examples of the contribution of new molecular methods in developing environmental biotechnology for hazardous wastes. Collectively, these and other developing methods such as highly sensitive biochemical detection techniques will provide revolutionary new ways to analyze the complexity, activity, and performance of biodegradative processes in the environment and in waste treatment technology.

System Analysis and Identification. When the bioanalytical approaches already discussed can be integrated into a dynamic testing and experimental approach using systems analysis techniques, even greater power to investigate system structure and activity is realized (22,23). As Figure 7 schematically shows, a microbial process of unknown internal structure (both from mathematical and physical/chemical/biological viewpoints) is operated in some type of a reactor system (including both engineered and "natural" reactors). Either a randomly-changing environmental disturbance or an induced perturbation is generated in one or more of the important system parameters (e.g., substrate concentration, flowrate, temperature). Using available and emerging biosensing technology in a time-series sampling mode, a related response in system activity or other "output" parameters can be measured. With mathematical tools long developed for this use, the known perturbation or disturbance signal may be related to the measured output signal and the nature and activity of the unknown "system" may be probed.

This approach offers the potential for (24):

1) A standardized approach for testing microbial systems to identify critical or dominant mechanisms for further targeted research,
2) A predemonstration testing and evaluation protocol to classify microbial processes for reliability and effectiveness under environmental disturbances,
3) A protocol leading to the identification of regimes of stability and robustness to disturbances that offer the potential for optimization by managing the nature of the disturbances,
4) A protocol to assess system adaptability and its importance to structure and activity, and
5) A dynamic experimental protocol that provides an experimental platform for testing and evaluation of new monitoring approaches and process improvement schemes.

Central to systems analysis is the attempt to develop a predictive model for the system's response to a given disturbance or perturbation. If a model can be proposed in advance, then experimental input/output measurement and signal analysis are used to verify that the proposed model is correct. This approach is called the "direct problem" of systems analysis.

For example, biotransformation of naphthalene in an operating activated sludge treatment system (after correction for abiotic processes) was modelled a priori by an elementary first-order (in naphthalene concentration) rate equation (24). The complex activated sludge system was perturbed by induction of sinusoidal naphthalene feed concentrations for eight sinusoidal frequencies while the naphthalene in the reactor offgas was measured every ten minutes. Abiotic fates (stripping, and sorption) were accounted for and

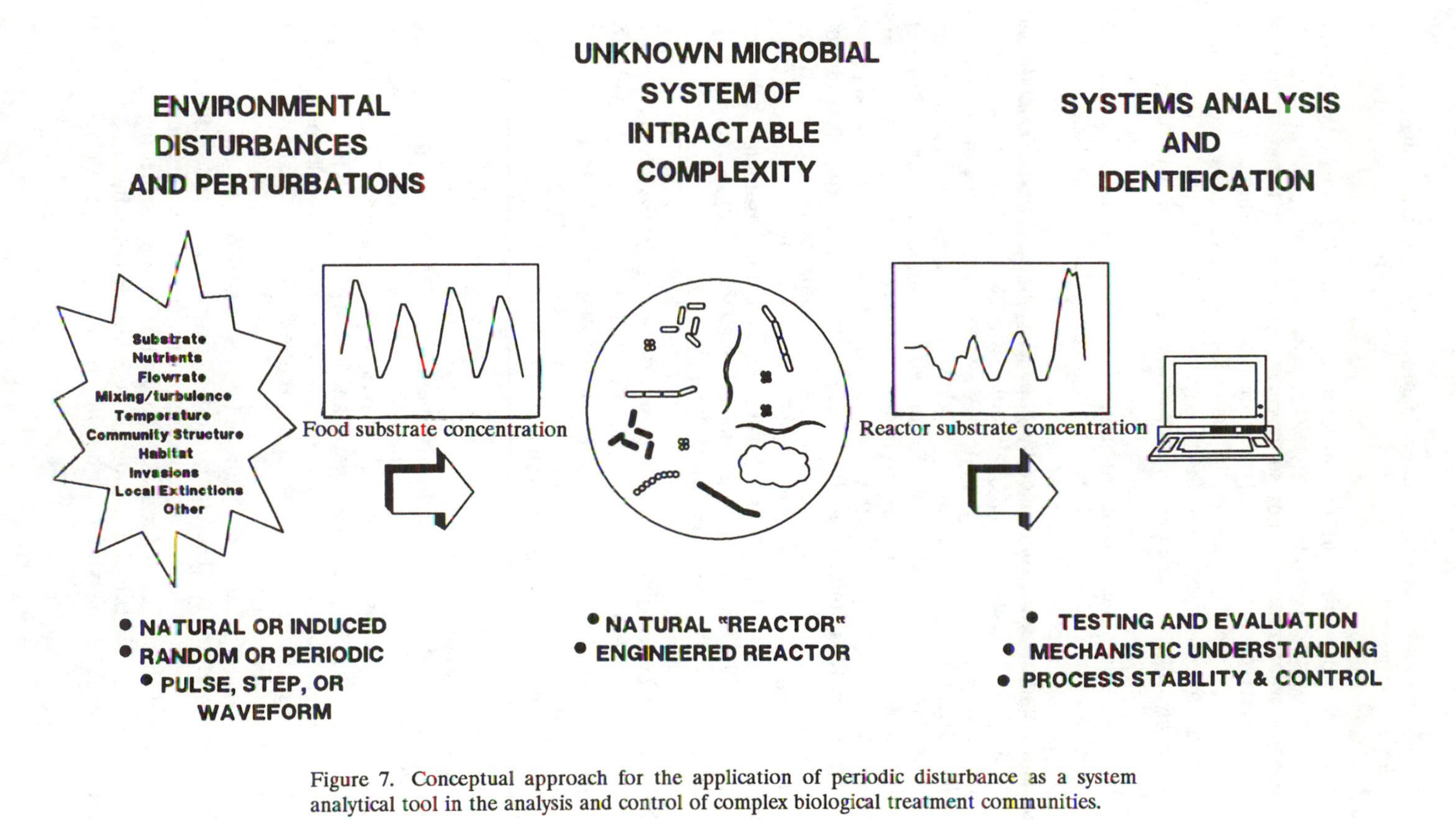

Figure 7. Conceptual approach for the application of periodic disturbance as a system analytical tool in the analysis and control of complex biological treatment communities.

the system's unsteady-state naphthalene material balance was solved for the first-order biotransformation rate "constant" arising from the assumed rate model (Figure 8).

The observation that the first order rate "constant" is not constant for the perturbation and relaxation test intervals leads to the conclusion that a simple first- order model is not sufficient to explain the behavior in this dynamically-perturbed system and another model should be proposed and tested. Often, the new models are in themselves more complex and require more information for verification than was originally collected. Thus, the direct approach may require the iteration of new experiments with additional sampling.

The second, or indirect systems analysis problem is also called system identification because it attempts to identify the model by knowledge of the input disturbances/perturbations and the measured response. Here, the input and response are experimentally known or measured and a model is computed with the capability of generating the observed response to the disturbance or perturbation. A system was identified in the naphthalene activated sludge work through use of Bode's graphical method (25). Here, the nature of the model was determined through the application of numerous perturbations to the experimental system, each at a predetermined frequency. This type of model is not based on the material balance or other deterministic relationships, but is made up of linear combinations of time derivatives (or integrals) with time as the independent variable. If the structure of the system does not change, such models are sufficient for predicting responses. Many other parametric and non-parametric schemes for system identification are also available for analysis of perturbation/response model development (26,27,28).

Attempts have been made to expand the technique to include the analysis of soil biotransformations (23,29). While the hydrodynamic nature and physical structure of soil systems vary widely and are difficult to establish with certainty, two limiting conditions may be specified. The first is where the soil particles are suspended and all phases are well-mixed. This case is not typically found in nature, but is found in various types of engineered soil-slurry reactors. The reactors currently used in our systems experiments include continuous stirred tank reactors (CSTRs) operated to minimize soil washout.

A second limiting physical/hydrodynamic case is the soil as a porous bed. Often others simulate undisturbed soils in the lab with soil columns, however we have chosen to use a slice of such a column--a differential volume reactor (DVR)--as the experimental design (29). This approach offers advantages in the ability to develop a more spatially homogeneous system and also contributes to the perturbation/response analysis needed for systems identification.

These reactors may be modelled with precision based on past engineering experience. Therefore, more obscure biotic and abiotic fate processes (e.g., biotransformation and sorption) may be studied in context of the structure of these obscure processes in context of the engineering model. For instance, a CSTR containing 400 g/L of contaminated MGP soil was fed a sinusoidal concentration of naphthalene in the liquid reactor feed (0-14

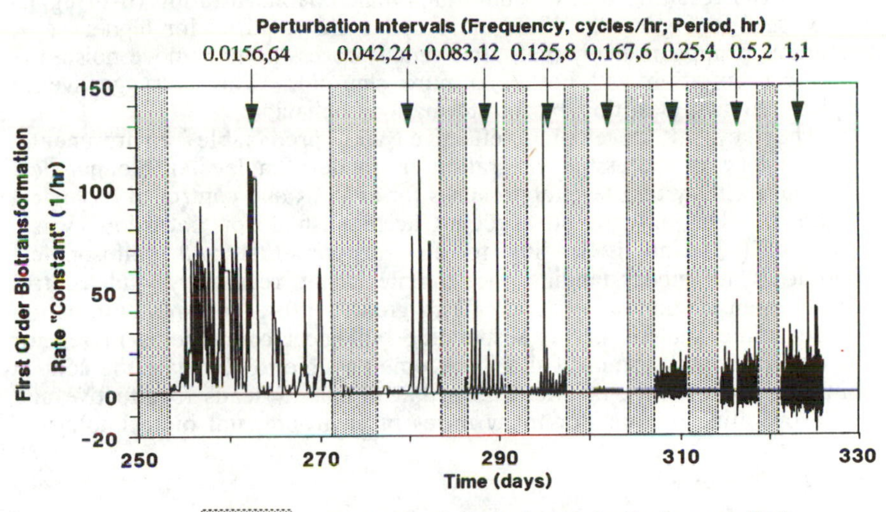

Figure 8. Response of first order biotransformation rate constants for naphthalene oxidation to naphthalene perturbation frequency in a continuous activated sludge biotreatment reactor.

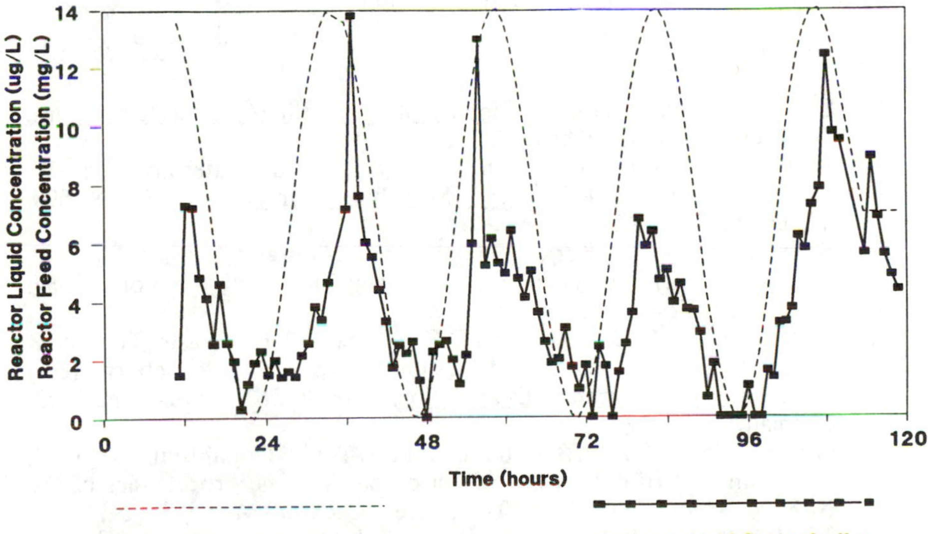

Figure 9. Comparative dynamic feed and liquid phage naphthalene in a continuous stirred soil slurry treatment reactor for contaminated Manufactured Gas Plant (MGP) soils.

mg/L). The resulting reactor liquid naphthalene concentration (0-14 ug/L) can be seen in Figure 9. While more cycles are required for higher resolution, it appears likely that, after signal processing to remove noise, the liquid concentration will not be a pure sinusoid. This is suggestive of atypical removal (sorption/biotransformation) behavior.

The key to developing effective and predictable environmental biotechnology is successful integration of modern molecular bioanalytical techniques with system level approaches for analysis and control of biological processes. This integration is being accomplished for hazardous waste treatment (15) and is leading to new experimental and philosophical approaches for understanding the complexity of regulatory and control circuits in environmental systems. With greater efforts towards automation and real time analysis of the responding biological components in reactor and environmental systems it should become possible to elucidate the control points for the network of interactions that ultimately leads to effective and stable operating regimes for many forms of environmental biotechnology.

Acknowledgments

The authors thank Henry King, Paul Dunbar, and Phil DiGrazia for contributions to this manuscript and to Mary James for word processing. The work was supported in part by EPRI contract #RP-3015-1, GRI grant #5087-253-1490, the U.S.G.S. award #14-08-0001-G1482, the U.S. Air Force contract #F49620-89-C-0023 and the University of Tennessee, Waste Management Research and Education Institute.

Literature Cited

1. Schonborn, W. 1986. Biotechnology: Microbial Degradations. Weinheim, West Germany: VCH.
2. Seal, K.J. and L.H. G. Aorton. 1986. Chemical Materials. In: W. Schonborn (ed.) Biotechnology: Microbial Degradation. Vol. 8: 590. Weinheim, West Germany: VCH.
3. Schmidt, O. and W. Kerner-Gang. 1986. Natural Materials. In: W. Schonborn (ed.) Biotechnology: Microbial Degradation. Vol. 8: 557. Weinheim, West Germany: VCH.
4. Finstein, M.S., G.C. Miller and P.F. Storm. 1986. Waste Treatment Composting as a Controlled System. In: W. Schonborn (ed.) Biotechnology: Microbial Degradation. Vol. 8: 563. Weinheim, West Germany: VCH.
5. G.S. Sayler, J.W. Blackburn, and T.A. Donaldson. 1988. Environmental Biotechnology of Hazardous Wastes, Proceedings of the NSF Workshop, Gatlinburg, Tennessee. ORNL/TM-10853. NTIS.
6. J.S. Hirschorn. 1983. Technologies and Management Strategies for Hazardous Waste Control. OTA-M-196, Office of Technology Assessment. Washington, D.C.
7. G.S. Sayler and J.W. Blackburn. 1989. Modern Biology: Application of Biotechnology. In: M. Huntley ed. "Applications of Biotechnology in Agricultural Wastewater". CRC Press Inc. pp. 53-71.

8. R.K. Jain and G.S. Sayler. 1987. Problems and potential for *in situ* treatment of environmental pollutants by engineered microorganisms. Microbiol. Sci. 4:59-63.
9. R.K. Jain, R. Burlage, and G.S. Sayler. 1988. Methods for detecting recombinant DNA in the environment. In: G.G. Stewart and I. Russell, CRC Critical Reviews in Biotechnology. Vol. 8, p. 33-84.
10. S.E. Lindow, N.J. Panopoulos, and B.C. McFarland. 1988. Genetic Engineering of Bacteria From Managed and Natural Habitats. Science. 244:1300-1307.
11. Sayler, G.S., Hooper, S.W., A. Layton, and J.M.H. King. 1989. Catabolic plasmids for environmental applications: A review. Microbial Ecology. Vol. 19, No.1.
12. G.S. Sayler, M.S. Shields, A. Breen, E.T. Tedford, S. Hooper, K.M. Sirotkin and J.W. Davis. 1985. Application of DNA:DNA colony hybridization to the detection of catabolic genotypes in environmental samples. Appl. Environ. Microbiol. 49:1295-1303.
13. G.S. Sayler and G. Stacey. 1986. Methods for Evaluation of Microorganism Properties. In: Biotechnology Risk Assessment, (J.R. Fiksel and V.T. Covello eds.) Pergamon Press, N.Y.
14. G.S. Sayler, R.K. Jain, L. Houston, A. Ogram, C.A. Pettigrew, J.W. Blackburn, and W.S. Riggsby. 1986. Applications for DNA probes in biodegradation research. In: Perspective in Microbial Ecology. (F. Megusar & M. Gantar eds.) Slovene Society for Microbiology. p. 499-508.
15. Blackburn, J.W., R.K. Jain and G.S. Sayler. 1987. The molecular microbial ecology of a Naphthalene-degrading genotype in activated sludge. Environ. Sci. Technol. 21: 884-890.
16. A.V. Ogram and G.S. Sayler. 1988. The use of gene probes in the rapid analysis of natural microbial communities. J. Indust. Microbiol. 3:281-292.
17. A.V. Ogram, G.S. Sayler, and T. Barkay. 1987. DNA extraction and purification from sediments. J. Microbiol. Methods. 7:57-66.
18. C.A. Pettigrew and G.S. Sayler. 1986. Application of DNA colony hybridization to the rapid isolation of 4-chlorobiphenyl catabolic phenotypes. J. Microbiol. Methods. 5:205-213.
19. Steffan, R.J. and R.M. Atlas. 1988. DNA amplification to enhance detection of genetically engineered bacteria in environmental samples. Appl. Environ. Microbiol. 54: 2185-2191.
20. O.A. Carmi, A.B. Stewart, S. Ulitzer, and J. Kuhn. 1987. Use of bacterial luciferase to establish a promoter probe vehicle capable of nondestructive real-time analysis of gene expression in *Bacillus* spp. J. Bacteriol. 169:2165-2170.
21. P.M. Rogowsky, T.J. Close, J.A. Chimera, J.J. Shaw, and C.I. Kado. 1987. Regulation of the *vir* genes of *Agrobacterium tumefaciens* plasmid pTiC58. J. Bacteriol. 169:5101-5112.
22. J. W. Blackburn. 1988. Problems in and potential for biological treatment of hazardous wastes. Proc. 81st Annual AWMA Meeting. Air and Waste Management Association. Pittsburgh.

23. J. W. Blackburn. 1988. Microbial systems analysis for increased reliability in polyaromatic hydrocarbon bioremediations. Proc. Sympos. on Gas, Oil, and Coal Biotechnology. Institute of Gas Technology. Chicago.
24. J. W. Blackburn. 1989. Improved understanding and application of hazardous waste biological treatment processes using microbial systems analysis techniques. Hazard. Waste & Hazard. Mat. 6(2):173-193.
25. J. W. Blackburn. 1989. Frequency response analysis of naphthalene biotransformation activity. Biochemical Engineering VI, Annals N. Y. Acad. Scien. In Press.
26. P. Eykhoff. 1974. "System Identification, Parameter and State Estimation". John Wiley, New York.
27. H. H. Kagiwada. 1974. "System Identification, Methods and Applications". Addison-Westley, Reading.
28. L. Ljung. 1987. "System Identification: Theory for the User". Prentice-Hall, Englewood Cliffs.
29. P. M. DiGrazia, J. W. Blackburn, P. R. Bienkowski, B. Hilton, G. D. Reed, J. M. H. King, and G. S. Sayler. 1989. Development of a systems analysis approach for resolving the structure of biodegrading soil systems. Appl. Biochem. and Biotechnol. In Press.

RECEIVED January 18, 1990

Chapter 3

Insect Symbionts

A Promising Source of Detoxifying Enzymes

Samuel K. Shen and Patrick F. Dowd

Mycotoxin Research Unit, Northern Regional Research Center, Agricultural Research Service, U.S. Department of Agriculture, Peoria, IL 61604

Insects are the most common organisms on earth in terms of numbers of species. Many are exposed to a diversity of toxins, including plant polymers such as lignins. In feeding on natural hosts such as plants or fungi, insects must detoxify these chemicals to survive. Many insects contain symbiotic microorganisms that are known to provide nutrients for their host, but also may contribute to detoxification. Detailed work with the cigarette beetle and its symbiotic yeast indicated that activity towards a representative substrate (1-naphthyl acetate) is significantly reduced in symbiont-free insects. Larvae were also more susceptible to plant toxins when symbionts were absent. Cultures of the symbiont were able to utilize (and apparently detoxify) representative plant toxins, mycotoxins, and insecticides. Intact cells hydrolyzed parathion, and cell-free extracts dechlorinated 1-chloro, 2,4-dinitrobenzene. The 1-naphthyl acetate esterase produced in culture was relatively resistant to inhibition by paraoxon, was relatively stable in organic solvents, and could be induced by exposing the cells to toxins. The tremendous variety of symbionts in other insects suggests that these microorganisms are a promising source of detoxifying and other novel enzymes.

Insects are the most successful group of organisms on earth; an estimated 726,000 species have been identified (1), and an equal or greater uncategorized number are thought to exist in the tropics and other areas.

Due to their feeding habits, insects may potentially encounter a diversity of toxins. Both plants (2) and microorganisms such as fungi (3) make a diversity of chemicals that act as agents to defend against insects and other predators, including plant polymers such as lignins that interfere with digestion. Those insects that feed on plants or fungi must be able to detoxify these toxins, whether by excretion, sequestration, or metabolic conversion in order to successfully feed on their hosts. The diversity of defensive compounds includes classes such as alkaloids, cardiac glycosides, cyanogenic compounds, flavonoids, phenolic compounds, and terpenoids in higher plants, and such mycotoxins as aflatoxin B_1, trichothecenes (e.g. T-2 toxin), and tremor-inducing compounds in fungi. Insects are known to detoxify enzymatically representatives of all major categories of plant toxins (4, 5). Detoxification of mycotoxins by insects can be inferred, based on apparent resistance of insects feeding on mycotoxin-contaminated materials, or documented by actual detoxification studies (6).

Exposure to natural toxins probably enables insects to adapt more readily to man-made materials such as insecticides (7). Insects have developed resistance to cyanide, chlorinated hydrocarbons, organophosphates, carbamates, synthetic pyrethroids, and other insecticides (8). This is not surprising when considering the same complex of detoxifying enzymes, mainly represented by hydrolytic, conjugative, and oxidative enzymes (9) is capable of detoxifying natural toxins as well as man-made materials. This ability is due to appropriate enzymes and/or isozymes that results in broad-substrate capabilities. For insects that feed on a wide variety of hosts (polyphagy), the spectrum of toxins that can be dealt with is truly remarkable.

Detoxification by Insect Symbionts

In an inciteful discussion of insect-microbe relationships, Jones (10) postulated that insect-microbial associations, known to involve catabolic (e.g. cellulose-degrading) and anabolic (e.g. biosynthesis of vitamins, sterols, and amino acids) processes necessary to the survival of the host, could also include detoxification abilities. Most investigations in this area have been limited (11). Nevertheless, some studies indicate detoxification of terpenoids (12, 13), lignin (14), and insecticides (15) by apparent symbionts. The roles other microbial symbionts of insects play in detoxification are discussed below.

Detoxification by Symbionts of the Cigarette Beetle

The cigarette beetle feeds on a wide variety of plant material (16), indicating a broad-spectrum ability to deal with plant toxins, including mycotoxin ochratoxin A (17). When the symbionts, *Symbiotaphrina kochii* Jurzitza ex. W. Gams and v. Arx c, were removed from the insect by hatching surface-sterilized eggs (the means of generation to generation transfer of the yeast) or by treatment with fungicides, the resulting symbiont-free insects became more susceptible to representative plant toxins.

Qualitative screening procedures were used to test toxins as sole carbon sources for growth of the cigarette beetle symbionts. By this method, we found the symbionts could utilize (and apparently detoxify) a diversity of plant flavonoids and phenolics including rutin, quercetin, caffeic acid, tannic acid, and gallic acid (11), as well as insecticides such as parathion, diazinon, and malathion; herbicides such as dinoseb, glyphosate, and 2,4-D; mycotoxins such as sterigmatocystin, (an aflatoxin precusor), ochratoxin A (a carcinogen), deoxynivalenol (a trichothecene), mycophenolic acid and citrinin; and meal toxins such as amygdalin, phytic acid and stachyose (18, Shen and Dowd, unpublished data). Presently we are working on demonstrating actual detoxification of representative toxins and on identifying and characterizing enzymes of interest. For example, ca. 4 pmole of ^{14}C parathion can be metabolized (to a 4-nitrophenol hydrolysis product) in one hour by a suspension of 10^6 cells (Shen and Dowd unpublished data). Dechlorination of 1-chloro, 2,4-dinitrobenzene, by a cell-free extract (1000 g supernatant) occurred at a rate of 20 nmole/min/mg protein (Shen and Dowd unpublished data). This dechlorination reaction appears to be performed by a glutathione transferase-like enzyme, since the reaction is glutathione dependent. Both α-glucosidase and β-glucosidase activity have also been detected (Shen and Dowd unpublished data).

Most of our work on enzyme properties has concentrated on 1-naphthyl acetate esterase. The enzyme responsible for most of the activity has a specific activity of ca. 1 μmole/min/mg protein and a molecular weight of ca. 38,000 when partially purified by gel filtration chromatography (Shen and Dowd unpublished data). Isoelectric focusing indicated a pI of 4.6-4.8 for the major source of enzyme activity (Shen and Dowd unpublished data). This enzyme is interesting in that it apppears to have relatively few external charges, because it moves relatively slowly by conventional polyacrylamide gel electrophoresis (Rf of ca. 0.33 in a 7.5% gel, with bromophenol blue as a tracking dye), in spite of its relatively low molecular weight (Shen and Dowd unpublished data). Although it is not particularly thermostable (inactivated at ca. 40° C.), it is stable when refrigerated or frozen (at least 1 week) (Shen and Dowd unpublished data). Another interesting property of the enzyme is that the activity is fairly stable in organic solvents; the rate of hydrolysis in a 50:50 acetone:buffer solution (after 1 hour preincubation) is still ca. 30% of control (Shen and Dowd unpublished data). The activity is somewhat resistant to inhibition by organophosphorous pesticide derivitives such as the extremely active paraoxon (ca 50% inhibition at 10^{-4} M), but is more susceptible to inhibition by heavy metal ions, such as Hg^{+2} (ca. 80% inhibition at 10^{-4} M) (Shen and Dowd unpublished data). Thus, the 1-naphthyl acetate activity from the cigarette beetle yeast does have properties, especially the solvent stability, that are likely to promote use for decontamination of lipophilic toxins. In addition, this activity (and in some cases additional molecular forms detected by gel electrophoresis) can be stimulated (based on higher activity relative to solvent controls) by malathion, β-pinene, griseofulvin, and flavone (1.8 fold); a relatively high level for a hydrolytic enzyme (19).

Potential for use of Insect Symbionts in Detoxification

The cigarette beetle symbiont has many desirable properties for use in decontamination. As indicated earlier, it has a broad-spectrum ability to detoxify, although we do not yet know if these reactions will occur with mixed substrates. The evidence of stimulation of 1-naphthyl acetate hydrolysis, however, suggests appropriate enzyme activity will be present with substrate mixtures as well. Since it is an obligate symbiont, it is not likely to persist in the environment when used apart from its host. Histochemical work indicates the symbiont appears to secrete or, at least, have externally bound hydrolytic enzymes, which overcomes some potential problems when detoxifying enzymes are located internally, and toxins are not taken up. However, the organism also appears to take up phenolic acids, (20). It is not known to produce any toxins and, in fact, it is known to produce B-vitamins, sterols and amino acids for its insect host (21). In many situations, however, its relatively slow growth compared to other microorganisms is likely to be a disadvantage since it can be overgrown by other organisms. In collaboration with Dr. Nancy J. Alexander at our lab, we are attempting to obtain faster growing strains or recombinant organisms that retain detoxifying capabilities.

Literature Cited

1. Borror, D. J.; DeLong, D. M.; Triplehorn, C. A.; In An Introduction to the Study of Insects; Holt, Rinehart and Winston: New York, 1976; p 138.
2. Whittaker, R. H.; Feeny, P. P. Science 1971, 171, 757.
3. Wicklow, D. T. In Coevolution of Fungi with Plants and Animals; Hawsworth, D. L.; Pirozynski, K. A. Eds.; Academic: New York, 1988; p 173.
4. Dowd, P. F.; Smith, C. M.; Sparks, T. C. Insect Biochem. 1983, 13, 453.
5. Dowd, P. F. In Natural Pesticides Volume V: Insect Attractants and Repellants; Morgan, E. D.; Mandava, L. B. Eds.; CRC Press: Boca Raton, in press.
6. Dowd, P. F.; VanMiddlesworth, F. L. Experientia. 1989, 45, 393.
7. Gordon, H. T. Annu. Rev. Entomol. 1961, 6, 27.
8. Georghiou, G. P. In Pesticide Resistance: Strategies and Tactics for Management; National Academy of Science: Washington, D.C., 1986; p 14.
9. Ahmad, S.; Brattsten, L. B.; Mullin, C. A.; Yu, S. J. In Molecular Aspects of Insect-Plant Associations; Brattsten, L. B.; Ahmad, S. Eds.; Plenum: New York, 1986; p 73.
10. Jones, C.G. In A New Ecology: Novel Approahces to Interactive Systems; Price, P. W.; Slobodchikoff, C. N.; Gaud, W. S. Eds.; Wiley: New York, 1984; p 53.
11. Dowd, P. F.; In Multitrophic Interactions Among Microorganisms, Plants, and Herbivores; Barbosa, P.; Krischik, V. A.; Jones, C. G. Eds.; Wiley: New York, in press.
12. Brand, J. M.; Bracke, J. W.; Britton, L. N.; Markovetz, A. J.; Barras, S. J. J. Chem. Ecol. 1976, 2, 195.
13 Brand, J. M.; Bracke, J. W.; Markovetz, A. J.;, Wood, D. L.; Browne, L. E. Nature 1975, 254, 136.
14. Nolte, D. J. J. Insect Physiol. 1977, 23, 899.
15. Boush, G. M.; Matsumura, F. J. Econ. Entomol. 1967, 60, 918.

16. Metcalf, C. L.; Flint, W. P.; Metcalf, R. L. Destructive and Useful Insects: Their Habits and Control; McGraw-Hill: New York, 1962; 888 pp.
17. Wright, V. F.; De Las Casas, E.; Harein, P. K. Environ. Entomol. 1980, 9, 127.
18. Shen, S. K.; Dowd, P. F. Amer. Chem. Soc. Abstr. 1989, 197, #BTEC 4.
19. Shen, S. K.; Dowd, P. F. Entomol. Exper. Appl. 1989, 52, 179.
20. Dowd, P. F. J. Econ. Entomol. 1989, 82, 396.
21. Jurzitza, G. In Insect-Fungus Symbiosis; Batra, L. R. Ed.; Allenheld and Osmun: Montclaire, 1979; p 65.

RECEIVED January 30, 1990

Chapter 4

Plastics Recycling Efforts Spurred by Concerns About Solid Waste

Ann M. Thayer

***Chemical and Engineering News*, Northeast News Bureau, 379 Thornall Street, Edison, NJ 08837**

With plastics taking up 20% of volume of municipal solid waste and costs of landfilling and incineration rising rapidly, waste reduction efforts focus on development of plastics recycling technology.

The management of solid waste is reaching a crisis in the U.S. and worldwide. This is illustrated by wandering garbage barges with no permission to land and by wastes washing up on beaches.

Every year the U.S. generates about 320 billion lb of what is called municipal solid waste, or postconsumer waste. About 85% of this trash is currently disposed of in landfills. Yet as the amount of solid waste increases---and the Environmental Protection Agency estimates that it will reach 380 billion lb by 2000---a third of the landfills are expected to close in the next five years. Many people are concerned that efforts to deal with the growing quantity of garbage are not moving fast enough.

Plastics make up only about 7% of solid wastes, according to figures prepared by Franklin Associates for EPA, but plastics and paper are the fastest-growing segments of such wastes. In 1987, the plastics industry produced a total of 55 billion lb of plastics, and the public discarded about 22 billion lb of that. By 2000, plastics production may reach 75 billion lb, and the public is expected to throw away about 38 billion lb of that, accounting for 10% of solid municipal wastes that year. Although not a significant proportion by weight, plastics have been targeted as a waste problem since, by volume, they are a very visible 20% of municipal solid waste.

Rapidly rising costs of landfilling, as high or higher than $100 per ton in some parts of the U.S., and of alternative waste treatment methods, such as incineration, are making it prohibitively expensive to deal with large amounts of waste. When coupled with the desire to conserve resources and have an environmentally acceptable means of reducing solid waste, recycling is often considered as an alternative. Recycling, although

NOTE: Reprinted from *Chem. Eng. News*, **1989**, *67*,(5), 7–15.

0097–6156/90/0433–0038$06.00/0

not inexpensive itself, can be less costly than other disposal methods because there is both a return on the reused materials and an offset of landfill charges. Of the more than 300 billion lb of solid waste produced, only about 10% is recycled annually. Aluminum and paper are the materials recycled most, with about 30% of the aluminum and 20% of the paper reprocessed. Only 1% of the plastics is recovered.

The apparent lack of movement in plastics recycling has helped to make plastics a scapegoat in the battle on solid waste. In the past decade, plastics have become increasingly prevalent because of their light weight, versatility in a range of applications, and convenience. But many plastic consumer items have short life spans and quickly make their way into the waste stream. More than half of the discarded plastics are found in the form of packaging, an area frequently targeted for recycling. Packaging, about one third by weight of consumer waste, consists primarily of paper (48%), glass (27%), and plastics (11%).

Much of the public believes that plastics cannot be recycled or safely incinerated and, because they do not decompose, that they should be removed entirely from the waste stream. One can argue as to whether plastics are in fact a major cause of the solid waste crisis, but the answer is almost irrelevant since plastics are generally perceived to be a significant problem.

The result has been the rapid introduction of a tremendous amount of local, state, and federal legislation that attempts to deal, one way or another, with the problems of solid waste. Much of the legislation has been aimed directly at plastics. On the federal level, former President Reagan signed the Plastic Pollution Control Act on Oct. 28, 1988, requiring that all six-pack ring carriers be made of biodegradable materials. The act will go into effect unless, in the next two years, EPA determines that it is either not feasible or that the by-products of the biodegradable materials are not environmentally safe. Efforts to deal with solid waste issues are also expected to continue with a reauthorization of the Resource Conservation & Recovery Act, which at this time promotes recycling only through government procurement guidelines.

The most comprehensive federal legislation proposed to date was reintroduced this month by Rep. George J. Hochbrueckner (D.-N.Y.). The bill, H.R. 500, entitled the "Recyclable Materials Science & Technology Development Act," has been described as the first measure introduced to authorize a comprehensive consumer product recycling program on a national level.

The bill stipulates that after a four-year joint review by EPA and the Commerce Department, all consumer items deemed recyclable would be identified and listed. Items determined to be recyclable must then be recycled, and all remaining items must be made biodegradable. After an additional year, the sale of certain nonrecyclable or nonbiodegradable consumer goods would be prohibited. The bill would establish an Office of Recycling Research & Information to distribute information and funds for research. Commerce would be directed to look into ensuring technology development, expanding the end-use markets for recycled materials, and encouraging the development of biodegradable products.

On the state level, the interplay of plastics and legislation began in 1978 with the introduction of the polyethylene terephthalate (PET) beverage bottle. In a relatively short time, nine states introduced "bottle bills," or deposit laws on the return of plastic bottles. In these states, the deposit laws account for collection of an estimated 80 to 95% of the PET bottles sold. At present, at least 30 states are looking into some sort of waste management legislation involving plastics.

The existing or proposed legislation runs the gamut of possible approaches, from the relatively benign sponsoring of feasibility studies for recycling to outright bans on the use of plastics. Other approaches include limited bans on certain items, such as polystyrene food containers, taxes on bottles or other containers, deposit laws, enforced usage of biodegradable materials, and restrictions on government procurement of recyclable or degradable items. In addition, voluntary or mandatory recycling programs are starting in many states. Susan Vadney of the Society of the Plastics Industry (SPI) Council on Solid Waste Solutions indicates that at least nine states have recycling programs and six more are likely to follow.

A few states such as Rhode Island, New Jersey, and California have become models for recycling programs, with Rhode Island and New Jersey having mandatory programs. In both states, several communities have had successful pilot programs for recycling that have included plastics. In October 1987, California enacted a mandatory beverage container redemption law (AB2020) with the condition that plastics manufacturers guarantee end-use markets for recycled plastics.

Massachusetts, in a joint study with Rhode Island, has recently released a Plastics Recycling Action Plan. The plan includes plastics as a significant part of a comprehensive recycling program aimed at reducing municipal waste levels 25%. The plan promotes pilot programs and further R&D into all aspects of recycling. Added incentives to recycling sometimes take the form of potential legislated penalties, such as taxes or deposits, that would be imposed if recycling goals, or those for certain materials, are not met. These incentives are often directed at bringing the recycling of plastics up to a level comparable to aluminum cans and glass bottles. Some states, like New Jersey, are also offering business loans, R&D funds, tax credits on equipment purchases, and market stimulation as incentives.

On a local level, some of the legislation already passed has made people take notice, since its approach has been more drastic. Suffolk County, New York, was one of the first counties to ban plastic food packaging–the apparent target being fast-food packages made of polystyrene. Similarly, Berkeley, California, banned all polystyrene packages made with chlorofluorocarbon (CFC) blowing agents. The Berkeley law requires that all fast-food restaurant containers be biodegradable by 1990.

Although the reaction by government has been fairly rapid and directed, the issue of plastics in the waste stream is complex. The number of legislative moves to ban plastics or replace them with biodegradable materials is increasing. Critics say that these approaches are based on misinformation and misconceptions, and are only a politically expedient response to the public's desire for solutions that appear to be environmentally safe. But biodegradable materials and plastics may not be the answer.

"There are still a lot of unanswered questions about biodegradable plastics," says Mary Sheil, deputy director of New Jersey's Division of Solid Waste Management and one of the drafters of that state's mandatory recycling law. Among these questions are concerns about the degradation products and how they will affect the environment. In addition, the introduction of degradable materials is often considered at cross purposes to recycling, because the degradable items can become contaminants of the more stable recyclable materials. One of the most startling facts that is coming to light is that even materials that are expected to decompose, such as paper and foodstuffs, are not breaking down in landfills. Therefore, producing more biodegradable materials would not necessarily ease the burden on landfills.

Public perception and the resulting trends in legislation have led manufacturers of plastics and plastic products to realize that they either have to act and get involved with the issue or be prepared to shoulder much of the blame for the solid waste crisis and face tremendous restrictions on their industry. The companies' initial responses included the formation of both internal divisions and cooperative industry associations to approach the issues of plastics recycling and solid waste management.

"Fortunately, we had some visionary people," says Wayne E. Pearson, executive director of the Plastics Recycling Foundation (PRF), an independent, nonprofit organization started through the Society of the Plastics Industry in 1985 by 20 members of the plastics and allied industries. "This group recognized, before it was as apparent to most of us, that there was a solid waste crisis, that recycling was important and that [group members] needed to know more about it." The foundation now has 45 member companies, ranging from resin producers to container manufacturers to users of the containers, who are concerned about the attitudes driving today's legislation and would like to see it take a new direction. "The legislation is coming from a quick fix," says Pearson. "And it's too early to make simplifying decisions, unless you know the impact on society and the environment.

"But," he says, "what we can't let it do is cause us to panic. We are trying to get a message out there about recycling, but get it out there in a calm and rational way. We want to get to various legislators and let them know that when they have the facts, they will see that banning plastics is not the answer."

But how and where does a group like this get its facts? In 1985, PRF made its primary objective the establishment and support of the Center for Plastics Recycling Research (CPRR) at Rutgers University. Funds totaling nearly $2.3 million in 1988-89 to support research come from PRF, Rutgers University, and other organizations such as SPI, the National

Science Foundation, the New Jersey Commission on Science & Technology (which has established CPRR as one of its Advanced Technology Centers), and several state governments and universities. In the near future, Rep. Jim Courter (R.–N.J.) is expected to reintroduce a proposal to fund the center as the National Center for Plastics Recycling, the hub of a national effort with other universities.

Starting up a research program in recycling is not an unlikely thing for plastics manufacturers to do, Pearson points out. "One of the things I don't think the public at all understands is that the plastics industry recycles every plastic, every thermoplastic, every day." He explains that the reprocessing of in-house scrap from the manufacture of plastic products has been a common practice for as long as the industry has existed. Thus, much of the technology for reusing plastics already exists. But, he notes, the use of inhouse scrap is relatively straightforward, since collecting and sorting the materials is not necessary. So the question CPRR was given regarding plastics recycling was, Pearson says: "How do you gather it back from 240 million Americans and do something with it that makes any sense?"

Under a charter to advance R&D in all aspects of plastics recycling and reuse, the job has been given to Darrell R. Morrow, director of CPRR, and his staff. The center operates with three divisions-research, process development, and information services-that work on covering the scientific, technological, economic, environmental, and practical sides of recycling plastics. This includes defining and developing the necessary infrastructure to recycle plastics. This framework consists of four areas: collection, sorting, reclamation, and reuse. Unlike other materials such as paper, metal, and glass, which have been recycled for 40 years or more, the infrastructure for plastics recycling is not yet in place.

The center initially concentrated on reclamation and reuse and developed a technology for the reclamation of postconsumer waste plastics. After only two years, CPRR published nearly 25 reports based on its research and held conferences to provide for transferring the technology. "We want to disseminate the information as widely and in as timely a fashion as we can to help develop the infrastructure that will cause plastics to be included in the recycling stream," says Morrow.

In its first two years, CPRR also constructed and began operating what is potentially a commercial-scale pilot plant facility. "To be able to reduce the applied research to practice in a timely manner," says Morrow, "we realized that we had to have a process development and demonstration facility that was significantly larger than you would ordinarily expect to find at a university, or even in industry." He adds that this has allowed them to move along faster toward a commercially viable design.

Through this focused effort, CPRR has produced a technology transfer manual that describes in detail the setup and operation of a plastics recycling plant. The hope of CPRR is that making the technology available will foster the commercialization of plastics recycling. There are currently 14 licensees of the CPRR process worldwide; two of them are looking into building plants in the near future.

The process developed at CPRR is said to be similar in its layout to those used in private industry. Most plastic reclamation systems are designed to work with rigid containers, such as PET beverage bottles, and HDPE milk or household product containers, because they are currently the easiest postconsumer items to collect and sort. PET beverage bottles are actually not one, but several materials: a PET body (clear or green), a pigmented high-density polyethylene (HDPE) base cup, aluminum cap, label, and adhesives. To separate these components, either a dry or wet separation method based on one or more of the different physical properties of the materials can be used.

Frank W. Dittman, project manager at the CPRR plant, describes its operation after the containers have been collected and delivered to the plant. The bottles are sorted by hand to separate out the more valuable resins. "We are looking at mechanical sorting based on optical and electrical principles," says Dittman, "but whether that's going to beat hand sorting at cost, I don't know. You can sort by hand pretty quickly." After sorting, the bottles are shredded and ground into chips for processing. The chips are then air classified, which, Dittman says, means that the lighter weight paper and other contaminants are separated in an air cyclone from the heavier plastic chips.

To remove the labels still attached to the chips, a cost-effective and efficient means of removing a waterproof adhesive had to be devised. After finally finding a detergent wash that would work, he explains, it had to be optimized with respect to temperature, washing time, concentration of detergent, power of mixing, and, of course, concentration of chips.

In the case of the multicomponent PET beverage bottles, once the labels are removed, the lighter HDPE fraction is separated from the heavier PET and aluminum pieces in a hydrocyclone system. It is at this stage that other reclamation methods often differ, using a separation step based on flotation methods; Dittman suggests that the hydrocyclone system will be an improvement over the flotation methods. The HDPE chips are then dried and collected.

The PET and aluminum chips require further drying so that they can be electrostatically separated. This is the most expensive part of the entire process. "The aluminum cap is only 1% by weight of the bottle," explains Dittman, "but the equipment to remove it represents about 30% of the investment in the plant. It's out of balance, and we are trying to remove the aluminum more cheaply." The process–which at the CPRR plant can handle about 600 lb per hour, or a potential capacity of about 5 million lb per year–generates clean, well-separated (99.9%) granulated plastic chips that can be sold to a manufacturer who uses the resins.

The chips sell for about 25 to 35 cents per lb for PET (with less than 100 ppm of aluminum), and 17 to 25 cents per lb for HDPE. In comparison, virgin PET costs about 50 to 60 cents per lb, and HDPE resins about 40 cents per lb. The center combines these and other factors in evaluating the financial aspects of operating a plant. "In our engineering manual we analyzed the economics and showed that a profitable plant should be able to process 20 million lb a year or more," says Dittman. He notes that the investment for a-plant of that size would be about $2.5 million and it would pay for itself in three to five years.

Other available reclamation technologies for rigid HDPE plastics are made by Transplastek of Canada and A.K.W. of West Germany and are similar in design to the CPRR process with granulating, washing, and separation steps. Transplastek uses a proprietary flotation tank method to separate plastics by density, whereas the A.K.W. method is based on the hydrocycloning technology.

Much of the resin reclamation and reprocessing technology comes from overseas, as the Europeans and Japanese have been involved in recycling for a longer period of time. There are other reclamation processes for different resins, which have been developed by or licensed to private companies, but because these are usually proprietary the details of their operations are not known.

Morrow makes a point of calling the system designed at CPRR a "resin recovery system," to stress the fact that it will work on about five different resins. Currently PET and HDPE can be processed together and are easily separated because of their different densities. The center hopes to add polystyrene and polyvinyl chloride to the mixed plastics stream they process. But, with the present technology, PET and PVC cannot be run together since they have similar densities.

"We have a very ambitious objective," says Morrow. "What we have challenged ourselves to is to make our resin recovery system a universal reclamation facility." Ideally, he explains, one would like to be able to feed in mixed, even granulated, plastics and have pure, generic resins come out the other end. This would remove the need for hand sorting and, for example, the separate processing of pigmented and unpigmented HDPE containers. Morrow notes that economics are driving this stage of the reclamation process: Colored HDPE is less valuable than uncolored HDPE, which has greater flexibility in end uses.

The value of recycled materials is present only if there are available markets, and plastics have become the second most valuable recyclable material after aluminum. The greatest return on recycled plastics is gained when the resins are reused in their original or similar applications, or by producing materials with added value, such as engineering plastics. If the properties of the resins are not compromised by the reclamation process, and if free enough of contaminants, generic resins can be sold to the manufacturers of plastic products as a replacement for virgin material. But Food & Drug Administration restrictions prohibit use of recycled resins in food applications, making one majaor plastics market inaccessible to recycled resins.

Mixed, or commingled, plastics must be dealt with differently to find end uses. Several technologies exist for mixed plastics, such as the ET/1 extrusion process from Advanced Recycling Technology of Belgium and a compression-molding process from Recycloplast of West Germany. At

CPRR, another pilot plant with an ET/1 extruder has been installed to make plastic lumber out of mixed plastics. Lower-melting plastics in the mixture act as a matrix that carries other plastics, and even up to 40% contaminants, such as paper, metal, glass, and dirt, into the mold. The resulting product can be treated and used like wood in nonconstruction applications such as decks, picnic tables, and park benches. Reclaimed HDPE resins are also used to make flower pots, pipe, bottle base cups, pails, and drums.

A few companies make mixed plastic items commercially. About six years ago, Eaglebrook Plastics of Chicago started recycling HDPE plant scrap, developing its own technology for cleaning and reclaiming the resin to be sold back to manufacturers. Eaglebrook since has extended its technology to deal with postconsumer waste and now buys bales of bottles from community collection centers. An offshoot company, Eaglebrook Profiles, was recently opened to make plastic lumber profiles using a proprietary, continuous-extrusion process developed in-house.

Similarly, Polymer Products, a division of Plastic Recycling Inc. of Iowa, is a young company that has started buying industrial scrap and postconsumer plastics, generally commingled, to make car stops, speed bumps, traffic bollards, and plastic lumber. Floyd Hammer, chairman of Plastics Recycling Inc., says that when he started the company he had in mind to manufacture value-added products with their own end-use markets, rather than deal in the more volatile commodity resins. He noted that two years ago there wasn't a market for mixed plastics products, but that the company has been building its market slowly, but successfully, as one of the few small companies making these types of products.

In contrast, PET end-use markets have been developing for at least 10 years. When the PET bottle was introduced in 1978 and was collected in states with deposit laws, the industry for recycling them and reusing them began to develop. PET resins can be reused to make polyols for insulation and unsaturated resins for bathtubs, shower stalls, boat hulls, and auto panels. Reclaimed PET is also used for strapping, paint brushes, geotextiles, fibers for fiberfill and carpets, and other textile applications.

Wellman Inc. of Clark, N.J., the largest consumer of reclaimed PET, recycles PET beverage bottles and scrap to manufacture fibers, fiberfill, geotextiles, and other products. Dennis Sabourin, vice president of Wellman, indicates that the company used about two thirds of the PET bottles collected in the U.S., or about 100 million lb, in 1988. About 150 million lb of PET bottles were collected nationwide and recycled in 1988, or about 20% of those manufactured. This is a substantial increase over the 8 million lb collected in 1979 at a time when the return of plastic bottles was just beginning. CPRR estimates a potential market now for at least 500 million lb of reclaimed PET, increasing to more than 900 million lb in 1993. Similarly, 55 million lb of HDPE containers were recycled in 1988 and the potential market is estimated at 440 million lb, growing to 660 million lb in 1993. By most estimates the market for plastics reuse is growing.

PET resins also can be reused to produce new unsaturated polyester resins. This takes a two-step reaction. The first reaction is similar to that used to make the polyols but, instead of ethylene glycol, propylene glycol is used in the glycol exchange. The end products of this reaction include bis(hydroxyethyl)terephthalate and bis(hydroxypropyl)terephthalate diesters, mixed ethylene glycol and propylene glycol terephthalate diesters, and free ethylene glycol and propylene glycol. Since all of the products are hydroxyl-terminated, reacting them with a dibasic acid like maleic anhydride produces new unsaturated polyester condensates.

This synthetic method has been successfully tested on PET from beverage bottles (a 2-L bottle, for instance, consists of about 50 to 60 g of PET). Producing unsaturated resins from reclaimed PET takes about half the time as other methods and produces materials of equal molecular weight, acid value, and viscosity. Glass fiber-reinforced laminates and unreinforced castings prepared from the reclaimed PET material test favorably with other materials.

Although Eastman Chemicals does not currently use the unsaturated resin synthesis from reclaimed PET in its production, it does frequently use a methanolysis reaction, which breaks down PET into its original starting materials, to reprocess large amounts (tens of millions of pounds) of Kodak film scrap. As David Cornell, manager of technology and manufacturing in Eastman Chemicals' polyester recycling business unit, says, "It makes good sense for us at Eastman Kodak to recycle our scrap as it helps treat the solid waste problem and saves on our disposal cost."

The methanolysis begins with reasonably clean PET scrap to which a catalyst and methanol are added. This mixture is heated under pressure to force the PET to depolymerize. The end products are ethylene glycol and dimethyl terephthalate (DMT). Pure DMT and ethylene glycol are obtained through recrystallization and distillation, respectively. These pure materials can be used as feedstocks for the synthesis of new polyesters.

A few special considerations must be dealt with when reusing PET from consumer sources in chemical recycling. For example, the PET must be clean and free of contaminating materials. Using plant scrap assures the manufacturers of both the identity of the resin and its purity. Using mixed colored resins may pose a problem, although when they are used in products that are not visible to the consumer, such as foam insulation, color is not a factor.

Whether an economic advantage is realized when using reclaimed materials depends on several factors, ranging from the relative costs of obtaining and reprocessing the materials to the cost of virgin materials or feedstocks. In addition, the supply of materials and the scale of the reprocessing effort may play a significant role.

According to an analysis of markets and economics of recycling by R. M. Kossoff and Associates of New York City, although the market for plastic resins can fluctuate widely, economic incentives are present in all aspects of the recycling chain. The firm sees the business as highly segmented, with opportunities at a variety of levels. These include collection, reclamation, selling of reclaimed resins, end-use products, and equipment manufacture. But as Richard M. Kossoff, director of the study, says, "In the long run, the markets will have to lead."

Increasing costs of virgin resins and viable end-use markets are currently making the recycling business attractive, and more and more companies, both large and small, are entering the business. The 1988 Plastic Bottle Recycling Directory and Reference Guide published by the Plastic Bottle Institute of SPI currently lists 67 companies that buy bulk bottles for recycling, 47 companies that sell reprocessed plastics, 50 companies that purchase reprocessed plastic, and 54 companies that manufacture equipment for recyclers.

Most people in the plastics recycling business will tell you they can always use more plastic than is currently available, but that the bottleneck is in getting the materials collected from consumers. Sabourin indicates that Wellman could use more than twice the amount currently available and as much as three and a half times that in the next two years. That more than 200 million lb of plastic containers were recycled with no organized recycling program, other than deposit laws in nine states, suggests that the level of plastics recycling and the requisite end-use markets will probably rise with increased collection efforts.

Having produced a working reclamation plant, CPRR is shifting its R&D emphasis to the areas of collecting and sorting plastics. The center is in the process of publishing another manual to help communities and industry establish the infrastructure for collecting and sorting plastics as a part of a comprehensive recycling program. In following the progress and analyzing the results of several community pilot programs, the center has found that the success of these programs depends on many factors. The most important factor is the careful design and planning of a program to fit a given community.

Even though including plastics in recycling programs is a new concept, participation rates can be as high as 70 to 90%. "If you want to get serious about collecting, you make it very simple for the consumer to participate," says Pearson. This usually means curbside pickup programs for trash and recyclable materials in suburban communities. In comparison, drop-off programs only have a 10 to 30% participation rate. Although mandatory recycling laws do help to directly communicate the need for recycling and enforce it, Pearson feels that with sufficient education about the solid waste problem and the value of recycling, most people are very willing to participate. And the value of education is not to be underestimated, because the most difficult initial obstacles to overcome are the previous misconceptions people hold about plastics not being recyclable.

The extent to which a community gets involved in recycling depends on the economics of collection, transportation, handling, reprocessing, the revenues from sales of recycled materials, and the offset of landfill or incineration fees. The large number of variables makes the overall economics of recycling programs complicated. Based on current costs and market prices, the CPRR economic model predicts that an average suburban community with landfill charges of $45 per ton or more cannot afford not to recycle. Although this should encourage some communities to become involved in recycling, this does not mean that local governments must become entrepreneurs in the plastics end-use market. However, government subsidies to start programs and stimulate market growth might be necessary.

CPRR suggests that communities work toward developing regional Material Recovery Facilities (MRF) to handle all the recyclable materials collected. Several MRF facilities exist across the U.S., and the number being looked at is growing rapidly. These larger facilities, serving several communities, can handle sorting and densification (baling or shredding) in addition to sale and transport of recyclables to reprocessors.

Promoting recycling and collecting plastic containers has led to the development of other organizations. One of these is NAPCOR, the National Association for Plastic Container Recovery, a nonprofit association of 12 major PET resin and container manufacturers. "We work with communities in recycling efforts to either encourage including PET in existing programs or to have it included in programs under development," says Luke B. Schmidt, president of NAPCOR. NAPCOR has a goal of achieving a 50% recycling rate for PET bottles by 1992.

NAPCOR works on a project-by-project basis to encourage collection, Schmidt says. These projects are usually in demonstration programs. Currently, they are working in at least seven states giving financial support, technical assistance, and public education and promotion programs. In addition, the member companies of NAPCOR have pledged to use a voluntary coding system developed by SPI to identify resins. The code consists of a number and letter symbol that can be imprinted on containers in the molding process and should help in the sorting of resins.

The Plastic Bottle Institute of SPI and the Council on Plastics & Packaging in the Environment (COPPE) work to educate and provide information on plastics recycling and waste issues. The Plastic Bottle Institute, with 23 industrial members, publishes newsletters in addition to its extensive plastics recycling directory. Along with other organizations, it hopes to inform the public about the recyclability of plastics and the existence of viable end-use markets. COPPE works to communicate information to industry, the public, and government about plastics recycling and waste management.

"We want to educate the public as to what can and should be done so that decisions can be made on an informed and not an emotional basis," says Connie Pitt, communications coordinator for COPPE. Although not a lobbying organization, it does interact on matters of public policy as an information source and in a peer review capacity. There are 42 companies in COPPE that cover a broad base in the plastics industry. Noel H. Malone, manager of marketing in Eastman Chemicals' polyester recycling division, says that the large number of groups spanning the plastics industry is beneficial. "The industry is deep and wide, and specific groups can focus in on specific problems," he says.

New Jersey Division of Solid Waste Management deputy director Sheil is encouraged that the plastics industry appears to be joining together and moving forward on recycling. "Unlike other industries [glass and paper], the plastics industry is a very diverse group, ...especially by the time you get to the end product," she says. "If the states hadn't put the pressure on, it wouldn't have happened." Although the industry has come a long way toward recycling plastics, comments Sheil, recycling plastics hasn't yet become real on a large scale, and it will remain to be seen what happens in the next few years.

In addition to getting involved in recycling through organizations such as PRF, NAPCOR, and COPPE, the plastics industry is expanding its own internal efforts. Many of the major resin manufacturers, such as Du Pont, Dow Chemical, Mobil Chemical, and Eastman Chemicals, have taken positions on and responsibility for solid waste issues and established corporate divisions to focus on plastics recycling and/or solid waste management. These divisions look into the governmental, environmental, and business aspects.

Of course, companies see the support and promotion of plastics recycling as vital to their positions in the marketplace and the growth of their companies. As happened with the aluminum can, the promotion of a product as recyclable lets the public perceive it in a positive light and helps to maintain a market. And joint ventures, acquisitions, and the building of recycling businesses offer the plastics industry new markets and promotional opportunities.

Mobil Chemical is a producer of polyethylene and polystyrene resins and a major manufacturer of polystyrene products. In addition to recycling all in-house scrap, the company purchases 100 million lb of polyethylene film scrap to use in the production of institutional trash bags. Robert J. Barrett, general manager of Mobil's Solid Waste Management Solutions unit, expects that business opportunities will continue to arise from the solutions to solid waste problems. An example is Mobil's $4 million joint venture company, Plastics Again, with Genpak of Glens Falls, N.Y.

Genpak, a plastic products manufacturer, and Mobil are in the process of opening and operating one of the first plants to recycle polystyrene foam items such as food containers, cups, and cutlery. The materials are being collected from Massachusetts schools and institutions by New England CRInc, a major reclamation firm and recycled materials end-use manufacturer. The plant has a capacity to recycle 3 million lb per year of polystyrene resin, which will be reused by the companies or sold to producers of insulation, fence posts, and flower pots. The new company is expecting a profit by 1992.

Similarly, Dow has entered into a joint venture agreement with Domtar of Canada to operate what the companies hope will be by 1990 a self-sustaining PET and HDPE recycling business. The North American company is expected to take postconsumer plastics and, using a proprietary Dow process, convert them into resins for use in Domtar's manufacturing or to be sold by Dow.

Other industry efforts to "close the loop" of product manufacture, recycling, and resin reuse include Procter & Gamble's marketing of a plastic household product container made from recycled PET. Johnson Controls, a major manufacturer of PET beverage bottles and containers, is also involved in promoting recycling efforts. Internally, the company recycles inhouse scrap and is working on proprietary R&D efforts in recycling. As a container manufacturer, Johnson Controls is also taking an interest in the design of packaging to make it more recyclable. A company spokesman indicates that approaches are being worked on to begin marketing nonfood containers from recycled materials. Manufacturers looking to use recycled resins are concerned with getting consistent quantities and quality of the materials to make these efforts economically feasible.

Although plastics have been targeted in the battle on solid waste, and both government and industry have responded, plastics are only one part of the growing solid waste problem. The importance of investigating alternatives in waste disposal is becoming obvious to all involved. And new organizations, such as the Council on Solid Waste Solutions of the SPI, are growing out of the plastics and other industries. Unlike PRF, which emphasizes plastics recycling, the Council looks into all aspects of the solid waste problem, focusing on reaching local, state, and federal governments to influence legislation and on evaluating waste treatment technologies.

Alternative solid waste treatment methods are being explored and plastics recycling will probably continue to be among them, because it doesn't seem likely that our society will entirely give up their plastic products. Morrow makes the important point that plastics should be integrated slowly into the recycling scheme, starting with beverage bottles and moving on to other types of packaging, to provide time for the collection, reclamation, and end-use infrastructure to grow. Forcing too much, too soon, into the existing structure could do more harm than good in terms of a collapse of the effort and backlash from the apparent failure to recycle plastics. The outlook on biodegradable plastics, as generally expressed by the plastics industry, is that they may only find limited applications in agriculture or in dealing with litter.

EPA's Office of Solid Waste has proposed that an integrated system of waste reduction, recycling, incineration, and landfilling be established. This view seems to be supported by most industries and many legislative and environmental groups, but with different emphasis on the different components. EPA has also set a goal for recycling 25% of solid waste nationwide.

Although the details and issues of solid waste management are both complex and controversial, it appears that after reducing the amount of waste through recycling and burning that which can be safely incinerated in waste-to-energy conversion facilities, the levels of materials left for landfills can be greatly reduced. Still left to be dealt with are the issues of safe and cost-effective incineration, air and ash toxicity levels and disposal, establishment of a nationwide recycling infrastructure, and development of efficient and environmentally safe landfilling methods.

Five resins are in common use in consumer packaging

Thermoplastics, which account for 87% of plastics sold, are the most recyclable form of plastics because they can be remelted and reprocessed usually with only minor changes in their properties. Five resins are commonly used in consumer packaging applications.

Polyethylene is the most widely used resin. High density polyethylene (HDPE) is used for rigid containers such as dairy and water jugs, household product containers, and motor oil bottles. Low density polyethylene (LDPE) is often used in films and bags.

Polyethylene terephthalate (PET) is used extensively in rigid containers, particularly beverage bottles for carbonated beverages.

Polystyrene (PS) is best known as a foam in the form of cups, trays, and food containers. In its rigid form it is used in cutlery.

Polyvinyl chloride (PVC) is a tough plastic often used in construction and plumbing. It is also used in some food, shampoo, oil, and household product containers.

Polypropylene (PP) is used in a variety of areas, from snack food packaging to battery cases to disposable diaper linings. It is frequently interchanged for polyethylene or polystyrene.

Chemical recycling of plastics is alternative to mechanical methods

In addition to mechanical methods of reusing plastics, such as molding or extruding, polyester resins also can be reclaimed through chemical means and used to produce new polymer compositions. Eastman Chemicals (a division of Eastman Kodak) and Freeman Chemical Corp. have led in developing the chemical recycling of polyester resins. By one of three reaction paths, polyethylene terephthalate (PET) can be used to produce polyols (for polyurethane foams) and unsaturated resins, or it can be broken down into its original starting components and repolymerized. Eastman Chemicals, through the Plastic Bottle Institute of the Society of the Plastics Industry, has provided detailed discussions of these reactions.

PET consists of repeating ethylene glycol and terephthalic acid (TPA) units joined together by ester linkages. Aromatic polyols, hydroxyl-terminated short-chain molecules consisting of TPA and glycol units, can be produced from PET in a one-step glycolysis reaction. The idea behind the reaction is very simple: High-molecular-weight PET is broken down into short-chain pieces through a catalytic reaction of PET and diethylene glycol. By heating a mixture of PET and diethylene glycol in the presence of a manganese catalyst, the polymer chain is cut into short fragments, and the ethylene glycol portions undergo exchange with the free diethylene glycol. This reaction is allowed to proceed until an equilibrium between the two glycols is reached and the PET has been reduced to short-chain polyols.

The aromatic polyols resulting from the reaction can be mixed with commercial polyols, blowing agents, surfactants, catalysts, and polymeric isocyanates to produce a rigid polyurethane foam. @n compared w@ control foams produced from commercially available polyester polyols, the foams produced from reclaimed materials were found to have essentially the same properties.

Freeman Chemical uses about 25 million lb of postconsumer PET bottles and film scrap to make polyols for the production of rigid foams. The company estimates that more than 50% of the laminate foam insulation used for construction is made from recycled material.

RECEIVED , 1990

Chapter 5

Plastic Degradability and Agricultural Product Utilization

J. Edward Glass

Polymers and Coatings Department, North Dakota State University, Fargo, ND 58105

Factors important in the commercialization of a process and in product acceptance in the areas of waste product disposal and increased use of agricultural materials are discussed. The relationships of market volumes to cost, achievement of cost performance through blending and the paradox of good compatibility and film properties with domain accessibility in the scenario of achieving degradability in commodity packaging are discussed. Greater use of agricultural products in which the thrust is not based on low cost secondary products is suggested, and an example of the role of noneconomic factors in the success of a new process and product is given in the concluding section.

Several years ago during the "oil shortage", when serious efforts to resolve the complexities of applying enhanced oil recovery processes were being made, a crude oil sample recovered from Chevron's Red Wash basin in Utah was analyzed as ca. 5% low molecular weight polyethylene. That initial analysis was hard for many on the project to believe, but several well equipped and staffed analytical departments confirmed its accuracy. Polyethylene is but one of several large volume commodity polymers produced above ground in billion pound quantities each year. It is very unlikely that the subterranean Utah polymer arose from pollution, but it does highlight the resistance of an all carbon backbone polymer to environmental degradation. The stability of commodity polymers is now recognized as an environmental problem in "advanced societies" and is one of the problems addressed in this text.

The second topic addressed is that of the American farmer. The remarks of one farmer at the Corn Utilization Conference in November, 1988 (Columbus, Ohio) defined the situation: "The soil and rainfall on my farm in Colorado are not good enough to compete with those in South America; we have to find new applications for agricultural products". Perhaps because he was not a traditional farmer (he had entered farming as a result of land acquisitions and resigned from a marketing position with an electronics company), he clearly saw the marketing need. One of the objectives of the annual Corn Utilization

0097–6156/90/0433–0052$06.00/0

Conference is to review progress in sponsored research for solution of both pollution and the farmer's problems by blending corn starch polymers with polyethylene (*1,2*). There are several problems in blending these two polymers of very different polarities.

Degradable Plastics

Market Restrictions. To evaluate the viability of commercializing carbohydrate/synthetic polymer blends, an understanding of the three laws of industrial polymer science must be appreciated. Any academic would find these laws, relative to the three laws of thermodynamics, repulsive; anyone with greater than five years of industrial experience knows their utility well. The three laws of industrial polymer science are:

1. The market volume of any polymer is inversely related to its price.
2. Avoid making a new polymer, keep blending the one with a long production history with other available materials, until the desired properties are obtained.
3. The way to improve the physical properties of any polymer is to decrease its price a few pennies a pound.

Blending of the lowest price commodity polymers from synthetic and carbohydrate polymer families [e.g., poly(ethylene) and starch] would appear to follow these laws. Although each polymer class is produced in large volume (first law), the production rate for corn starch/synthetic polymer blends is much lower than that for the synthetic polymer; this slower extrusion rate directly affects the final cost. Ignoring this limitation, the film properties of the blend are significantly poorer than those of the synthetic polymer film. Both deficiencies are related to the poor thermoplastic properties of water-soluble polymers such as corn-starch.

Technical Impositions. When chemically similar polymers of very low polarities, such as polyethylene and polypropylene, are mixed, there is no enthalpic interaction between the two nonpolar macromolecules, and phase separation occurs due to entropic restrictions (*3*). A compatibilizer, such as an ethylene/propylene block copolymer for the above blend, aids in compatibilization by residing at the interface of the two phases. The need for a compatibilizer was discovered early in the blending of corn starch and polyethylene (*2*); an ethylene/acrylic acid copolymer compatibilizing agent is generally used. The first alarm sounded in these studies was that this compatibilizer added too much to the cost of polyethylene/starch blends. The significant loss in production rate of such blends is mentioned infrequently, but the economics imposed by this parameter are significantly greater than the cost of the compatibilizer.

If economic restraints were not a limitation, there would still be the consumer acceptance factor to consider. A polyethylene/starch blend does not have the film properties (e.g., tensile strength) of a polyethylene film. One can approach biodegradable packaging through the inclusion of weak linkages (e.g., degradable to sunlight) in the hydrocarbon backbone, without the complication of adding starch. However, an ultraviolet unstable linkage serves no purpose if the material is buried in a landfill. The performance of polyolefins in the various consumer markets is dependent on their molecular weight and molecular weight distribution, which are determined in large part by the organometallic catalyst used in production. The incorporation of weak linkages such as one sensitive to ultraviolet radiation would necessitate definition of new catalysts and associated scaleup studies. The processability (primarily stability) at comparable times and temperatures of the "weak linked" polyolefin to the polyolefin also would need verification.

A second approach to biodegradable packaging is to blend polyethylene with a second synthetic polymer with polar repeating units that are capable of degradation, such as ester linkages (chapter 12). Poly(caprolactone) represents such a class of polymer, which has a long history of compatibility (*4*) with a variety of polymers and degradability (*5*); recently, improved miscibility and film properties have been reported when poly(caprolactone) is blended with commodity plastics (*6*).

Given a resolution of the major deficiencies, production rate and film properties, a third restriction with respect to polymer blends becomes evident: the film produced would degrade in proportion to the fraction of the poly(caprolactone) present and the extent of degradation would decrease with improving miscibility. The kinetics of degradation are influenced by the limited domain accessibility as compatibility of the polyolefin with the poly(caprolactone) increases. Domain accessibility is discussed for polyethylene/starch blends in chapter 8.

There are no realistic answers in this area nor any promising approaches on the horizons. The definition of an optimum polymer blend, however, is only the beginning. Whatever the degrading moieties, they function in a highly complex system. A proper study of the matrix interactions must be defined before a viable scenario is realized for resolution of the environmental problems. Such a matrix approach, in the absence of a polymer blend, is addressed in chapter 2. This type of system approach is not a standard technique in classical academic studies, but it is the standard operational approach in most industrial research efforts. The contributions of insect symbionts (chapter 3) and recycling of plastics (chapter 4) will make an impact, but without a breakthrough improvement in the degradability of plastic packaging it is unlikely that long-standing progress on the waste disposal problem will be realized. The breakthrough improvement, noting the deficiencies cited above, would seem to dictate a totally degradable plastic. The blending of degradable natural polymers with nondegradable polyolefins is not just objectionable in packaging. The ecological threat of such blends has been debated in disposable diapers and recently in a new total package, the "juice box," which is a natural/commodity plastic blend that also includes a metal layer.

Utilization of Agricultural Products

Historical Perspective. During the 1930s each new synthetic polymer found a ready application and the manufacturer profited from its production. In the late forties studies in the copolymerization of different monomers accelerated and in the fifties, the remaining unpolymerizable, low cost alpha-olefins were converted to macromolecules by Ziegler-Natta organometallic catalysts. In the late fifties, the realities of the three laws of industrial polymer science began to surface, which were reflected in the middle sixties by significant reductions in industrial staffs. Intermingled with synthetic polymer acceptance, a few carbohydrate polymers were accepted in commodity applications.

Nonnutrition Application Areas. With food consumption demands approached through increased agricultural productivity worldwide, the utilization of agricultural products in plastic applications is considered by many as a future avenue for American farm products. In the "shortage situations" of the seventies, the Japanese very competently demonstrated the usefulness of a variety of fermentation carbohydrate polymers as commodity plastic substitutes, but agricultural products outside of nutritional areas have not received wide acceptance. In many cases one encounters the biodegradation goal in plastic waste a shortcoming; the commercial use of nature's products (e.g., Chapter 21) are often found too biodegradable in many applications.

In the third section of this book, it is evident that agricultural products are directed at low cost segments of a given market for two reasons. The thrust is to replace an established synthetic product, and the agricultural material is almost always a by-product of an established farm product grown for food applications. These conditions demand that the secondary farm products be low cost. There are good contributions from three USDA laboratories in this book and some of the past USDA studies have made significant contributions to the use of agricultural products. The use, however, is nowhere near its potential. If this trend is to be broken, the growth of a specific product for a specific market appears to be a requirement. The question of the biodegradability of agricultural products (compared to petroleum based products) and the reliability, both in terms of availability (again when compared to petroleum based products), and uniformity of compositions appear to be some of the primary determinants in agricultural product growth outside of nutritional areas.

Both degradable packaging and the greater use of agricultural material require a process for the product's production. This topic is considered in the next section.

Realization of a Commercial Process

Will environmental factors bring about the goals of the chapters addressed in this book? It is unlikely, for it is unwise to make business decisions on environmental factors. The following is an appropriate example. In the mid-sixties, emphasis was placed on the production of powder coatings by many industrial organizations, some of whom were major solvent suppliers to the coatings industry. They were concerned about the California Rule 66 pollution control law on solvent emissions. As the seventies were approached the research efforts in powder coatings became nonexistent among solvent suppliers. As we approach the mid-nineties, powder coatings is "the fastest growing segment of the coatings industry" but this segment is still a minor percentage of the total coatings business. It might be argued that the latex technology developed in the forties took 25 years of incubation to achieve commercial reality in latex coatings in the sixties, and that success will be realized in powder coatings in the nineties.

Is profitability an inducement to realizing commercialization? Not necessarily, as illustrated by another example of a reaction to environmental influences in the coatings industry. Lead was removed from alkyd coatings over a decade ago. In the seventies, removal of mercuric biocides from aqueous latex coatings was advocated. The use of phenylmercuric acetate inhibited the enzymic degradation of hydroxyethyl cellulose. A sequential process (*7,8*) to provide greater substitution at the O-2 carbon position of the repeating glucopyranosyl units of cellulose negated the need for mercuric biocides in latex coatings. The sequential process had a few drawbacks in that a threefold increase in reaction time was required and the increased degree of substitution produced a more hydrophilic surface, which created minor problems in product recovery and storage. These deficiencies were more than off-set by significant increases in adduct addition efficiencies and solvent and product recovery costs (*8*), but that did not result in a new process modification. Business team decisions are dominated by marketing and sales personnel. Their annual performance ratings are highlighted by new market penetrations, not necessarily from implementation of research contributions. The new process is now a commercial realization (*9*), not directly related to a greater profit and a more environmentally sound process, but because it regained market share lost from not producing the less biodegradable cellulose ether.

The Non-Cost Factor

There are many examples where commercial success of a new product and process are not achieved even when all the elements of an industrial business team are pulling together. The latest appears to be Group Transfer Polymerization. This is unfortunate for it is a beautiful piece of technology, but it is understandable from the laws of industrial polymer science presented previously.

An example of the complexity in achieving a commercial success, more related to the environmental theme of this book is the Q-resin technology developed for the production of porous poly(vinyl chloride) [PVC] particles by a suspension process. This technology (*10*) was developed to produce polymer particles with a porous surface to aid rapid penetration of the particle by commercial plasticizers when "dry-blended". Particles by a conventional process contain pericellular skins that in most cases retarded plasticizer penetration in "dry-blend" (*11*) processes, and in the final film, nonplasticized "fisheyes" are observed. During the time of its commercial run, the carcinogenic nature of the vinyl chloride monomer was declared. Limits on residual monomer in a given production polymer were imposed; a lower limit that a more porous particle could help facilitate. The Q-resin process failed to achieve real commercial success because of the mind set of industrial formulators: fisheyes would be present if the particle size was above 200 microns. The Q-resin technology produced particles that were essentially fisheye free after application of the PVC/plasticizer dry-blend, but the resin's initial median size was slightly greater than 300 microns.

The best perspective on the efforts of both thrusts of this book are reflected in the TV response of a McDonalds representative. Under questioning he responded that McDonalds was going to fast food containers that were thinner. He then raised a question and volunteered an answer. "Is going to thinner packaging part of a public relations effort? Maybe, but we are trying". The contributions in this text will not resolve the two major problems addressed, but the efforts are serious ones that hopefully will catalyze support for additional studies that may provide more substantive answers.

Conclusions

Many factors are involved in determining a commercial process and product's acceptance. The chapters that follow should be read with this understanding. The technical contributions and financial support in areas addressed by this book are much less than that expended in development of enhanced oil production in our "shortage years." If I had a choice for an illustration for this text, it would be a photo on display in the London Museum of Natural History in 1989 taken by a young American. A bald eagle is photographed atop a mound of garbage bags covered with a second type of transparent commodity plastic. The mountains of Alaska are in the background. Plastic waste and agricultural product utilization are not seen as major crises. In the perspective of other social problems, they are not.

The crisis efforts in the shortage years of the seventies did not lead to a technology that would have delivered us if the oil crisis had really existed. It is hoped that the topics covered in this text will stimulate a sustained effort in the areas covered so that progress can be realized for a critical situation that may develop in the future.

Literature Cited

1. *Proceedings of the First Annual Corn Utilization Conference*, June 11 and 12, 1987, National Corn Growers Association, St. Louis, MO.
2. *Proceedings of the Corn Utilization Conference II*, November 17 and 18, 1988, National Corn Growers Association, St. Louis, MO.
3. Olabisi, O.; Robeson, L. M.; Shaw, M.T. *Polymer-Polymer Miscibility*, Academic Press: New York, NY, 1979.
4. Paul, D.R.; Barlow, J. W. *J. Macromol.Chem.* **1980,** *C18*(1), 109-168.
5. Potts, J. E. *Plastics Engr. Tex. Pap.* **1975,** *217*.
6. Private communication from one of the chapter reviewers of this text.
7. Glass, J. E. ; Buettner, A. M. ; Lowther, R. W. ; Young, C. S. ; Cosby, L. A. *Carbohyd. Res.* **1980,** *84*, 245-263.
8. Glass, J. E. ; Lowther,R. G., U.S. Patent 4 084 060, 1978.
9. Cellosize Enzyme, Resistant Hydroxyethyl Cellulose, 1984 Product Bulletin, Union Carbide Corp., Old Ridgebury Road, Danbury, CT 06817.
10. Nelson, A. R.; Floria, V. E. Br. Patent 1 195 478, 1970.
11. Glass, J. E.; Fields, J. W. *J. Appl. Polym. Sci.* **1972,** *16*(9), 2269-2290.

RECEIVED March 29, 1990

DEGRADABILITY OF COMMODITY PLASTICS AND SPECIALTY POLYMERS

Chapter 6

Polyethylene Degradation and Degradation Products

Ann-Christine Albertsson and Sigbritt Karlsson

Department of Polymer Technology, The Royal Institute of Technology, S–100 44, Stockholm, Sweden

The increasing use of different polymers in materials such as packaging etc. means increasing problems of disposing of garbage. Plastic waste is a considerable part of the garbage and demands are put on using degradable materials as well as increasing the possibility of recycling.

Predicting the changes in long-term properties of new polymers is therefore even more important than before. Parameters which must be controlled are the rate of degradation and the evolution of low molecular weight degradation products. The degradation rate is important in case of using the plastic waste as landfills. It is somewhat contradictory to use degradable waste as landfill, it will be quite difficult using land for example to build on if the plastic waste is degradable.

Various techniques are used for studying the degradation of polymers. Some of the more important ones are size-exclusion chromatography (SEC) for measuring of the molecular weight changes during degradation, infra-red spectroscopy (IR) for following changes in for example carbonyl-index and liquid scintillation countings (LS) following evolution of carbon-14-dioxide ($^{14}CO_2$), a measurement of degradation. LS is a very convenient and sensitive method which has been developed and described in a series of papers (1, 2, 3). Even small degradation rates can be monitored, below 0.1%, which is about 100 times better than traditional methods.

Chromatography is a relevant technique when studying the evolution of low molecular weight degradation products. Using gas chromatograpgy (GC) and liquid chromatography (LC) it was possible to monitor the biodegradation of casein (biopolymer) used as additive in some cement products (4, 5, 6).

0097–6156/90/0433–0060$06.00/0

Arbin et al. (7) have studied the contamination of intravenous solutions from polyvinylchloride (PVC) bags using LC-diode-array detection. Examples of products identified in the solutions were different phthalates used as plasticizers. This is one example of the enormous importance of controlling what type of degradation products evolved during degradation of polymers. Phthalates are known to be carcinogenic and the levels of phthalates in for example blood-bags were often too high. Nowadays the manufacturers have managed to lower this level. It is equally important to monitor the degradation products of polymers used as packaging. Different additives are incorporated in otherwise inert polymers in order to make them degradable (in abiotic as well as biotic environments).

Biodegradable additives

Griffin (8, 9, 10) introduced the idea of increasing the biodegradability by adding a biodegradable additive to the polymer material. By mixing biodegradable biopolymers such as starch with PE and studying the degradation of LDPE film in compost he found that the autooxidation was enhanced.

When a biodegradable additive is employed, microorganisms can easily utilize the additive. The porosity of the material is thereby increased and a mechanically weakened film is obtained. The surface area will be increased, and this film will be more susceptible than the original film to all degradation factors including biodegradation.

Mixing 10% $C_{32}H_{56}$ with high density polyethylene (HDPE) powder gave a film containing an additive that will be degraded by fungi (11). Over a period of two years then degradation of this film was followed and compared with pure PE film containing ^{14}C. Using the $^{14}CO_2$ LS method it was concluded that the liberation of $^{14}CO_2$ was slightly increased by the additive during the first years, but somewhat retarded thereafter. In combination with other degradation factors, the biodegradable additive presumably has a positive effect on the degradation factors.

Molecular weight and degradation

Paraffins may not only be additive in PE but can also be regarded as the low molecular counter part of synthetic polyolefins. Several groups have performed studies on the biodegradation of alkanes. Jen-Hao and Schwartz (12) were probably the first to claim that the number of bacteria that PE was able to support was dependent on the molecular weight of the polymer.

Initial photooxidation and several other factors will diminish the molecular weight of polymers thereby releasing low molecular weight portions of polymeric chains which eventually can be biodegraded. The total environmental effect on polymers is therefore put together by several separate degradation factors working synergistically towards deterioration of polymeric materials outdoors (2).

Degradation products of PE

In connection with an ongoing project aiming at evaluating and developing new methods for studying the low molecular weight degradation of polymers a preliminary abiotic degradation of LDPE was performed. LDPE with a melt index (MI_2) of 45 g/10 minutes and a thickness of 150 μm was kept in hot water (95°C) in three different pH (0.5, 6 and 12.5). LDPE samples were withdrawn at regular intervals and studied in fourier transform infra-red spectroscopy (FTIR) and in differential scanning calorimetry (DSC) (a complete description of treatments, instrumental, etc. is described in a paper under preparation) (13). The water fraction was also withdrawn and extracted with diethylether.

The ether extracts were analysed in GC showing increasing amount of degradation products as the degradation times were increased. Series of low molecular weight compounds were evolved at the same time as the crystallinity increased from about 32 % to 43 % during 40 weeks of degradation (13). Initially the amorphous part of the PE chain is degraded thereby giving an increase in the crystallinity. It is also possible that water diffuses into the polymer leading to a more ordered structure and a higher crystallinity.

During the degradation the keto-carbonyl groups increased in amount as obtained by attenuated total reflectance (ATR)-FTIR. In the basic environment a growing peak corresponding to carboxylate anion is observed (13).

After 15 days of degradation only low amounts of degradation products have evolved, the amount being lower than the detection limit of the GC system. After 17 weeks, however, 2-butanol, propionic acid, 1-pentanol, butyric acid, valeric acid and caproic acid were detected in pH 6 water fraction. After another 20 weeks several alkanes could also be detected: n-octane, n-nonane, n-decane, n-dodecane, n-tridecane and n-tetradecane (13).

Carlsson and Wiles have in an early work (14) discussed the ketonic oxidation products of PP films. The volatile products were analysed in GC with a flame ionization detector (FID) and a thermal conductivity detector (TCD) giving the major oxidation products carbon monoxide and acetone. Other products detected were water, formaldehyde, formic acid, propane, acetic acid and iso-propylalcohol.

The monomeric yield from thermally degraded PE is very low, less than one percent. The mechanism is a random scission type giving many volatile degradation products with a very complex pattern. This makes analyses and identification of low molecular weight degradation products of PE very difficult and demands skilful and careful emthod developments.

Important areas for the study of PE and its degradation products are in particular the packaging industry. Demands are put that all packaging materials should be degradable, *i.e.* its life-time should be controlled and in this context it is important to show degradation products.

Degradation mechanisms

Photooxidation increases the low molecular weight material by breaking bonds and increasing the surface area through embrittlement. The hydrophicility is also increased. The formation of carbonyl groups, usually expressed as carbonyl index is one way of monitoring the photooxidation effect. Thermolytic degradation of PE starts with initiation: scission of primary bonds randomly or at chain ends. The depropagation occurs mainly through intermolecular or intramolecular hydrogen abstraction and the termination proceeds by combination or disproportionation.

The accelerated degradation briefly described in this work is a combination of different degradation mechanisms. A photooxidation is unavoidable and by keeping a moderately high temperature (95°C) a thermolytic degradation can be accounted for. The influence of different pH will primarily affect the low molecular weight compounds giving rise to several organic acids besides the hydrocarbons to be expected from the thermolytic degradation mechanism.

Conclusions

A brief description of evolved low molecular weight degradation products of LDPE has been given. The experiment was conducted in order to evaluate the usefulness of GC for detecting degradation products of polymers. Further studies will be performed in order to improve the possibility of studying degradation products of degradable polymers and also discuss how different biodegradable additives will affect the degradation product pattern.

References

1. A-C. Albertsson J. Appl. Polym. Sci., 22, 3419 (1978).
2. A-C. Albertsson: Advances in Stabilization and Degradation of Polymers, Volume 1, A. Patsis (ed.), Technomic Publishing Co., Lancaster, Pennsylvania, 1989.
3. A-C. Albertsson and S. Karlsson: J. Appl. Polym. Sci., 35, 1289 (1988).
4. S. Karlsson, Z.G. Banhidi and A-C. Albertsson: J. Chrom., 442, 267 (1988).

5. S. Karlsson, Z.G. Banhidi and A-C. Albertsson: Materials and Structures, 22, 163 (1989).
6. S. Karlsson and A-C. Albertsson: Materials and Structures, (in press) (1990).
7. A. Arbin, S. Jacobsson, K. Männinen, A. Hagman and J. Östelius, Int. J. Pharmaceut., 28, 211 (1986).
8. G.J.L. Griffin, British Pat. Appln. No. 55 195/73 (1973).
9. G.J.L. Griffin, J. Polym. Sci., Poly. Symp., 57, 281 (1976).
10. G.J.L. Griffin, Pure Appl. Chem., 52, 399 (1980).
11. A-C. Albertsson and B. Rånby: in J.M. Sharpley and K.M. Kaplan (eds.) Proc. of the 3rd Int. Biodegrad. Symp., Appl. Science, London, 1970, p. 743.
12. L. Jen-Hao and A. Schwartz, Kunststoffe, 51, 317 (1961).
13. S. Karlsson and A-C. Albertsson, Manuscript under preparation.

RECEIVED February 16, 1990

Chapter 7

Biodegradation of Starch-Containing Plastics

J. Michael Gould, S. H. Gordon, L. B. Dexter[1], and C. L. Swanson

Biopolymer Research Unit, Northern Regional Research Center, Agricultural Research Service, U.S. Department of Agriculture, Peoria, IL 61604

Plastic composites containing high levels of gelatinized corn starch (20-40% dry weight basis) in combination with petrochemical-based, hydrocarbon polymers such as polyethylene (PE) and poly(ethylene-co-acrylic acid) (EAA) are being developed in an effort to facilitate the breakdown, or metabolism, of the composites by living organisms. Starch in films composed of starch/EAA and starch/PE/EAA was resistant to degradation by some amylolytic bacteria tested, but was readily metabolized by others. Spectroscopic analysis of starch-containing plastic films indicated that, in some cases, almost all of the starch component could be removed from the film within 20-30 days by these organisms. PE and EAA were not affected by the microorganisms tested after up to 60 days incubation. Tensile strength of the films decreased as the starch component was metabolized, facilitating physical disintegration of the films by mechanical actions.

The persistence of petrochemical-based plastic materials in the environment beyond their functional life has resulted in a broad range of pollution, litter, and waste disposal problems for modern society. Research to alleviate these problems includes efforts to develop plastics that degrade more rapidly in the environment. Plastic formulations have already been developed that degrade chemically as the result of exposure to sunlight (1). Recently, plastic formulations containing from 6% to 50% starch from corn have been studied as part of an effort to increase the susceptibility of plastic products to biological degradative processes (2-6). There is presently very little published information on the rates, extents and mechanisms by which plastics containing starch are degraded by living systems. In fact, there is some confusion about the definition of the term "biodegradable" as it applies to these plastics. Clarification of this situation has become imperative since governmental and legislative initiatives are already underway to regulate the use of "biodegradable" and "non-biodegradable" plastics.

[1]Current address: Canadian Harvest USA, 1001 S. Cleveland Street, P.O. Box 212, Cambridge, MN 55008

For purposes of this paper, the term "biodegradable" is defined simply as the ability of the molecular components of a material to be broken down into smaller molecules by living organisms, so that the carbon contained in that material can ultimately be returned to the biosphere. Thus, starch is biodegradable because it can be readily metabolized by a wide array of organisms. Polyethylene, on the other hand, is not readily metabolized by living organisms, and so must be considered to be, for all practical purposes, essentially non-biodegradable. Plastic formulations containing both starch and polyethylene can therefore be considered to be composites or mixtures containing both biodegradable and non-biodegradable components. Confusion can arise when these starch-plastic composites are referred to as "biodegradable plastics". The use of such terminology is based upon the observation that microbial removal of starch from starch-plastic composites can cause severe reduction in the mechanical strength of the remaining, non-biodegradable portion, so that the manufactured plastic product disintegrates readily into smaller pieces (Figure 1). If appropriate chemical catalysts (e.g. divalent metal ions) are present, however, even the non-biodegradable polymers can be chemically oxidized to lower molecular weight compounds, some or all of which may be further metabolizable by living organisms. However, the chemical nature of these lower molecular weight chemical degradation products has not yet been determined. Consequently, their degree of biodegradability is at this point largely unknown.

Two different technologies for the production of starch-plastic composites are currently being studied. Griffin (6) developed a process for incorporating granular starch particles into plastic films, producing a textured product with improved printing properties (Figure 2). Films produced with this technology typically contain 6-12% starch by weight. Otey (2-5) developed a process for incorporating gelatinized starch into plastic films, yielding a more uniform distribution of starch molecules throughout the starch-plastic matrix (Figure 3). Films produced with this technology typically contain 20-50% starch by weight.

Starch-plastic composites produced by both technologies have been claimed to be biodegradable, although there is currently no acceptable standard laboratory method for determining the relative rates and extents by which the starch component in these composites is metabolized by microorganisms (7). There is a growing need for this type of information by legislative bodies and regulatory agencies, as well as by those involved in the development and marketing of starch-plastic composites. For example, the U.S. Food and Drug Administration (7) has indicated the need for more information in each of the following four areas: 1) whether a polymer is degradable and, if so, under what conditions and in what time-frame; 2) potential for increased environmental introduction of degradation products and additives from a degrading polymer; 3) potential effects of small pieces of the degrading plastic in terrestrial and aquatic ecosystems; and 4) the impact of degradable plastics on recycling programs. The studies reported here are intended to address only the first of these areas. We have been developing a simple laboratory assay for biological degradation of starch-containing plastics using the following criteria: a) the assay system is self-contained and employs a defined media, allowing for complete quantification and analysis of degradation products; and b) the assay utilizes natural biological processes and organisms, optimized specifically for the assay. In this paper some preliminary results are reported on the use of this assay to characterize starch-plastic films produced by the Otey technology.

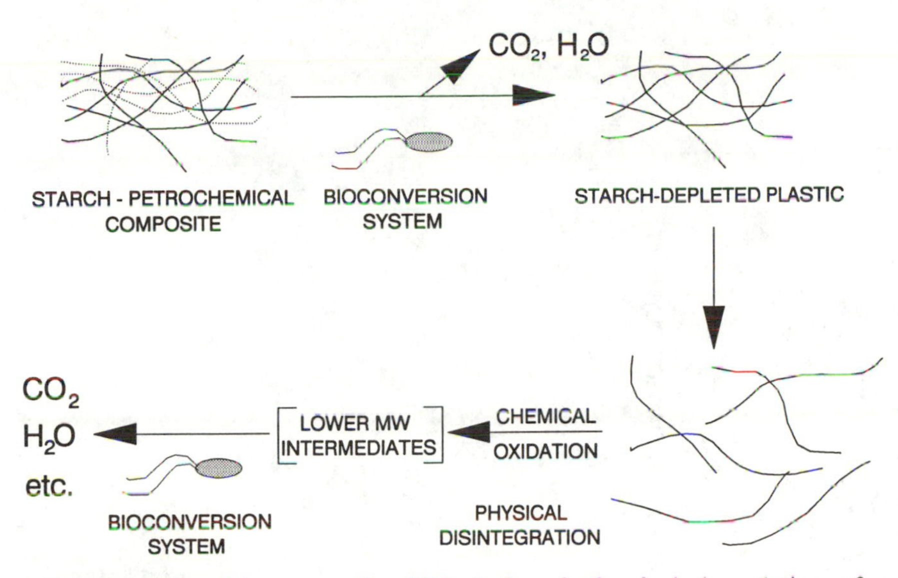

Figure 1. Possible routes for biological and chemical degradation of starch-plastic composites. Note that direct biological degradation of petrochemical-based polymers does not occur. Rather, these polymers must first undergo chemical degradation to form as yet uncharacterized, lower molecular weight intermediates.

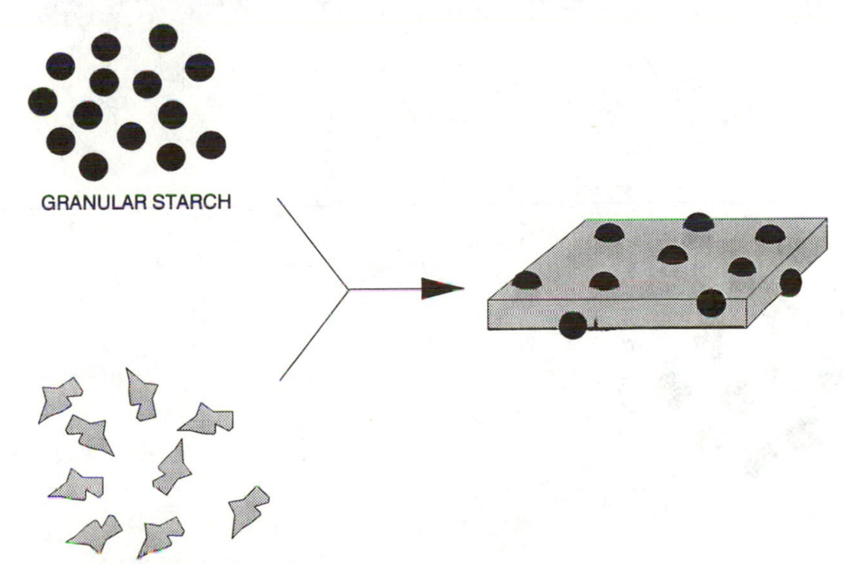

Figure 2. Method for producing plastic composites containing granular starch. See text for details.

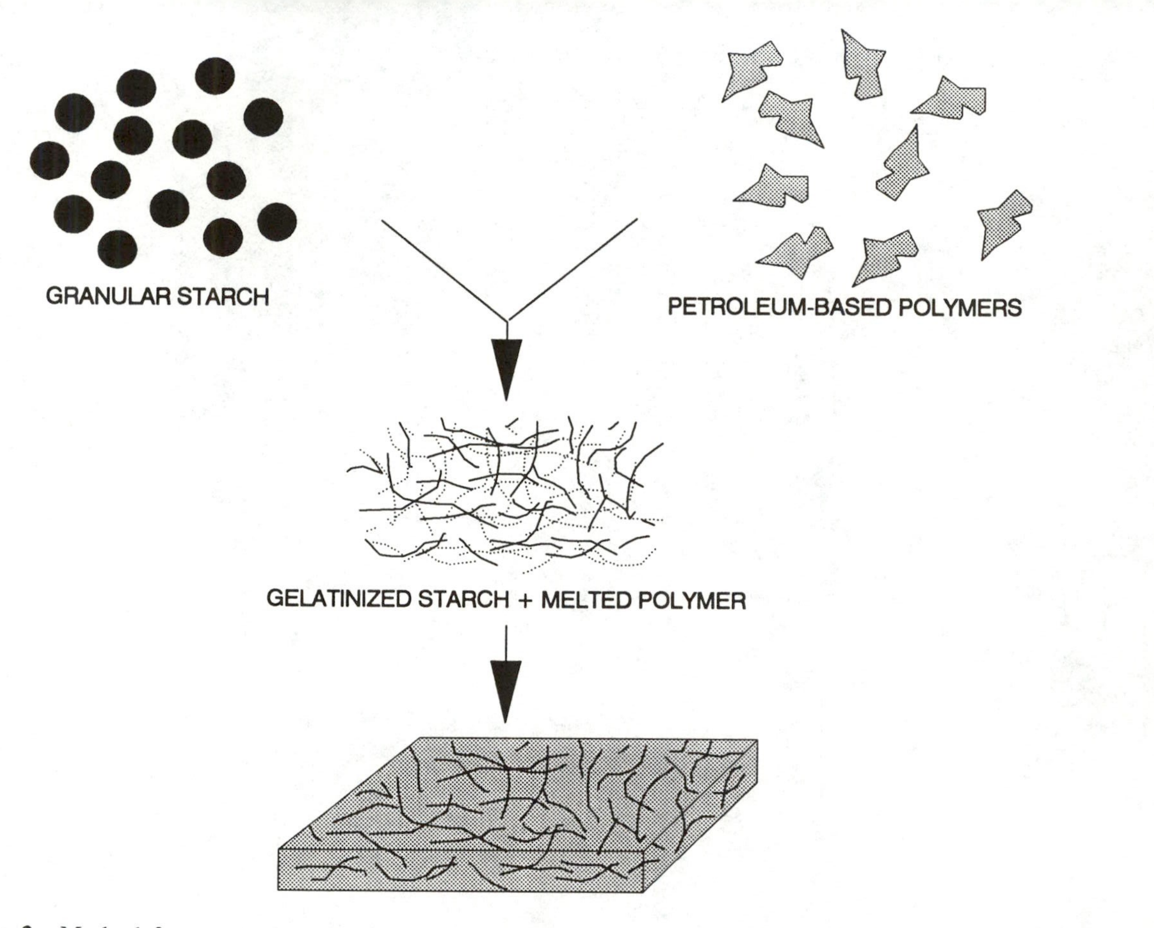

Figure 3. Method for producing plastic composites containing gelatinized starch. See text for details.

Materials and Methods

Plastics. The plastic films used in this study were prepared by the Otey semi-dry method (3), and were formulated to contain (dry weight basis) 40% starch, 45% poly(ethylene-co-acrylic acid) (EAA), 15% urea (starch/EAA plastic) or 40% starch, 25% EAA, 25% low density polyethylene (PE), 10% urea (starch/PE/EAA plastic). Films were blown with a Brabender Plasticorder extruder as described in detail elsewhere (3).

Microorganisms. Some of the bacteria used to degrade plastic films in these studies were obtained from selected environmental sites by employing a proprietary selection and enrichment process. Three different mixed cultures of these proprietary bacteria were used, and are identified as cultures LD54, LD58 and LD76. These mixed cultures contained 7 (LD54, LD76) or 8 (LD58) individual isolates. These isolates have not been fully characterized taxonomically. Preliminary studies indicate that many of the isolates are Gram-negative rods, although Gram-positive rods typical of the genus *Bacillus* are also present. All three mixed cultures used here are stable, and have been maintained in the laboratory for more than 2 years. Bacteria were grown in sterile 250 mL Erlenmeyer flasks containing 100 mL of a defined medium. Plastic films were cut into strips 1.2-1.3 cm wide and about 20 cm long. Each strip was numbered, and its average thickness determined with a micrometer for subsequent tensile strength measurements. Strips equivalent to 0.75 g (4-6 strips) were added to each culture flask after being sterilized by soaking for 30 minutes in a solution of 3% H_2O_2, followed by several rinses in sterile distilled water. Flasks were inoculated with 2 mL of an actively growing stock culture, and were incubated at 28°C with gentle agitation provided by a gyrorotary shaker (90 rpm).

Analytical Methods. Plastic strips were recovered from the culture flasks after the desired period of incubation, rinsed carefully in distilled water, and air-dried at constant temperature and humidity (25°C, 50% humidity) for 5 days (final moisture content = 5%). The air-dry strips were weighed, and their tensile strength determined with an Instron Universal Testing Machine (3). Small pieces of the strips were also subjected to FTIR spectroscopic analysis in an Analect RFX-75 spectrometer. KBr pellets of the plastic strips were prepared by freezing under liquid N_2 a 1.33 mg plastic sample in a stainless steel vial (Wig-L-Bug) containing two stainless steel balls. The vial was shaken rapidly for 20 seconds, refrozen in liquid N_2, and shaken again for 20 seconds. This procedure was repeated as needed to convert the film into a fine powder. The powdered film sample was mixed for 10 seconds with 400 mg powdered KBr in the same vial, and a 300-mg portion of the mixture was pressed to form the KBr disk. Spectra with 4 cm^{-1} resolution were obtained by averaging 32 interferometer scans by means of Happ-Genzel apodization.

Results

Starch-plastic composites contain a mixture of two very different types of materials: (*i*) hydrophobic, petrochemical-derived polymers (PE, EAA) known to be highly resistant to degradation by living organisms, and (*ii*) a hydrophilic, natural polymer (starch) that is easily broken down by a wide array of organisms. In the process developed by Otey (3), these fundamentally incompatible materials are forced into an intimate mixture during production of the plastic film. Since starch does not occur in nature embedded in a hydrophobic matrix, it is perhaps not too surprising that some highly amylolytic microorganisms, such as *Azotobacter chroococcum* and *Bacillus subtilis*, did not remove significant amounts of starch from the starch-plastic composites, even after several weeks incubation (data not shown). The films tested were also found to be resistant to degradation by even high levels of commercial amylase enzymes. To address these problems, we developed in our laboratory a number of proprietary bacterial cultures that are capable of rapid, sustained growth and metabolism using starch-plastic films as carbon source.

The effects of three different cultures of these bacteria on strips of starch/PE/EAA plastic film are shown in Figures 4-6. During the first 20-30 days of incubation the weight of the plastic strips declined steadily, until about 40% of the original weight was lost. After that point, no further weight loss was observed for up to at least 60 days of incubation. Although the initial decrease in film weight was accompanied by a concomitant reduction in the average tensile strength of the plastic strips (Figure 5), the gross physical appearance of the film strips was essentially unchanged for the first 20-30 days. Similar results were also obtained with starch/EAA plastic films. Scanning electron microscopic examination of the films after exposure to bacterial cultures for >30 days revealed the presence of small pits and other areas of deterioration on the surface of the film (8) and the presence of numerous small pieces of plastic, ranging in size from $<1\mu$ to $>10\mu$, in the growth medium.

Changes in composition of the plastic films resulting from bacterial action were investigated by Fourier transform infrared (FTIR) spectroscopy. FTIR spectra of the individual components used in film production are shown in Figure 6. The spectrum of low density PE exhibited characteristic C-H stretching bands at 2851 cm^{-1} and 2921 cm^{-1}, and a weaker C-H bending absorbance at 1468 cm^{-1}. There were no absorption bands in the fingerprint region of the spectrum (1400-900 cm^{-1}). A small band at 721 cm^{-1} was also present. The spectrum for EAA was similar to PE, with the addition of a C=O stretching band at 1705 cm^{-1} and some very minor bands in the fingerprint region. In contrast, corn starch exhibited a broad O-H stretching absorbance centered around 3400 cm^{-1}, a minor C-H stretching band at 2921 cm^{-1}, and a characteristic set of strong C-O stretching bands between 960 and 1190 cm^{-1}.

The FTIR spectrum for starch/PE/EAA film after 24 hours incubation in sterile medium (Figure 7) was essentially a composite of spectral features from each of the individual film components. Urea used during film preparation (3) was not apparent in the film's FTIR spectrum because it was extracted from the film during the aqueous sterilization process (not shown). Even after several weeks in sterile liquid medium, no apparent alteration in the film's FTIR spectrum was observed. When the starch/PE/EAA films were incubated in the presence of the proprietary culture of amylolytic bacteria, however, there was a dramatic loss of O-H and C-O absorption bands, indicating removal of starch from the plastic film. Absorption bands attributable to PE and EAA remained essentially unchanged. Because starch was the only component of the plastic that exhibited significant absorption bands in the 960-1190 cm^{-1} region, it was possible to quantitatively estimate the relative amount of starch present in the films by integrating total IR absorption over this spectral range. The data presented in Figure 8 show the approximate time-course for starch disappearance from the plastic films during their incubation with a bacterial culture. Essentially all of the starch was removed within 20 to 30 days incubation with the bacteria. These spectral data are in good agreement with the data presented in Figure 4 on the loss of weight from the plastic.

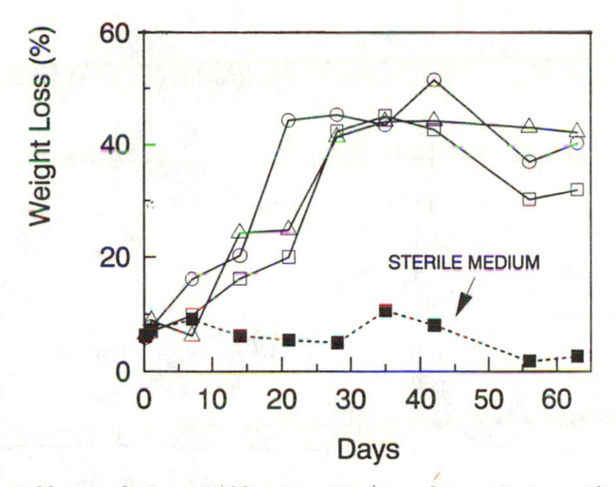

Figure 4. Effects of three different cultures of starch-degrading bacteria on the weight of starch/PE/EAA plastic films incubated in liquid culture. The bacterial cultures used (identified as LD54 (○), LD58 (□) and LD76 (△)) are composed of proprietary starch-degrading bacteria isolated by the USDA.

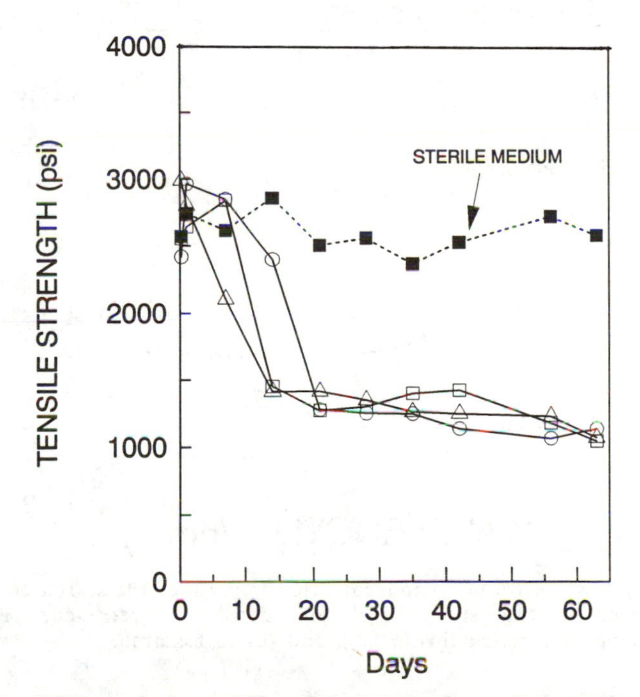

Figure 5. Effects of three different cultures of starch-degrading bacteria on the tensile strength of starch/PE/EAA plastic films. See legend to Figure 4 for details.

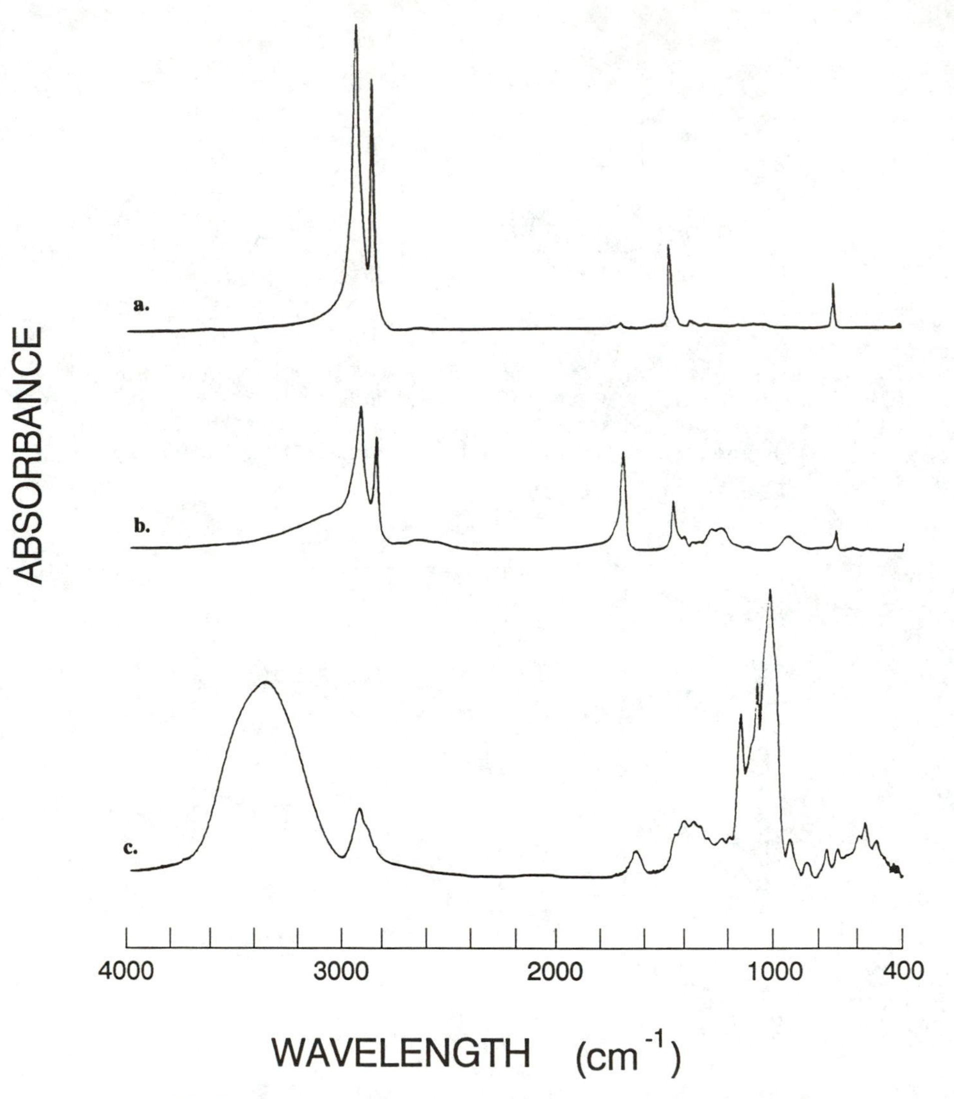

Figure 6. FTIR spectra of components used to produce the starch-containing plastics used in this study: (a) low density polyethylene (PE); (b) poly(ethylene-co-acrylic acid) (EAA); and (c) corn starch.

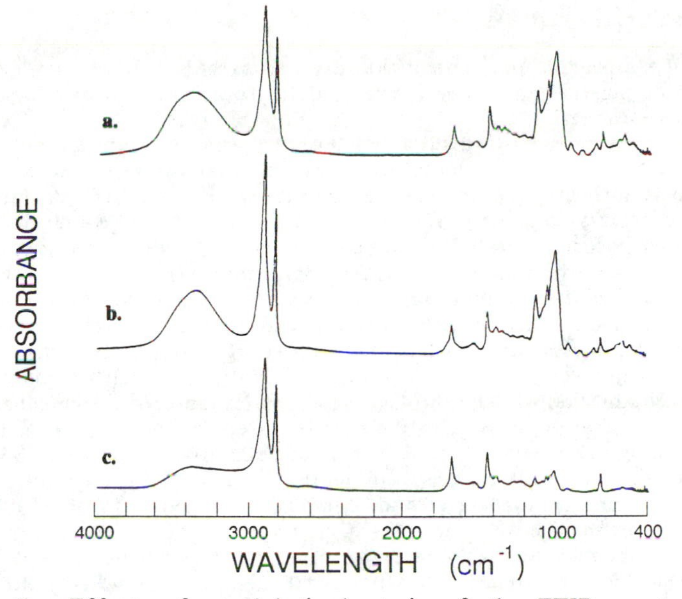

Figure 7. Effects of amylolytic bacteria of the FTIR spectrum of starch/PE/EAA plastic films: (a) film incubated for 24 hours in sterile medium; (b) film incubated for 28 days in sterile medium; and (c) film incubated for 28 days in medium inoculated with LD76.

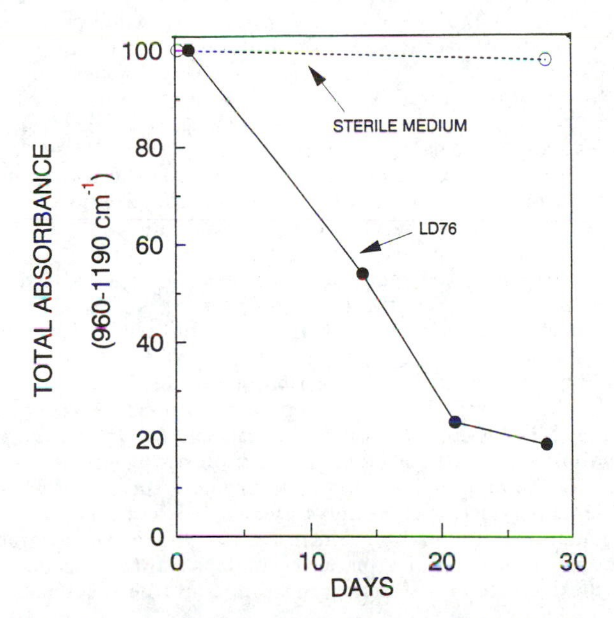

Figure 8. Rate of starch disappearance from starch/PE/EAA plastic films incubated with a culture of starch-degrading bacteria (LD76). The proportion of starch remaining in the film was estimated from FTIR spectra as the integrated absorbance over the range 960-1190 cm^{-1}.

Discussion

The ability to predict and control the rate and extent of starch biodegradation from starch-plastic composites is prerequisite for successful development and commercial production of these products. For this reason we have focused our attention on development of a suitable laboratory assay procedure for determining the susceptibility of starch-containing plastics to microbial processes. Such an assay should ultimately be standardized, so that work from different laboratories can be accurately compared, and should be based upon microbial degradation processes that occur naturally in the environment. However, it is also desirable that the time required for an individual determination be as short as possible. For the starch-plastic formulations we have studied to date, the rate of starch breakdown by many common starch-degrading enzymes and bacteria we tested was too slow to be of practical use in a laboratory assay. The starch-degrading bacterial cultures we employed in this study were specifically developed for their ability to rapidly degrade starch in starch-plastic composites, and so can be considered to simply compress in the laboratory the time required for such degradation to occur in the natural environment. Therefore, the time required for degradation of starch in a particular starch-plastic composite must be taken only as an indicator of the relative rate of degradation of that composite compared to other composites subjected to the same assay. The actual time required for biodegradation of starch in any starch-containing plastic in the environment depends upon a host of highly variable ecological, climatic and biological factors, and will be difficult to predict accurately.

The data presented here provide a direct demonstration that the starch portion of a starch-plastic composite can be metabolized by microorganisms. We have found no evidence that the presence of starch in the plastic matrix increased the susceptibility of hydrophobic polymer moieties (PE or EAA) to attack by the bacterial strains tested, within the 60-day maximum incubation period used in these studies. Given the known resistance of these polymers to microbial attack, it is reasonable to conclude that starch-plastic composites consist of two distinct milieus, only one of which (starch) can be considered to be biodegradable within a reasonably short period of time. Nevertheless, removal of starch from the polymer matrix did reduce the tensile strength of the plastic dramatically. Such alterations in physical properties of a plastic material resulting from biological actions on the starch component can be beneficial, for example by facilitating disintegration of the plastic by mechanical actions. This could be a very useful property for non-recycled plastic products such as bags for yard waste, for products that contribute to environmental litter, and for specialized applications such as planter pots and agricultural mulch films.

Our preliminary studies suggest that the amount of starch present in a starch-plastic composite may influence the rate and extent of its physical disintegration in the environment. Production of plastics in which petrochemicals, derived in part from imported oil, are replaced with starch from domestically produced, surplus corn is a highly worthwhile national goal. However, caution must be used in claiming that all starch-containing plastic products are completely biodegradable. In the absence of catalysts to promote the chemical oxidation of non-biodegradable polymers such as polyethylene, it is conceivable that a plastic formulation containing 10% starch could in fact be 90% non-biodegradable. More accurately, the actual percent of a plastic formulation that is biodegradable, and perhaps the relative rate of its biodegradation, could be specified.

At this point little is known about the interrelationships between composition, structure, starch-degradation and physical disintegration properties of starch-plastic composites. Continued work towards development of a laboratory assay for biodegradability will eventually result in the establishment of a sufficient database to elucidate these relationships, allowing development of a host of starch-containing plastic products for both existing and new markets.

Literature Cited

1. Johnson, R. *Proc. Symp. on Degradable Plastics, Soc. Plastics Industry, Inc.*, 1987, p 6-12.
2. Otey, F. H.; Westoff, R. P.; Doane, W. M. *Ind. Eng. Chem., Prod. Res. Dev.* 1980, **19**, 592-595.
3. Otey, F. H.; Westoff, R. P.; Doane, W. M. *Ind. Eng. Chem. Res.* 1987, **26**, 1659-1663.
4. Otey, F. H.; Doane, W. M. *Proc. Symp. on Degradable Plastics, Soc. Plastics Industry, Inc.*, 1987, p 39-40.
5. Otey, F. H.; Westoff, R. P.; Russell, C. P. *Ind. Eng. Chem., Prod. Res. Dev.* 1977, **16**, 305-308.
6. Griffin, G. J. L. U.S. Patent No. 4 016 117, 1977.
7. *Degradable Plastics: Standards, Research and Development.* U.S. General Accounting Office. Report No. GAO/RCED-88-208, U.S. G.A.O.: Gaithersburg, MD, 1988.
8. Gould, J. M.; Gordon, S. H.; Dexter, L. B.; Swanson, C. L. *Proc. Corn Utilization Conf. II, National Corn Growers Assn.*, 1988.

The mention of firm names or trade products does not imply that they are endorsed or recommended by the U.S. Department of Agriculture over other firms or similar products not mentioned.

RECEIVED January 18, 1990

Chapter 8

Constraints on Decay of Polysaccharide–Plastic Blends

Michael A. Cole

Department of Agronomy, University of Illinois, Urbana, IL 61801

Embedding biodegradable polymers like starch and cellulose in a plastic matrix restricts microbial and enzymatic attack on the polysaccharide. These limitations apply even when the biodegradable component exists as a continuum within the plastic. Microbial invasion and degradation is limited by low rates of diffusion of oxygen, nutrients, and hydrolytic enzymes into the blends and product release from them. Boundary conditions can be modified to enhance or retard biodegradation of blends. Many commonly used plastic additives and processing aids are likely to retard decay of biodegradable materials and a careful selection of these compounds is necessary if biodegradability is desired.

Numerous polysaccharide-plastic blends have been described with the majority of them containing starch, cellulose, or their derivatives as the polysaccharide. The primary goal of that work was to replace plastic with inexpensive and renewable resource materials derived from plants. More recently, there has been a substantial interest in polysaccharide-plastic blends that are somewhat biodegradable. These blends have been suggested as desirable alternatives to non-degradable plastics, for which a number of solid waste disposal and environmental problems exist. The principal degradable materials are mixtures of starch and polyethylene (1) or starch, polyethylene (LDPE), and ethylene-*co*-acrylic acid (EAA)(2). Although most currently-available materials are starch-plastic blends, there is no good scientific reason why other plastics or polysaccharides cannot be used to prepare blends. A variety of societal and economic incentives for developing polysaccharide materials were presented by U.S. Senator Conrad (3). Briefly, he indicated that these materials are desirable if they are biodegradable, if they will help reduce the grain surplus while increasing commodity prices, and if they

0097–6156/90/0433–0076$06.00/0

stimulate development of new products that are not wholly dependent on imported, non-renewable petrochemical resources for their manufacture. It should be noted that not all these incentives require that blends be biodegradable.

The work described in this chapter was initiated to determine the kinetics of enzymatic and microbial attack on blends containing starch and LDPE or starch, LDPE, and EAA. From the initial studies came a more general goal of identifying structural and compositional characteristics of blends (in general) that facilitate or retard degradation of the polysaccharide component.

Volumetric Content of Degradable Component and Degradability

Packaging plastics like LDPE are widely used because they are good barriers for aqueous-based solutes. Rapid penetration of LDPE by enzymes and low-molecular weight products of enzymatic digestion (maltose, for example) does not occur rapidly enough to allow rapid or extensive degradation of starch surrounded by LDPE. Therefore, facile access of hydrolytic enzymes and removal of water-soluble products is restricted to pathways containing the enzymatically susceptible component. Minimum percentages of degradable component that are required for enzymatic access can be estimated from data obtained with inorganic filler-plastic blends. In these blends, connectivity of spherical filler particles in plastic occurs at approximately 30% filler by volume (4). When microbial growth on blends of LDPE and granular cornstarch is assessed by ASTM procedures (5), a threshold for extensive microbial growth is seen between 25% and 30%(v/v) cornstarch (Cole, M.A., unpublished data). This value would also be applicable to commercially available LDPE-starch blends since they currently contain granular cornstarch.

According to filler theory, connectivity can be achieved at lower values when the filler form is plates rather than spheres. Depending on the proportions of the plates and whether or not an inactive phase is included in the blend, connectivity can be achieved at 8 to 16% (v/v) filler (4). The starch-plastic blends developed by Otey (2) have a laminate structure when the starch content is under 30% by volume (Figure 1) and the threshold for microbial attack on these materials is under 13% starch by volume (Figure 2). This low threshold value can be explained by considering the LDPE as a non-conductive (enzyme-impermeable) phase combined with a conductive phase of starch-EAA complex.

Wool and Cole (6) described a simulation model based on percolation theory for predicting accessibility of starch in LDPE to microbial attack and acid hydrolysis. This model predicted a percolation threshold at 30% (v/v) starch irrespective of component geometry, but the predicted values are not in accordance with results of enzymatic or microbial attack on these materials (Cole, M.A., unpublished data). Since a model that incorporates component geometry provides a better fit to experimental data than a geometry-independent model does, development of advanced models should be based on material geometry and composition, rather than on composition alone.

Figure 1. Scanning electron micrograph of cryofractured film containing 40% starch, 60% LDPE + EAA after 64 days of soil burial. Top of photo shows film surface; bottom shows film interior. Bar = 20 μm.

Kinetic Aspects of Film Degradation

When comparing the decay of films containing these blends with decay of natural materials such as wood when buried in soil, two major differences between natural products and the blends were noted. First, decay of natural materials typically give complex curves with plots of (carbon lost) versus time or $time^{0.5}$. In contrast, carbon loss from blends was linear in plots of (carbon lost) versus $t^{0.5}$ (Figure 2). Carbon dioxide evolution is a measure of complete degradation of organic compounds, with higher values indicating a greater extent of decomposition (6,7). Among materials discussed in this chapter, starch is the primary degradable component in the blends. Decay rate was not not well-correlated with starch content, since a number of films containing 40% (w/w) cornstarch decayed at quite different rates and large differences in the extent of decay were found (7). The decay kinetics of the blends suggested that diffusion rather than environmental and biological attributes was the controlling factor in decay of these materials. Second, microbial growth on these blends was restricted to the film surface, in contrast to penetration of the entire volume seen with natural products (8), and films made from cellophane or amylose (9,10). These observations suggested two questions: (1) What was the identity of the diffusible component whose mobility was controlling the decay rates shown in Figure 2 and (2) What was the barrier to microbial intrusion into the films? Diffusion-controlled microbial growth could be attributed to either slow movement of some nutrient (nitrogen or carbon source, for example), limited mobility of the amylase required for starch degradation, or to insufficient O_2 supply. Since soil fungi are generally aerobic organisms, poor O_2 supply was the most likely explanation for surface growth (see below), but would not be the controlling factor for overall decay rate. Diffusion of water-soluble nutrients like nitrogen into the films was eliminated because microbial intrusion did not occur. Eliminating inward nutrient diffusion left either diffusion of enzymes into the films or diffusion of products from the films as candidates for the rate-controlling process.

Non-invasive Microbial Growth on Starch-plastic Blends

Decay of heterogeneous natural materials such as wood by fungi begins with surface colonization followed by intrusion into the material via pre-existing pores and subsequent digestion of insoluble components (11). Invasion sites are also generated by degradation of intact, relatively non-porous surfaces. One would expect intrusion into starch-filled pores of polysaccharide- plastic blends if the components form an interpenetrating network (Figure 3), but these pores would be the only potential decay sites because the plastic is generally non-degradable. If the pore diameter is greater than hyphal diameter (typically 1-2 μm), there is no obvious physical impediment to intrusion as the starch is digested and one would expect the starch to be replaced by a reticulum of fungal hyphae. Engler and Carr had demonstrated fungal invasion of cellophane films (10), and Griffin (12) reported occasional fungal intrusion into polyvinyl acetate (PVA) and PVA-starch films. By extension, one would expect the same results with starch-PE materials. Contrary to expectations, fungal growth was seen only on the

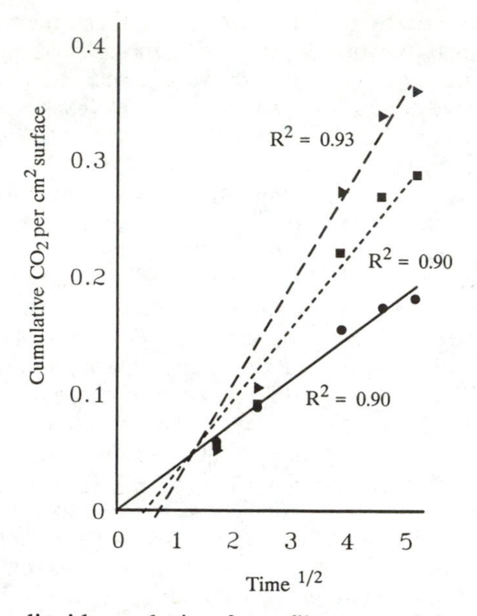

Figure 2. Carbon dioxide evolution from films containing cornstarch, LDPE, and EAA during soil burial. Circles: 13% (v/v) starch; squares: 20% starch; trianges: 28% starch. R^2 values were obtained by linear regression of time$^{1/2}$ *versus* cumulative CO_2 evolved.

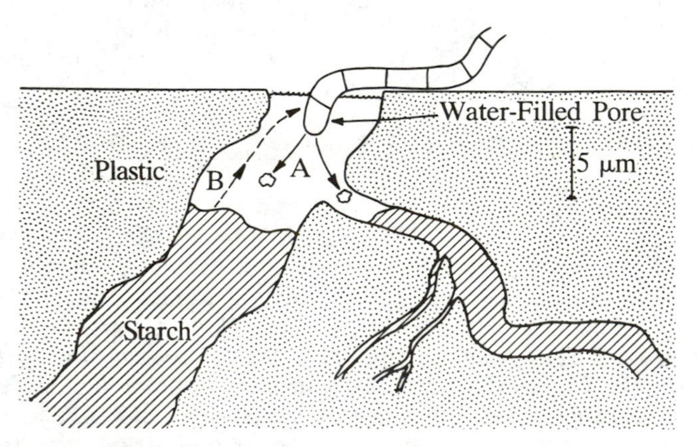

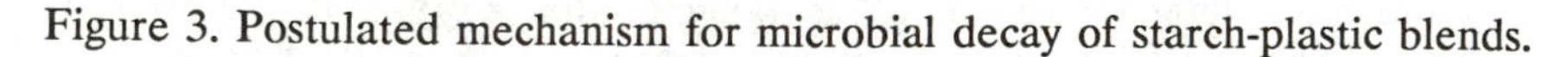

A: Diffusion of amylase to starch

B: Diffusion of digestion products to fungus

Figure 3. Postulated mechanism for microbial decay of starch-plastic blends.

surface of these materials even when extensive starch removal had occurred. The predominance of surface rather than interior growth was observed with several blends, including 6% (w/w) starch-LDPE films, 40% starch-60% LDPE+EAA films (Figure 1), and with 80% starch-20% LDPE+EAA injection-molded plaques. A few cases of hyphal intrusion were found, but these were a rarity. Many bacterial species also produce amylases, but bacterial growth was seen only on the surface as well.

Oxygen Availability in Degrading Films. A major difference between natural materials and starch-plastic or cellulose-plastic blends is that the hydrophilic and relatively permeable matrix of materials like wood and hydrated polysaccharide films allows diffusion of O_2 and release of nutrients from sites at a distance from the invasion site. As colonization proceeds, pore enlargement occurs when the pore walls are degraded (8) or as the polymer matrix of amylose or PVA films is hydrolyzed (10,12). In contrast, the LDPE matrix supplies no nutrients, hinders diffusion of water and O_2, and the pore diameter cannot be increased. The consequence of impermeability is that the sole means of obtaining O_2 and nutrients is by diffusion through water-filled pores.

Oxygen-limited growth of aerobic microorganisms is common in engineering applications and natural environments. In aquatic sediments, rapid metabolism and slow diffusion of O_2 through water-filled interstitial spaces results in O_2 depletion at short distances (< 1 mm) from O_2-saturated water (13); similar behavior has been proposed for soil aggregates (14) when pores are water-filled. Oxygen diffusion rate is a critical variable in various liquid waste degradation systems (15), as well as in the fermentation industry (16). Low permeability in relation to oxygen demand of soil organisms was documented by Bremner and Douglas (17) who found that the O_2 permeability of LDPE bags was not sufficient to maintain bulk soil in an aerobic state. Although LDPE is generally considered to be quite permeable to O_2 (18) when judged as a packaging material, it was not sufficiently O_2-permeable to allow aerobic microbial growth when metabolically active bulk soil was enclosed in LDPE. Based on these precedents, the inability of aerobic organisms to grow into starch-LDPE films seemed explicible and some simple modeling was done to assess the problem.

Based on reported metabolic rates for soil fungi (19), the estimated O_2 consumption for the fungus in Figure 2 was estimated to be about 2.1×10^{-9} $\mu g\ O_2/sec/\mu m$ of intruding length [due to a typographical error, a previously reported value (20) was given as 2.1×10^{-9} mg $O_2/sec/\mu m$]. This value is based solely on fungal metabolism of nutrients obtained from the surrounding environment and does not include an increase in metabolic rate from metabolism of starch digestion products. Oxygen flux in 5 μm diameter pores containing a 1.25 μm diameter fungus was calculated from Fick's law; O_2 flux through the plastic matrix was calculated to be <1 % of that contributed by water-filled pores and was not considered in these estimates. A comparison of respiration rate with O_2 supply indicated that pores would be O_2-deficient at depths of 2 μm or less from the film surface. Since filamentous fungi require O_2, but have a respiratory system with a very high O_2 affinity (21), colonization will cease only when $[O_2]$ approaches zero. The calculated values for maximum intrusion depth are sufficiently similar to observed intrusions of 3 to 4 μm (Cole, unpublished data) to suggest that fungal colonization into these materials is precluded by inadequate O_2 supply. Increasing pore diameter would not remedy the problem because intrusion of more than one fungal element into each pore would be likely, thereby maintaining the same fungal diameter/pore diameter relationship given above. Since many commercially available starch-plastic films are about 40 to 100 μm thick and since LDPE has a high O_2 permeability compared to most widely-used plastics, thorough penetration of the matrix of polysaccharide-plastic films by fungi appears to be improbable. Although intrusion would probably accelerate decay of the polysaccharide portion of the films, there is a possibly beneficial aspect of the lack of intrusion; namely, that contamination of packaged products by invading fungi would be limited by the inability of the organisms to grow through the films.

Limitations to Bacterial Colonization. Amylase and cellulase production is a widespread character among anaerobic bacteria and their growth and activity would be expected when aerobic bacteria and fungi fail. Several factors may limit bacterial attack on polymer blends. First, bacteria in soil and water are firmly attached to the solid phase (i.e., to soil and sediment components and debris) of terrestrial and aquatic ecosystems and facile movement from an attachment site to adjacent film is not readily done. Rai and Srivastava (22) found that fungi were major colonizers of plant debris buried in soil and Witkamp (23) indicated that fungi are primary agents for decay of above-ground cellulosic materials. Since plant debris and plastics are mainly water-insoluble, it is likely that the same microbial group--to whit, the fungi--should be the primary colonizer of insoluble plastics, whether the plastics are buried in soil or remain above ground. Second, Imam and Gould (24) found that bacteria they tested did not attach readily to starch-plastic films. Since attachment of cells to insoluble substrates precedes substrate decay in most cases, failure to attach would limit potential attack. The results obtained by Imam and Gould are not typical, since numerous investigators have reported rapid bonding of microorganisms ranging from bacteria to algae to plastic surfaces. Third, bacteria tend to be relatively poor colonizers of pores. Hattori (25) showed that when pore diameters were in the range of 100 to 200 μm, about 35% of all pores were not colonized and few pores were heavily colonized. Internal pore diameters of films examined by me have ranged between 5-25 μm. In general, smaller pores should present a greater impediment to bacterial invasion and therefore one would expect a higher percentage of vacant pores in polysaccharide-plastic blends than Hattori observed. Kohlmeyer (26) indicated that aerobic marine bacteria attacked cellophane films mainly from the surface and that cellulase diffusion into the films did not occur readily. He had previously reported (27) that bacterial attack on wood in the deep sea was seen only on the wood surface, even though wood is a naturally porous material. Fourth, most soil bacteria produce exopolysaccharides that would clog pores even when successful invasion occurred. Clogging would greatly reduce the degradation potential of the colonizing organism by limiting diffusion of nutrients, enzymes, and degradation products within the pore.

Principal Mechanism for Starch Degradation

For the reasons stated above, deep intrusion of degrading microbes into polysaccharide-plastic films is demonstrably and theoretically improbable. Since starch removal does occur when the films are buried in soil, the primary mechanism must be microbial production of amylase in or near a pore, diffusion of the enzyme into pores and diffusion of soluble digestion products back to the surface where they are metabolized (Figure 3). This mechanism would be the only choice when the pore diameter is too small to admit a microbial cell (i.e., at diameters $< 0.5\ \mu m$). An alternative mechanism could be diffusion of a water-soluble polysaccharide to the film surface, at which point degradation would occur. None of the materials used in these investigations showed loss of starch even when soaked in water for extended periods with microbial inhibitors present. Therefore, diffusion of amylase to the substrate rather than diffusion of the substrate to the film surface is the more likely mechanism.

The scheme proposed above requires microbial colonization of the material and excludes degradation by amylases and cellulases that are present in soils (28), but are not newly synthesized or associated with microbial cells. Active polysaccharide hydrolases are found in nearly all soils, but these enzymes are primarily bound to soil organic matter or mineral components; attachment is firm enough to severely limit migration of the enzymes from surrounding soil to the film.

Degradation of Moist or Submerged Films. A requirement for synthesis of hydrolases on or near the film surface has a large impact on the environmental degradability of films. When the films are moist, a thin water layer would be present on the film surface (29), but would not necessarily be continuous with the surrounding environment (Figure 4A). This scenario would be applicable to films buried in moist soil, lying on the soil surface, or the top surface of film lying on a solid culture medium (as in ASTM test procedures, reference 5). Under these conditions, enzymes produced by organisms on the films could only diffuse across the films and possibly into pores; all produced enzyme has the potential for contributing to film degradation. In contrast, complete immersion of films--as would occur in water-saturated soil, flooded landfills, aquatic sediments, and in liquid microbial cultures--creates a situation in which enzymes produced by adherent organisms might diffuse on the surface and into pores, but would be more likely to diffuse away from the film (Figure 4B). This situation would be analogous to sediments, in which decay of woody materials is much slower than observed in soils.

Specific Diffusion-based Limitations to Decay. If microbial colonization is confined to the surface of materials, the decay rate will inevitably be lower than seen where proximity between substrate and microbial cells is possible because enzymes produced by the cell and soluble products formed by enzymatic attack must diffuse a considerable distance. For example, if closer contact between the starch face and fungus were possible than seen in Figure 2, uptake of starch digestion products would occur at the growing tip and translocation within the mycelium by active transport would be possible. This

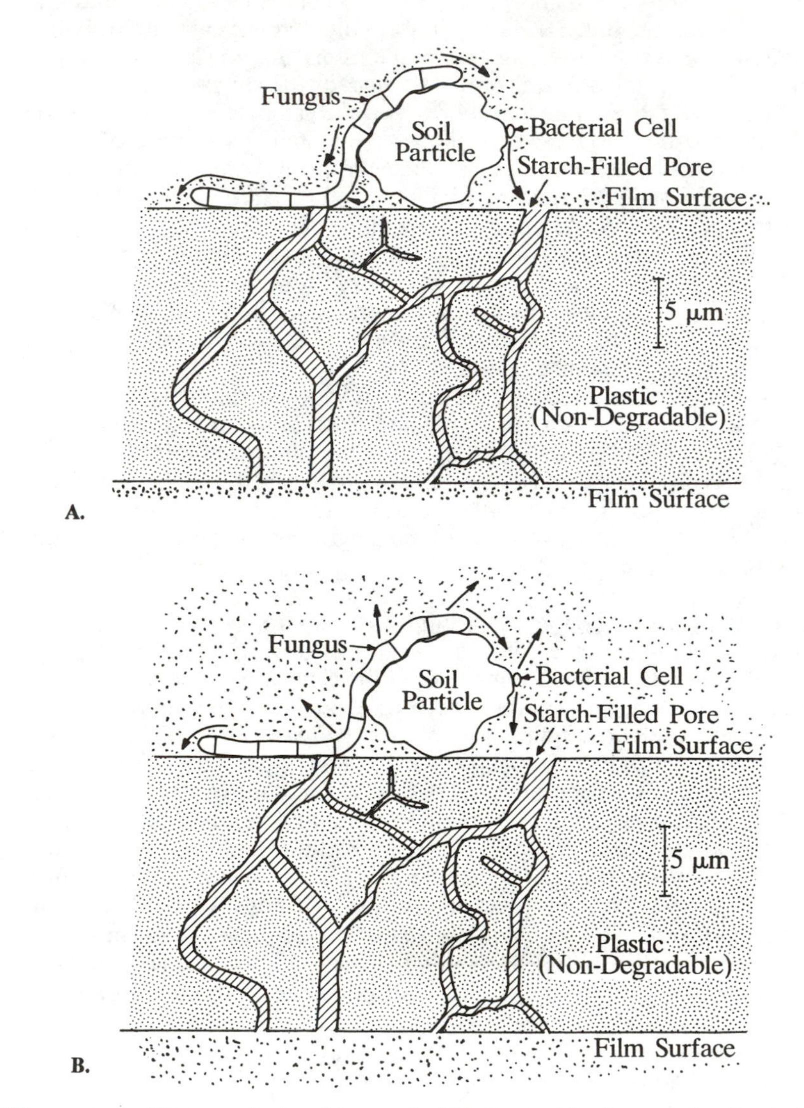

Figure 4. A. Water relationships and diffusion paths for starch-plastic film in a moist environment. Lightly shaded area indicates thin water layer on surface; arrows show diffusion paths for enzymes produced by microbes on film surface. B. Same as 4 A., except that film is immersed in water.

process would accelerate decay for two reasons: (1) Active transport of metabolites is much faster than diffusion, and (2) Accumulation of amylase digestion products, some of which are inhibitors of amylase (30), would not be so likely. Some products of cellulose hydrolysis are also inhibitory to cellulase activity (31), so the same impediment would apply to cellulose-plastic blends. An analysis of diffusion rates for soluble products indicated that the rate of removal of these products from the starch face is the most likely rate-limiting process in decay of these materials when excess enzyme is present, but enzyme migration may be the rate-limiting factor under natural conditions where enzyme quantities are low.

Degradation of Cellulose-containing Blends. No distinction was made in previous discussion between blends containing cellulose or starch. A process requiring enzyme diffusion will limit the possible microbes that would degrade cellulose-plastic blends because many cellulases are cell-associated (32). In such cases, intimate contact between cellulose and the microbial cell is required, but such contact would not be possible if the cellulose particles are smaller than microbial cell diameters. Some microbes produce diffusible cellulases (33), so cellulose hydrolysis would still occur as suggested previously for decay of starch-plastic blends. Degradation of a cellulose-plastic blend may require that a prospective colonizing organism not only be able to attach to the material but also must produce a diffusible cellulase. If so, environmental decay of cellulose-plastic blends would be much slower than starch-based blends because many microbes would not have the necessary qualifications and because cellulose is generally more slowly degraded than starch is. Based on these considerations, cellulose-plastic blends or composites may be a better choice than starch-based materials when a moderate degree of environmental persistance is desired.

Critical Film Properties that Affect Degradation Rate

A common feature to both intrusive and non-intrusive mechanisms of attack is that decay can be initiated only from the surface, a conclusion from which three correlaries can be derived: (1) The decay rate ultimately depends on the percentage of surface area consisting of degradable material; (2) Among materials of the same composition, the number of vulnerable sites per unit area depends on the dimensions of the degradable component, and (3) Decay will not occur if the dimensions of the degradable component are so small that neither microbial intrusion nor amylase diffusion into the material can take place.

Pore dimensions may have a more subtle effect on decay rate depending on component dimensions and production method of the manufactured material. Products made from pasted starch, LDPE, and EAA (2) typically appeared as laminates of starch and plastic when examined by scanning electron microscopy (Figure 1). The dimensions of inter-laminate channels (i.e., pores) were not uniform and ranged from about 50 to 325 μm^2 in cross-section (33). Since flux is dependent on diffusional path area, the smaller pores can be an impediment to movement of solutes from the interior to the surface of the films. Figure 5 illustrates two films in which the laminate units are the same thickness, but differ in length. When the starch is removed from the surface inward, channels are created through which amylase would diffuse inward and digestion products outward (B, D, Figure 5). In the degraded material (D, Figure 5), there are a few sites where constrictions occur (arrows), and these sites would control enzyme and product flux through the entire channel. In contrast, the degraded material (B, Figure 5) does not have the severe constrictions in channels. Therefore, material A would likely be more rapidly degraded than material C, even though the percentage composition of A and C is identical. In such cases, the decay rate is not governed by the average pore size, but by the minimum pore size. This type of restriction can occur when the interior is more porous than the surface (see discussion below) and smaller surface pores would further hinder product release from the film interior. Bottlenecks (arrows, Figure 5) may also have a localized effect on concentration gradients. With substantially higher solute concentrations on the "upstream" side of the path, there may be some compensation for the smaller path dimensions and the influence of path restrictions can be reduced. It should be evident from such considerations that predicting the decay rate of these materials requires more than information about the volumetric composition of blends.

Porosity characteristics also influence the degradation rate of blends containing intact starch grains. Amylase removal of starch from these films was not highly correlated with starch content, since films whose starch content was above possible "percolation thresholds" (6) were degraded at very different rates when starch content was not very different (Table I).

Microscopic analysis of the film surface indicated that total porosity of 50% and 60% starch (v/v) films was similar (Table I), but pore size distribution was much different (Figure 6). The internal matrix of these materials is very porous, with an entrained air volume about twice the starch volume, but surface pore area is low. The air is displaced by water when the material is submerged in water. Even when cut into small pieces to increase enzyme access to the porous edge of the film, and incubated with excess alpha-amylase [EC 3.2.1.1] (sufficient to digest the amount of starch present in the film in a few minutes if there were no hinderances), removal of 22% of total starch was not achieved until 3 days of amylase treatment, by which time the rate of product release was a constant 18 μg soluble product/h/cm^2 surface area. At this rate, complete starch removal would require about 10 days. When this material was buried in moist soil, removal of accessible starch required approximately 120 days of incubation. A comparison of these values strongly suggests that surface porosity and diffusion path characteristics within the films are major contributors to the potential decay rate. These data suggest that smaller pores, although they provide a larger number of access sites to the film interior, have a deleterious effect on decay rate even when the pore diameter is quite large in relation to amylase dimensions. Critical pore dimensions for cellulase attack on wood chips were determined by several workers (34,35,36) and were generally very similar to the diameter of the permeating enzymes. Substantially larger dimensions were found for starch-plastic blends and further work is in progress to determine why the size requirements for the blends are so much larger than found for natural cellulosic materials.

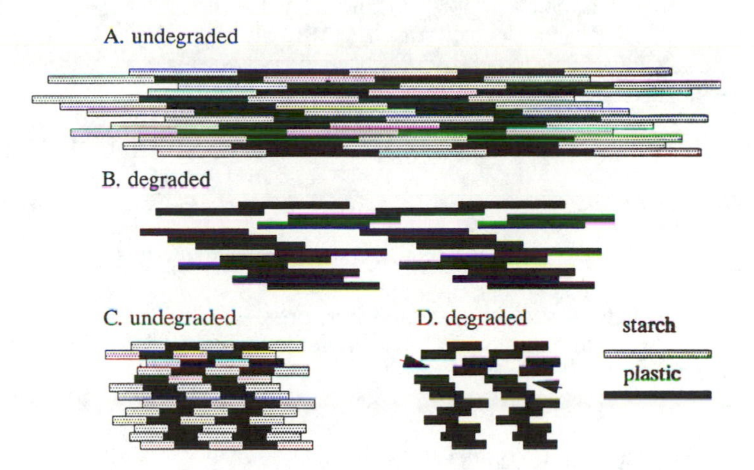

Figure 5. Influence of subunit size on diffusion paths in starch-plastic laminates containing equal volumes of starch and plastic (not to scale). A and C: intact film; B and D: after starch removal. Arrows indicate constrictions that would control diffusion processes.

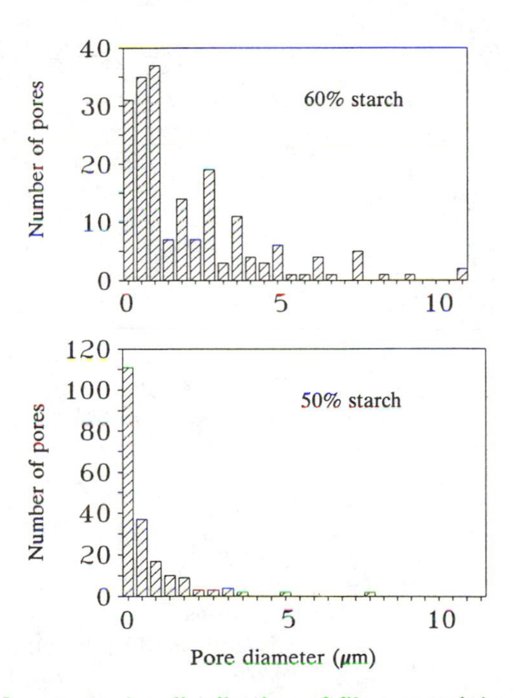

Figure 6. Surface pore size distribution of films containing 50% or 60% granular starch with the balance being LDPE.

Table I. Surface porosity and amylase digestion rate of starch-LDPE blends

% starch (v/v)	Surface porosity[a]	Amylase digestion rate[b] µg product/cm^2 film/hr		µg product/cm^{2}pore/hr	
		24 hr	48 hr	24 hr	48 hr
60	3.3%	18	12	509	353
50	3.6%	6	4	345	239

[a] Percentage of total surface occupied by pores.
[b] Release of reducing sugar by alpha-amylase during 0 to 24 hr or 24 to 48 hr interval.

Enzyme Exclusion and Non-degradability

In the examples discussed above, limitations on the rate of decay of the degradable component existed because of material structure, but the material was ultimately degradable. However, it is possible to construct blends in which the matrix dimensions are smaller than the dimensions of degradative enzymes. Enzyme diffusion cannot occur in these materials if the pore diameter is less than the diameter of the degrading enzyme, irrespective of the inherent degradability of the component. For example, the biodegradability of Otey-type films during soil burial was found to vary widely among different lots of similar starch content (7) and in all cases, the starch content was well above calculated connectivity thresholds. Starch in the films stained rapidly and intensely with aqueous dyes, indicating that the film surface was porous and that the starch formed a continuum within the plastic. The simplest explanation was that the starch was inaccessible to enzymatic attack because it was distributed in the plastic matrix in channels that were too small for microbial intrusion or enzymatic attack. Restrictions of this type have been discussed at length for cellulose hydrolysis (34,37) and much of that work is conceptually applicable to starch-containing blends because of the similar molecular weights and dimensions of amylases and cellulases, whose molecular weights typically range between 12,000 and 60,000 (34). Molecular dimensions among cellulases average 59 Å for spherical enzymes and 33 X 200 Å for ellipsoidal enzymes (34). When considering movement of proteins within pores, hydraulic flow would not occur in a closed-end pore, and therefore, diffusion would be the primary means by which the enzyme would make contact with the starch surface (Figure 7). A diffusing enzyme probably cannot maintain the most hydrodynamically favorable configuration with the long axis oriented along the direction of movement because of rotation about the enzyme's central axis (indicated by dotted circles in Figure 7); therefore, the largest dimension of the enzyme is probably the appropriate size boundary for entry into a porous matrix. Since physical contact between enzyme and substrate is necessary for degradation, minimum diameters for substrate-filled pores based on enzyme size alone would be around 60 to 200 Å. In practice, pore diameters would have to be substantially larger than the enzyme alone to allow diffusion of products to the surface (Figure 7), to provide space for random lateral motion of the enzyme within a small pore, and to compensate for a hydrophobic layer near the plastic surface. Small pores would also result in high product concentrations near the degrading polysaccharide face, a situation in which product inhibition could severely limit enzyme activity. Based on a spherical protein of 60 Å diameter, oligosaccharide products ranging from 10 to 15 Å diameter, and allowing one enzyme diameter (60 Å) for movement of molecules past each other, the projected minimum diameter of a substrate-filled pore would have to be 130 Å or more. Minimum pore size would be over 400 Å for ellipsoidal proteins. In either case, the blends would have to be microscopic, not molecular blends if they were to be biodegradable. Models for cell wall ultrastructure of wood postulate that cellulose occurs as fibrils about 50 X 100 Å in cross-section (34) and surrounded by non-cellulosic materials (35). In this state, the cellulose is only slowly degraded and dramatic increases in cellulose digestion rate can be achieved by removal of non-cellulosic components (38) or by explosive steam treatment to increase pore diameter (37). With cellulose-plastic blends, pore diameter cannot be increased and slow degradation seems to be inevitable.

Effects of Plastic Additives on Biodegradability

The physical properties, durability and ease of processing of plastics can be improved by use of additives and processing aids. There are several potential impacts of these compounds on the utility and degradability of plastic-polysaccharide blends. The majority of commonly used additives are metal salts or low molecular weight alkanes, aromatics and their derivatives Many of the organics in this group are biodegradable, but are sufficiently toxic to inhibit microbial growth in culture at concentrations in the parts per thousand range. These compounds are weak inhibitors compared to antibiotics that are effective at parts per million concentrations and inhibition of microbial activity is an inadvertant secondary effect, not the primary reason for their inclusion in plastics. If these compounds migrate to the plastic surface during processing or are deliberately applied to the surface (lubricants, for example), concentrations may be high enough to prevent microbial growth on the material and thereby retard degradation of an otherwise biodegradable material. Where inhibitors are deliberately added to minimize marine fouling, the best control is achieved with additives that diffuse to the plastic surface and then dissipate into the surrounding water. An inhibitory surface concentration is maintained by continuous diffusion, but biocontrol is eventually lost upon exhaustion of the inhibitor (39,40). If one applies this concept to polysaccharide-plastic blends, then one would expect temporary inhibition of decay if the additive migrates, but there should be less effect if the additive has a high affinity for the plastic. Where inhibition does occur, microbial degradation would be delayed, but not prevented. There is virtually no published work on the partitioning of additives between the plastic and polysaccharide components in prepared blends and this area appears to be a critical deficiency in our ability to predict degradability based on material composition. It seems likely that judicious use of additives will be a simple means of controlling the lifespan of products. For example, starch loss from LDPE-cornstarch films is complete within 120 days of soil contact when starch connectivity exists (Cole, unpublished data), but this rate of loss is too rapid when the films are used as agricultural mulches with a desired residence time of 180 to 300 days. In this case, use of an inhibitory but diffusible additive would increase the useful life of the film but would not have a long-term effect on film stability.

Several additives are known microbial inhibitors, cannot readily migrate from the films and avoiding these compounds is recommended. For example, Otey *et al.* reported that carbon black inhibited biodegradation of starch-plastic blends (2), which indicates that use of carbon black or solvents added with it is not compatible with a biodegradable technology. Sparingly-soluble cadmium and lead salts are likely to inhibit degradation of starch-plastic blends since they have been shown to be solubilized by soil bacteria (41) and to inhibit starch degradation in soil (42).

Inhibitory Properties of Photodegradable LDPE. There is considerable interest in materials that utilize a combination of biological, chemical, and photolytic degradation mechanisms (43,44) to facilitate environmental destruction of plastics, but the available literature on the degradability of these materials suggests that some of these approaches may be mutually incompatible. The work of Nyqvist (45) is a good example of the potential problems. Films of LDPE with or without accelerants of photodecomposition were photodegraded and then buried in biologically active soil. Release of ^{14}C-CO_2 was used as a measure of biodegradability of the plastic and the results (Figure 8) indicated that small increases in biodegradability could be achieved following photodegradation. What one should find is either no change or an increase in ^{14}C-CO_2 released as the amount of added substrate is increased. In contrast, Nyqvist found that CO_2 evolution decreased dramatically as the amount of plastic added to soil was increased. His results indicate clearly that either photodegradation products or additives in LDPE severely inhibited degradation of the plastic. Long-term biodegradation studies of extensively photodegraded polyethylene (46) indicate that this material is slowly decayed and a question arises: Is the polymer intrinsically non-degradable or is microbial activity being prevented by photolysis products or additives remaining in the disintegrated particles? Complete inhibition of starch decay in some, but not all, 6% starch-LDPE films was reported by Cole (4) and the differences could not be attributed to film structure or gross composition; inhibition of microbial attack by additives is the most likely reason for this result.

Additive Release upon Degradation of Blends

A frequently cited merit of biodegradable plastics is their lack of persistance in an intact state; "environmentally friendly" is a widely used vernacular phrase, but there has also been speculation in the popular press that degradable plastics will release potentially harmful additives into the environment when the plastics degrade or disintegrate. An ideal biodegradable plastic will leave no undegraded polymeric residues, and for these materials, the persistance of additives as well as the polymers must be considered. In this case, the issue is not whether or not the additive

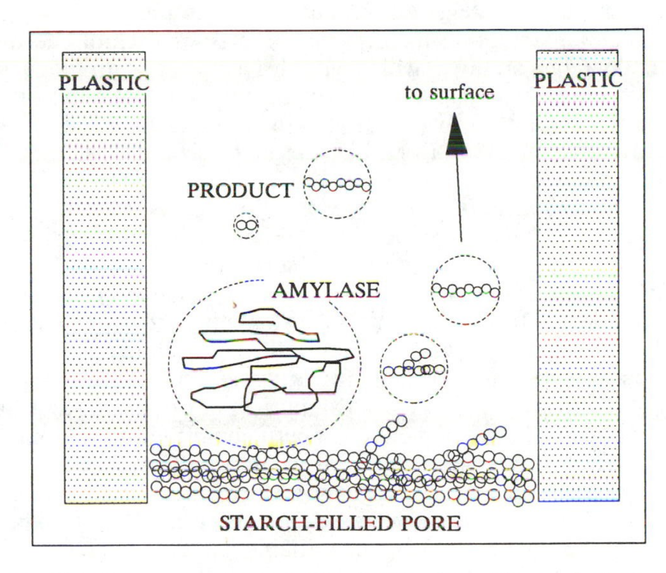

Figure 7. Enlargement of starch-filled pore (see Figure 3 also) showing space requirements for enzyme and product diffusion. Dotted circles indicate the effective radius due to molecular rotation.

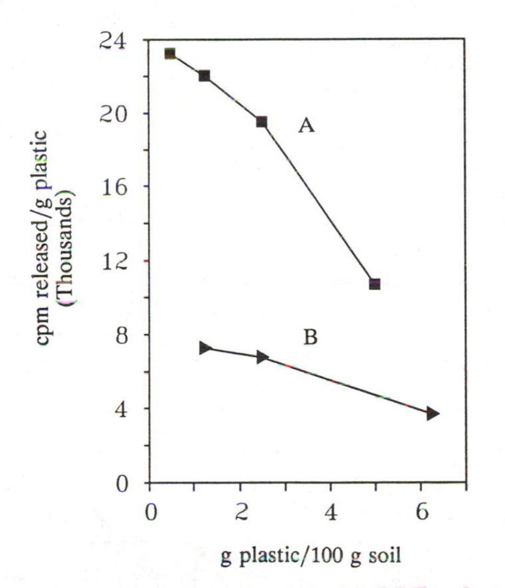

Figure 8. Biodegradation of photodegraded LDPE. Cpm released (as ^{14}C-CO_2) demonstrates decomposition of the low molecular weight fraction of the plastic. Line A: LDPE with accelerant of photodegradation; Line B: LDPE without accelerant. Maximum cpm released is less than 5% of total cpm added as LDPE. Data are from (45), Table 5.

interferes with polymer degradation, but addresses the question of overall environmental impact of using degradable plastics. Upon decay of the polymer, undegraded additives will inevitably be released into the surrounding soil or water and will contribute to environmental contamination if they are not degradable. There are two questions involved in this issue: First, is the additive harmful?, and second, is it likely to be persistant if released from the degraded plastic?

Ionnatti and coworkers (Proc. Internat. Conf. on Degradable Plastics, in press) indicated that commercial films did not generally contain priority pollutants (USEPA list), but their survey did not other lists of environmentally significant chemicals such as (47). Several "inerts" in these lists are compounds found in pesticide formulations that are not pesticidal (hence, the "inert" designation) and are added as stabilizers or emulsifiers, while some are by-products of manufacturing. A number of compounds on the "inerts" lists are also used as plastic additives or processing aids and a few have a long history of environmental concern about their use. Among these, phthalate ester plasticizers are the most widely cited (48,49).

If one goes from a simple comparison of chemical lists to consideration of why particular compounds appear on the lists, a number of additives (50) may be inappropriate for use in degradable materials. Historically, the organic compounds of greatest concern and with bans on their use have three characteristics in common: they are poorly biodegradable, lipophilic, and have adverse health effects of humans or wildlife. Halogenated alkanes and aromatics and substituted and polynuclear aromatics are good examples. Based on these criteria and analogies in chemical structure with already restricted compounds, several groups of additives or degradation products thereof fit at least two of the criteria cited above (Table II).

A general concern exists about the environmental persistance and possible health effects of several entire classes of organics because there is a high frequency of problem compounds within the groups. If a comprehensive view of improving environmental quality by using degradable plastics were taken, manufacturers of these products should consider replacing poorly degradable additives with less persistant ones where alternatives are available. This task would be simplified somewhat by the fact that many of the less desirable additives are stabilizers whose use would be incompatible with degradable technologies employing chemical or photodegradation mechanisms to disrupt the plastic matrix.

Conclusions

If one considers simultaneously all the boundary conditions of material composition and arrangement, it is evident that potentially incompatible requirements can exist between desirable properties from the engineering perspective and necessary characteristics imposed by biological systems that may decompose those materials. One consequence of this conflict is that material strength--particularly where interface bonding is weak--and biodegradability may have opposing requirements. The overall strength of weak interfaces is increased by decreasing component dimensions (thereby increasing contact area/volume values), but the dimensions of the degradable component cannot fall below the boundaries established above if

TABLE II. Chemicals of environmental concern and plastic additives with analogous structures

Known Concern	Analogous Additive	Use in plastics
Polynuclear aromatics	*bis*-phenols, triphenyl -*t*-butyl compounds	Stabilizers
Cresols	Substituted phenols and cresols	Antioxidants
Zinc, nickel, lead, and cadmium salts	Ba-Cd-Zn stabilizers Organo-nickel complexes CdS, CdSe, lead stearate Cadmium laurate	Stabilizers Pigments Lubricants
Polychlorinated biphenyls (PCB) Polybrominated biphenyls	Polybrominated biphenyl ethers	Flame retardants
Chlorinated alkanes and aromatics	Brominated neopentyl alcohol and derivatives Dibromophenol	Flame retardants

biodegradability is desired. For example, molecular-level composites of degradable and non-degradable components should not decay unless matrix disruption occurs first. This limitation may be used advantagously to control the decay rate if the plastic component is photodegradable or if destabilizers such as metal salts or fatty acids are added to promote disintegration of the plastic.

There is a definite need for further work on the compatibility of additives from the perspective of their interfering with biological or chemical processes that are needed to disrupt the plastic matrix. Although additives are valuable components when they improve the properties of a plastic, their use cannot be sanctioned if they prevent degradation of the plastic.

The discovery that several controlling factors in decay of polysaccharide -plastic composites are physico-chemical in nature (rather than environmental or biological) should make it possible to develop an accurate simulation model to define structural and biodegradable boundaries to achieve the best balance between the two conflicting requirements. This type of model would facilitate development of materials that are satisfactory by both engineering and biodegradability criteria; a model of this type is currently being developed in my laboratory.

Acknowledgment.

This work was supported in part by a grant from the Illinois Corn Marketing Board.

Literature Cited

1. Griffin, G.J.L. In Degradable Plastics; Society of the Plastics Industry: Washington, D.C., 1987; pp 47-50.
2. Otey, F.; Westhoff, R.P.; Doane, W.M. Ind. Eng. Chem. Prod. Res. Dev. 1980, 19, 592.
3. Conrad, K. In Agricultural Commodity-Based Plastics Development Act of 1988 (S. Hrg. 100-1003); U.S. Government Printing Office: Washington; 1988; pp 215-217.
4. Reboul, J.-P. Inorganic Fillers. In Lutz, J.T., Ed.; Marcel Dekker, Inc. NY; 1989; pp 255-280.
5. ASTM Procedure D1924-70 (reapproved 1985).
6. Wool, R.P.; Cole, M.A. In Engineered Materials Handbook, volume 2; ASM International: Metals Park, OH, 1988; pp 783-787.
7. Cole, M.A. Abstr. Annu. Mtgs. Amer. Soc. Microbiol. 1989, 89, 364.
8. Boddy, L. Soil Biol. Biochem. 1983, 15, 501.
9. Bradley, S.A.; Engler, P.; Carr, S.H. Appl. Polym. Symp. 1973, 22, 269.
10. Engler, P; Carr, S.H. J. Polym. Sci., Polym. Physics. 1973, 11, 313.
11. Cowling, E.B.; Brown, W. In Cellulases and Their Applications; Hajny, G.J., Reese, E.T.; Eds.; American Chemical Society: Washington, D.C., 1969; pp 152-187.
12. Griffin, G.J.L.; Mivetchi, H. Proc. Third Internat. Biodegrad. Symp.; Sharpley, J.M., Kaplan, A.M. Eds.; Applied Science Publishers, Ltd.: London, 1976; pp 807-813.
13. Jorgensen, B.B.; Revsbech, N.P. Appl. Environ. Microbiol. 1983, 45, 1261.
14. Greenwood, D.J.; Berry, G. Nature (London). 1962, 195, 161.
15. Nienow, A.W. In Environmental Biotechnology; Forster, C.F., Wase, D.A.J., Eds.; Ellis Horwood Ltd: Chichester, 1987; pp 377-401.
16. Metz, B.; Kossen, N.W.F. Biotechnol. Bioengin. 1977, 19, 781.
17. Bremner, J.M.; Douglas, L.A. Soil Biol. Biochem. 1971, 3, 289.
18. Steingiser, S.; Nemphos, S.P.; Salame, M. In Encyclopedia of Chemical Technology, 3rd ed., vol. 3; Grayson, M. Ed.; John Wiley & Sons: New York, 1978; pp 480-502.
19. Anderson, J.P.E.; Domsch, K.H. Soil Biol. Biochem. 1978, 10, 215.
20. Cole, M.A. Polym. Prepr. Am. Chem. Soc. Symp., American Chemical Society: Washington, D.C., 1989.
21. Griffin, D.M. Ecology of Soil Fungi; Syracuse University Press: Syracuse, 1972.
22. Rai, B; Srivastava, A.K. Soil Biol. Biochem. 1983, 15, 115.
23. Witkamp, M. Soil Sci. 1974, 118, 150.
24. Imam, S.H.; Gould, J.M. Abstr. Annu. Mtgs. Amer. Soc. Microbiol., 1989, p 356.
25. Hattori, T.J. Gen. Appl. Microbiol. 1981, 27, 43.
26. Kohlmeyer, J. In Biodeterioration; Oxley, T.A.; Becker, G. Eds.; Pitman Publishing Ltd: London, 1980; pp 187-192.
27. Kohlmeyer, J. ASTM. STP. 1969, 445, 20.
28. Pancholy, S.K.; Rice, E.L. Soil Sci. Soc. Amer. Proc. 1973, 37, 47.
29. Avissar, R.; Mahrer, Y.; Margulies, L.; Katan, J. Soil Sci. Soc. Amer. J. 1986, 50, 202

30. Robyt, J.; French, D. Arch. Biochem. Biophys. 1963, 100, 451.
31. Barras, D.R.; Moore, A.E.; Stone, B.A. In Cellulases and Their Applications, Hajny, G.J.; Reese, E.T. Eds.; American Chemical Society: Washington, D.C., 1969; pp 105-138.
32. Mandels, M. In Biotechnol. Bioengin. Symp. No. 5, Wilke, C.R. Ed.; John Wiley & Sons: New York, 1975; pp 81-105.
33. Cole, M.A.; Wool, R.P. Bull. Amer. Physical Soc. 1989, 34, 708.
34. Cowling, E.B. Biotechnol. Bioengin. Symp. No. 5, Wilke, C.R. Ed.; John Wiley & Sons: New York, 1975; pp 163-181.
35. Fan, L.T.; Gharpuran, M.M.; Lee, Y.-H. Cellulose Hydrolysis. Springer-Verlag: New York, 1987.
36. Grethlein, H.E. Bio/Technology, 1985, 3, 155.
37. Burns, D.S.; Ooshima, H.; Converse, A.O. Appl. Biochem. Biotechnol. 1989, 20, 79.
38. Converse, A.O.; Kwarteng, I.K.; Grethlein, H.E.; Ooshima, H. Appl. Biochem. Biotechnol. 1989, 20, 63.
39. Bowden, R.D.; Taylor, J.M. In Proc. Fourth Internat. Biodeter. Symp. Oxley, T.A.; Becker, G.; Allsopp, D. Eds.; Pitman Publishing, London, 1980; pp 291-296.
40. Bowden, R.D.; Heeson, I.E.; Taylor, J.M. In Proc. Fourth Internat. Biodeter. Symp.; Oxley, T.A.; Becker, G.; Allsopp, D. Eds.; Pitman Publishing, London, 1980; pp 297-306.
41. Cole, M.A. Soil Sci., 1979, 127, 313.
42. Cole, M.A. Appl. Environ. Microbiol. 1977, 33, 262.
43. Maddever, W.J.; Chapman, G.M. In Degradable Plastics; Society of the Plastics Industry: Washington, D.C., 1987; pp 41-44.
44. Gilead, D.; Ennis, R.; Degradable Plastics, *ibid*, pp 37-38.
45. Nyqvist, N.B. Plastics and Polymers. 1974, 42, 195.
46. Karlsson, S.; Ljungquist, O.; Albertsson, A.-C. Polym. Degrad. Stabil. 1988, 21, 237.
47. United States Environmental Protection Agency. Extremely hazardous substances list, Federal Register. 1987, 52, 308 .
48. Mayer, F.L.; Stalling, D.L.; Johnson, J.L. Nature. 1972, 238, 411.
49. Giam, C.S.; Chan, H.S.; Neff, G.S. Anal. Chem. 1975, 47, 2225.
50. Thuen, J. Ed.; Additives for Plastics. D.A.T.A., Inc.: San Diego, CA, 1987; 391 pp.

RECEIVED January 18, 1990

Chapter 9

Biodegradation Pathways of Nonionic Ethoxylates

Influence of the Hydrophobe Structure

L. Kravetz

Shell Development Company, Westhollow Research Center, P.O. Box 1380, Houston, TX 77251

The structure of the hydrophobe in alcohol ethoxylates (AE) and alkylphenol ethoxylates (APE) has a significant influence on the biodegradation pathways and biodegradation rates of these nonionic surfactants in bacterial inocula present in waste treatment plants. APE undergo initial microbial attack at the terminal part of the polyoxyethylene (POE) chain followed by a shortening of the POE chain to leave intermediates which are slower to biodegrade than intact APE. Ethoxylates of alcohols derived from propylene or butene oligomerization appear to degrade by a similar mechanistic pathway. In contrast, ethoxylates of alcohols derived from linear olefins undergo initial attack which results in scission of the linear alkyl chain from the polyoxyethylene chain with rapid biodegradation of the hydrophobe to CO_2 and H_2O accompanied by slower degradation of the polyoxyethylene chain. The use of double radiolabeled nonionic surfactants has shed significant light on their biodegradation pathways by providing detailed information on degradation of the hydrophobic and hydrophilic moieties.

Increasing quantities of surfactants are used in household products and such industrial and institutional applications as textile processing, pulp and paper deinking and deresination and hard surface cleaning (1). Of the two largest classes of surfactants, anionics and nonionics, the latter have shown the fastest growth rate over the past twenty years. Such environmental issues as foaming and aquatic toxicity shown by all the major classes of surfactants may be faced by selecting those whose structural features permit rapid biodegradation to intermediates which have non-foaming and non-toxic properties under practical waste treatment conditions. In an era of increasing environmental awareness, a detailed understanding of biodegradation pathways is essential in developing an overview of the potential impact surfactants usage and treatment practices can have on aquatic life.

Secondary waste treatment, employing aerobic biodegradation, is the most widely used approach to reducing the undesired properties of organic materials, like surfactants, which enter domestic waste treatment plants. In this approach, bacteria provide enzymatic systems which utilize oxygen to convert organics to CO_2, H_2O and cellular mass. The process is complex with biodegradation proceeding through a series

0097–6156/90/0433–0096$06.00/0

of intermediates produced by specific enzymes. Biodegradation rate is dependent upon the structure of the organic material assuming sufficient oxygen and viable bacteria are present. Those molecules having essentially linear structural features generally biodegrade in sewage treatment plants considerably faster than those having substantially branched carbon skeletons. A thorough review of the effect of structure on surfactant biodegradation has been written by Swisher (2).

This paper will review the biodegradation of nonionic surfactants. The major focus will be on alcohol ethoxylates and alkylphenol ethoxylates—the two largest volume nonionics. In this paper the effect of hydrophobe structure will be discussed, since hydrophobe structure is considered more critical than that of the hydrophile in biodegradability of the largest volume nonionics. The influence of the hydrophobe on the biodegradation pathway will be examined with an emphasis on the use of radiolabeled nonionics.

Structural Features of Nonionic Surfactants

Figure 1 summarizes the significant structural features of four classes of nonionic surfactants discussed in this paper. These are simplified representations of commercial surfactants which contain varying hydrophobe and hydrophile chain lengths. The hydrophiles are represented as polyoxyethylene chains, although such hydrophiles as polyglucosides and polyoxypropylene/polyoxyethylene chains are used to a lesser extent.

Linear Primary Alcohol Ethoxylates (LPAE). The hydrophobes of LPAE are made in several ways:

1. Catalytic addition of CO and H_2 to linear detergent range olefins yields alcohols which generally contain 15-50 percent alkyl branches at the 2-carbon position.
2. Oligomerization of ethylene using aluminum alkyl technology followed by hydrolysis to alcohols which contain less than 5% branching.
3. Conversion of triglycerides, found in animal fats or vegetable oils, to corresponding alcohols which contain less than 1% branching.

Ethoxylates derived from essentially linear alcohols produced by the above processes have been shown in many extensive studies (3-10) to biodegrade rapidly to products which do not foam or show toxicity to aquatic life.

Branched Primary Alcohol Ethoxylates (BPAE). The hydrophobes of BPAE are produced by oligomerization of propylene or butene followed by catalytic addition of CO and H_2 to yield highly branched alcohols. The ethoxylates of these alcohols biodegrade more slowly and less extensively than the linear alcohol ethoxylates (11,12).

Linear Secondary Alcohol Ethoxylates (LSAE). The hydrophobes of LSAE are made via borate-modified oxidation of n-paraffins to form inorganic esters followed by hydrolysis. The resulting alcohols contain secondary hydroxyl groups randomly located along the linear alkyl chain. LSAE biodegrade slightly slower than LPAE (13,14).

Alkylphenol Ethoxylates (APE). The hydrophobes of most commercial APE are made by reacting phenol with either propylene trimer or diisobutylene to form nonylphenol or octylphenol. These products contain an aromatic moiety and extensive branching in their alkyl chains. It has been shown that APE biodegrade more slowly and less extensively than LPAE (3,15-20). The difference is more pronounced when the treatment system is operating under stress conditions such as low temperatures and high surfactant loadings.

Primary and Ultimate Biodegradation

As shown in Figure 2, biodegradation of nonionic surfactants to intermediates having different physical and chemical properties is termed primary biodegradation. In the example shown, a surface-active, aquatically toxic alcohol ethoxylate has degraded to less surface active, less toxic intermediates. Surfactant intermediates may be relatively resistant to further microbial attack when they arise from nonionics which contain significant branching and/or aromaticity. These intermediates, shown in Figure 3, are more toxic to aquatic life than their intact precursors (21). The older methods used to measure primary biodegradation, such as foaming (3) and cobalt thiocyanate active substance (CTAS) (22), frequently do not determine the presence of these less biodegradable intermediates. Newer methods, such as high performance liquid chromatography (HPLC) (23-25) and gas chromatography (GC) coupled with mass spectroscopy (MS) (26,27), are useful for the determination of these intermediates.

The aerobic biodegradation of a linear alcohol ethoxylate to its ultimate products CO_2 and H_2O is shown schematically in Figure 4. Also shown are methods for determination of ultimate biodegradability. Even the most rapidly biodegradable substrates such as glucose, are not completely mineralized to CO_2 and H_2O since some organic material is converted to cellular mass. Generally, an oxygen uptake of 50% or greater of the theoretical amount as measured by biochemical oxygen demand (BOD) in respirometry tests is considered sufficient to categorize the substrate as having "ready biodegradability" (28).

The ultimate biodegradability of a substrate, such as is depicted in Figure 4, may, in addition to oxygen uptake, be measured by disappearance of organic carbon, CO_2 evolution and the formation of water. A radiotracer approach provides a more accurate determination and is the only feasible way of measuring the formation of water in the aqueous medium required for all metabolic processes.

Potential Points of Initial Attack for Nonionic Surfactants. Bacterial inocula have been shown to produce enzymes which can initiate degradation at any of three different locations (2) of the nonionic surfactant. As depicted in Figure 5, these are:

1. A central fission mechanism in which the hydrophobe is cleaved from the hydrophile. The well established β-oxidation mechanism then converts the linear chains to CO_2 and H_2O as indicated in Figure 6.
2. ω-hydrophobe attack in which the far end of the hydrophobe is first oxidized to a carboxylic acid. Biodegradation can then proceed inward, particularly in the case of linear alkyl chains, via β-oxidation.
3. ω-hydrophile attack in which the far end of the polyoxyethylene (POE) chain is oxidized initially to a carboxylic acid. Further conversion of the POE chain to CO_2 and H_2O then proceeds by a mechanism that has yet to be elucidated in detail.

Figure 6 depicts the β-oxidation process (2,29) which proceeds after initial oxidative attack by enzymes to split a linear alcohol ethoxylate into an alkyl carboxylic acid and polyethylene glycol (PEG). β-oxidation begins with esterification of the acid by a coenzyme called coenzyme A. The alkyl chain is then degraded to acetyl groups and another carboxylic acid with two less carbon atoms. This process continues two carbon atoms at a time to ultimately produce acetyl groups in the case of even carbon-numbered alcohols and propionyl groups in the case of odd carbon-numbered alcohols. The acetyl and propionyl groups can easily be incorporated into cellular mass or converted to CO_2. β-oxidation is an extremely fast process and does not leave any measurable quantities of biodegradation intermediates.

Ultimate Biodegradation of the Four Types of Nonionic Ethoxylates

In a Gledhill-modified version (30) of the Sturm CO_2 evolution shake flask test (31), the following nonionic ethoxylates were studied:

Linear Primary Alcohol Ethoxylates

$$RO(CH_2CH_2O)_nH$$

$$R = CH_3(CH_2)_m$$

Branched Primary Alcohol Ethoxylates

$$RO(CH_2CH_2O)_nH$$

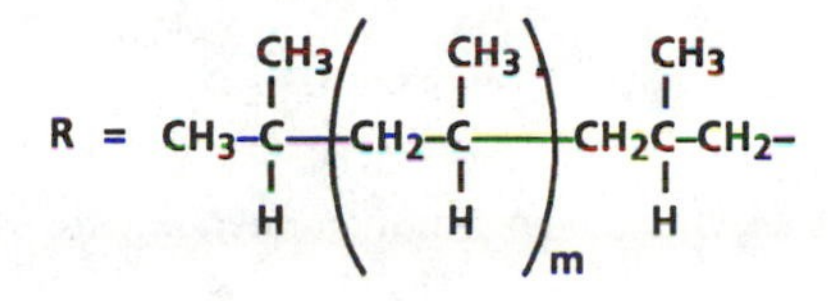

Linear Secondary Alcohol Ethoxylates

$$\begin{matrix} R_1 \\ R_2 \end{matrix} > CH{-}O(CH_2CH_2O)_nH$$

$$R_1 = CH_3(CH_2)_m \; ; \quad R_2 = CH_3(CH_2)_p$$

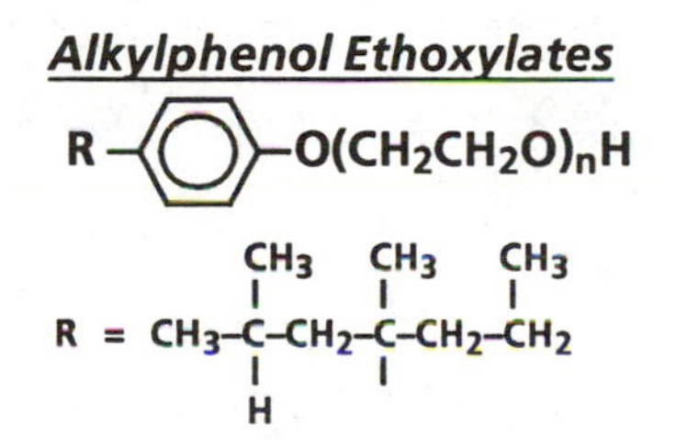

Figure 1. Structural features of nonionic classes.

$$RCH_2O(CH_2CH_2O)_nH$$

Surface Active
Toxic to Aquatic Life

↓ Bacteria + O_2

$$RCH_2OH + HO(CH_2CH_2O)_nH$$

Not Surface Active
Not Toxic to Aquatic Life

Figure 2. Primary biodegradation.

When R is Essentially Linear Alkyl

$$RCH_2O(CH_2CH_2O)_nH \longrightarrow RC(=O)OH + HO(CH_2CH_2O)_nH$$

When R is Aromatic or Highly Branched Alkyl

$$RO(CH_2CH_2O)_nH \longrightarrow RO(CH_2CH_2O)_{n-1}H + HOCH_2CH_2OH$$

$$HOCH_2CH_2OH \downarrow CO_2 + H_2O$$

$$RO(CH_2CH_2O)_{n-1}H \downarrow \downarrow RO(CH_2CH_2O)_2H$$

Relatively Bioresistant

Figure 3. Biodegradation intermediates.

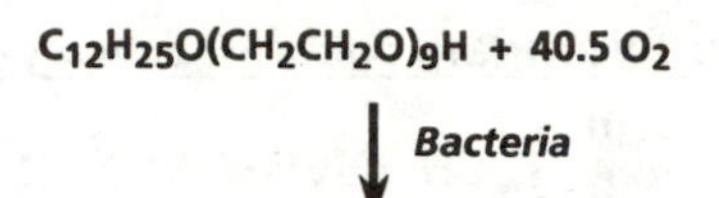

Determination	Method
• Oxygen Uptake	Respirometry
• Loss of Organic Carbon	DOC/^{14}C
• CO_2 Evolution	Titration/^{14}C
• Formation of H_2O	^3H Only

Figure 4. Ultimate biodegradation.

Central Fission

$$RCH_2O(CH_2CH_2O)_nH \xrightarrow{O_2} RC(=O)OH + HO(CH_2CH_2O)_nH$$

ω-Hydrophobe

$$CH_3(CH_2)_nO(CH_2CH_2O)_nH \xrightarrow{O_2} HO{-}C(=O){-}(CH_2)_nO(CH_2CH_2O)_nH$$

ω-Hydrophile

$$RO(CH_2CH_2O)_nH \xrightarrow{O_2} RO(CH_2CH_2O)_{n-1}CH_2C(=O)OH$$

Figure 5. Points of initial attack.

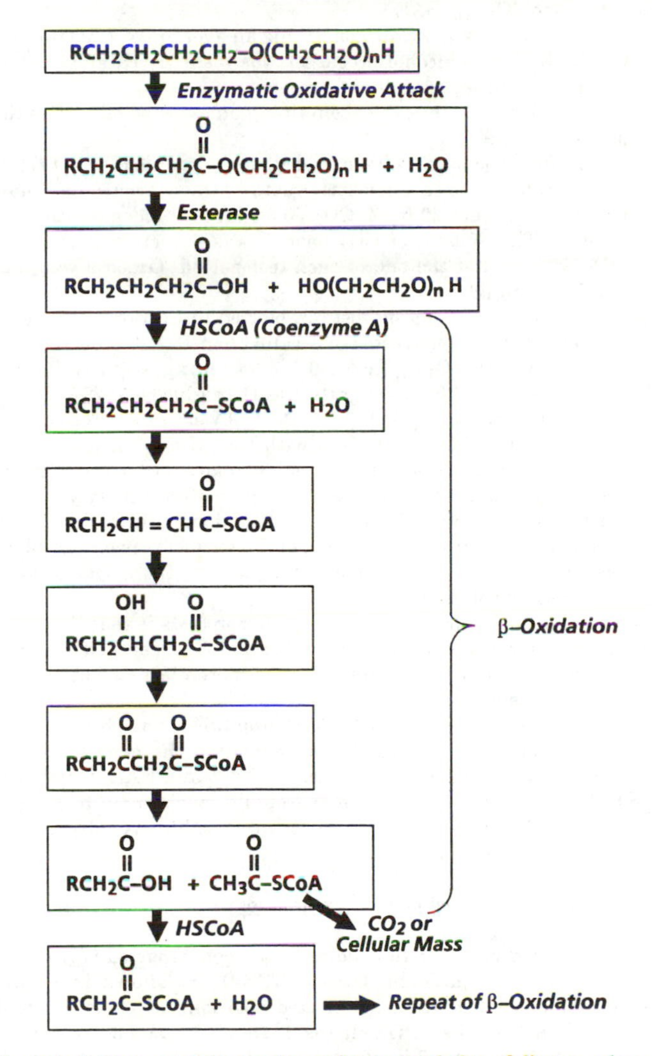

Figure 6. Suggested biodegradation pathway for hydrophobe of linear primary alcohol ethoxylates.

1. A C_{12-15} essentially linear primary alcohol ethoxylate having an average of 9 ethylene oxide (EO) units per mole of alcohol (C_{12-15}LPAE-9). The alcohol was prepared from C_{11-14} olefins using catalytic addition of CO and H_2. Approximately 80% of this alcohol contained linear alkyl chains. The 20% remaining contained 2-alkyl branches, mostly methyl.
2. A C_{11-16} linear secondary alcohol ethoxylate containing an average of 9 EO units per mole of alcohol (C_{11-16}LSAE-9).
3. A C_{13} branched alcohol ethoxylate containing an average of 7 EO units per mole of alcohol (C_{13}BAE-7). The alcohol precursor was made by catalytic addition of CO and H_2 to highly branched propylene tetramer.
4. A branched nonylphenol ethoxylate containing an average of 9 EO units per mole of nonylphenol (NPE-9).

Shake flasks were inoculated with mixed liquor suspended solids from activated sludge units in a Houston area domestic waste sewage treatment plant. Initial surfactant concentrations were 20 mg/ℓ. CO_2 formed from biodegradation was trapped in aqueous $Ba(OH)_2$. The amount of CO_2 formed was determined by back-titrating residual $Ba(OH)_2$ with HCl at the end of each test period. Glucose was included as a positive biodegradation standard.

The results of this CO_2 evolution test are plotted in Figure 7. As shown, the two linear alcohol ethoxylates were converted to greater than 75% of their theoretical yields of CO_2 in 30 days with the C_{12-15}LPAE-9 biodegrading slightly faster then the C_{11-16}LSAE-9 and to a level of 88%—slightly less than glucose (92%). Conversely, the two highly branched nonionics, C_{13}BAE-7 and NPE-9 were converted to less than 50% of their theoretical yield of CO_2 in 30 days with the NPE-9 biodegrading somewhat slower than the C_{13}BAE-7. In these studies a slight induction period was observed for all the surfactants but not for the glucose. This induction period was significantly greater for the branched nonionics than for linear nonionics. It is likely that induction represents sorption of the hydrophobic portion of the surfactants onto sludge with later release as less surface active intermediates desorb from the sludge and are enzymatically attacked in solution (13).

Another interesting feature of these CO_2 evolution tests is that CO_2 formation was still increasing for the linear alcohol ethoxylates but had reached a plateau for the branched nonionics. This suggests the formation of more bioresistant intermediates in the case of the branched surfactants.

Biodegradation of the linear products LPAE and LSAE have been shown to rapidly produce significant quantities of PEG (13,32) indicating the central fission pathway. However, biodegradation of the branched products NPE and BAE do not produce PEG, which suggests they biodegrade by the ω-hydrophile mechanism to give NPE's and BAE's with shorter POE chains (11,33,34). These intermediates biodegrade more slowly by a mechanism which has yet to be elucidated.

Biodegradation Pathways By Radiotracer Techniques

Somewhat more detailed mechanistic studies have been reported (13) using a double-labeled linear C_{14}AE-9, and a double-labeled NPE-9. As shown in Figure 8, these nonionics were labeled with tritium in selected positions of their hydrophobes and uniformly with carbon-14 in their POE chains. It should be noted that tritium labeling in the hydrophobe of the C_{14}AE-9 is concentrated in the alpha and gamma positions. Figure 9 shows the pattern of ultimate biodegradation for the hydrophobe of C_{14}AE-9 in a shake flask test similar to that used in the results discussed previously for the CO_2 evolution studies of the four types of nonionics. The alkyl CO_2 curve was obtained by subtracting $^{14}CO_2$ of the POE chain (determined by scintillation counting) from total

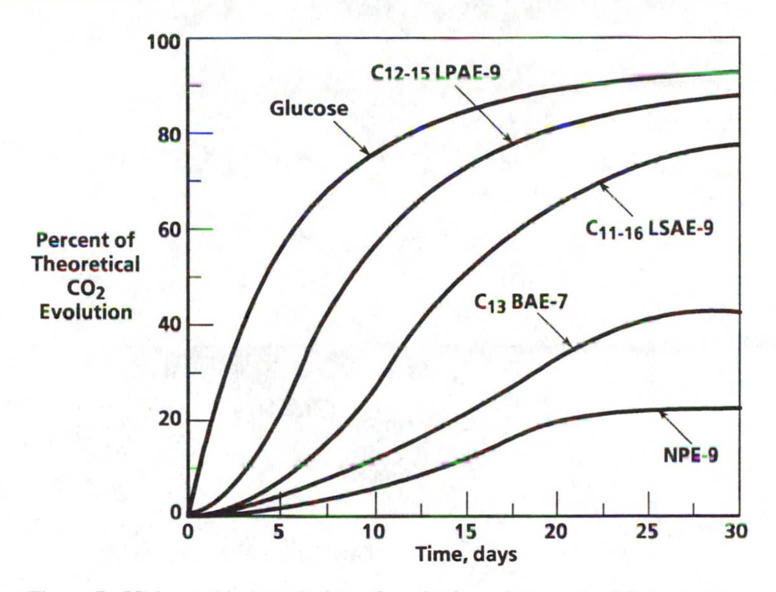

Figure 7. Ultimate biodegradation of nonionic surfactants by CO_2 evolution.

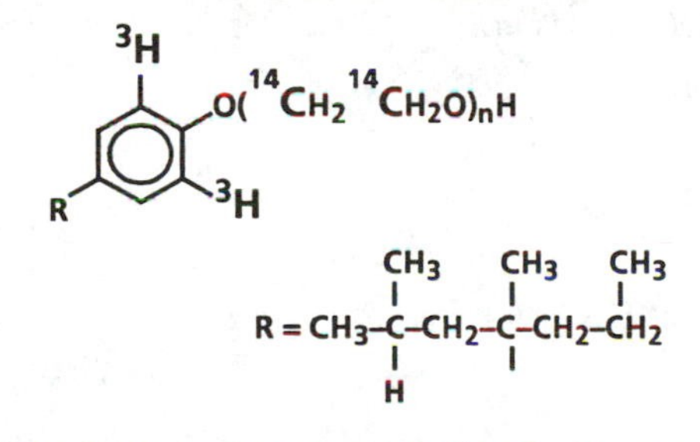

Figure 8. Double radiolabeled surfactants. (Reprinted with permission from Ref. 33. Copyright 1982 Rodman.)

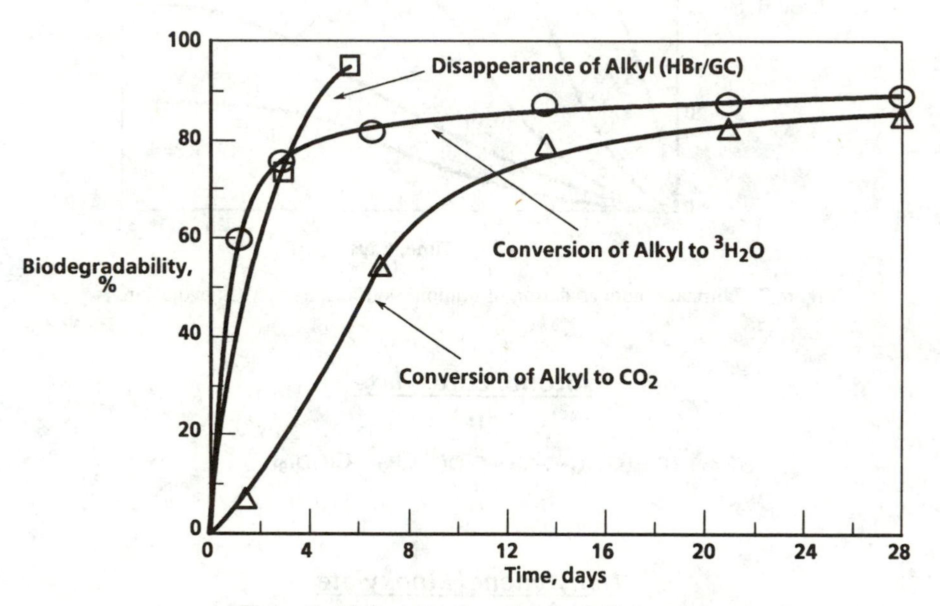

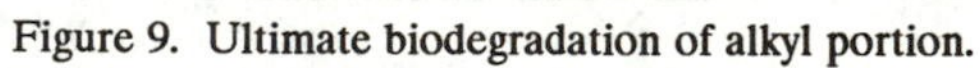
Figure 9. Ultimate biodegradation of alkyl portion.

CO_2 (determined by titration). The rapid appearance of ^3H_2O in solution was accompanied by very little CO_2 evolution. This indicates that the tritiated portion of the alkyl chain, located near the hydrophobe-hydrophile junction, was biodegrading at a much faster rate than the alkyl chain as a whole. Disappearance of the alkyl chain by HBr/GC is also shown and indicates that primary biodegradation proceeded at a rate which was parallel to that of ^3H_2O formation in the early stages of this study. These results are consistent with a biodegradation mechanism in which the initial step is cleavage of the molecule to form hydrophobic and hydrophilic products followed by alkyl degradation to CO_2 beginning at the functional group rather than at the terminal methyl group. The above mechanism is in line with studies by Patterson and co-workers (32) which indicate an initial cleavage of the alkyl material from the POE material followed by alkyl degradation. However, Patterson's studies did not indicate whether biodegradation proceeded from the functional portion of the alkyl chain or from the terminal methyl group. In contrast, Nooi and co-workers using ^{14}C-labeled AE at different locations in LPAE, indicate that the initial oxidation takes place at the terminal methyl group prior to hydrolytic cleavage of the alkyl group from the POE group (35). It appears that different bacterial strains may exist with selective capabilities to initiate biodegradation of alcohol ethoxylates by more than one mechanism. However, the major pathway with non-selected bacterial strains such as are found in domestic waste treatment plants appears to be central fission.

In a separate study (33), the double-labeled NPE-9, shown in Figure 8, was fed, along with a slipstream of a domestic waste treatment plant influent, to a bench-scale activated sludge unit. Table I lists data from this slipstream study showing the ratio of EO to hydrophobe in effluent samples rapidly went from 9 to approximately 2-3 within eight hours and remained at this ratio throughout the 28-day course of the study. These results are in line with APE biodegradation mechanisms proposed by other workers (11,34) who indicate that APE biodegradation is initiated by ω-attack at the far end of the POE chain with gradual shortening of the chain to a relatively bioresistant alkylphenoxy moeity.

Table I. Ratio EO/Hydrophobe During Biodegradation of NPE-9

Time, days	Ratio, EO/Hydrophobe*
0.08 (2 hr)	5.4
0.33 (8 hr)	2.4
1	2.2
18	2.4
24	2.7
28	2.2

SOURCE: Ref. 33. Copyright 1982 Rodman.
**Basis, $^{14}C\,/\,^{3}H$ Activity in Effluent*

Table II lists the radioactivity balance for the 28-day slipstream study (33) in which doublelabeled $C_{12\text{-}15}$LPAE-9 was also a substrate. The considerably greater amount of effluent ^3H_2O found at the completion of the study for $C_{12\text{-}15}$LPAE-9 (91.6%) compared to that found for NPE-9 (28.3%) indicates the ultimate biodegradation of the

LPAE hydrophobe was more extensive than that of the NPE. Also, the much higher level of tritium found in the biomass (activated sludge) for NPE suggests that the biomass had sorbed significant quantities of the NPE-2 biodegradation intermediate. The sorption of an NPE with a short EO chain onto activated sludge has recently been observed in our laboratories using environmental samples from a waste treatment plant which was experiencing normal high loadings of NPE-9 from industrial point sources (36). The results for influent, activated sludge and effluent samples from the plant are listed in Table III. CTAS values, basis dewatered sludge, were 1300 mg/ℓ. Since CTAS does not respond well to low EO-containing nonionics, the recently developed HPLC/MS procedure of Giger (23-25) was used and showed almost double the levels found by CTAS. In addition, the HPLC/MS procedure positively identified the sorbed material as an NPE mixture much more concentrated with NPE-2 than was present in the influent NPE-9. This sludge enrichment in shorter chain NPE's has also been observed by Giger and co-workers (37).

Table II. Radiochemical Product Distribution

Product	Distribution, %		
	$C_{12\text{-}15}$ LPAE-9	NPE-9	Glucose
^3H_2O	91.6	28.3	--
Soluble ^3H-Metabolites	1.5	36.7	--
^3H in Biomass	4.1	31.5	--
Total	97.2	96.5	
$^{14}CO_2$	65.4	54.0	42.8
Soluble ^{14}C-Metabolites	1.3	8.5	1.4
^{14}C in Biomass	24.7	27.6	52.0
Total	91.4	90.1	96.2

Table III. NPE Fate in Biotreatment

Influent, mg/ℓ		Activated Sludge, mg/ℓ 2)		Effluent, mg/ℓ	
CTAS	HPLC	CTAS	HPLC	CTAS	HPLC
63	55 1)	1300	2340 3)	5.4	4.6 1)

SOURCE: Ref. 33. Copyright 1982 Rodman.

1) *Identified as NPE-9*

2) *Basis, Dewatered Sludge*

3) *Identified as NPE Heavily Enriched with NPE-2*

The use of these radiolabeled $C_{12\text{-}15}$LPAE-9 and NPE-9 in continuous activated sludge benchscale units simulating winter conditions has been reported (38). The results for biodegradation of the hydrophobes to ^3H_2O are shown in Figure 10 and suggest that under winter conditions biodegradation of the hydrophobe of linear alcohol ethoxylates is essentially unaffected while biodegradation of the hydrophobe of NPE is slowed significantly with decreasing temperature.

Aquatic Toxicity of Effluents

In a recent study (39), bench-scale activated sludge units were fed up to 100 mg/ℓ of LPAE and NPE to simulate influent concentrations from industrial sources. The effluents of these units were subjected to aquatic toxicity tests. The results, summarized in Figure 11, show the effluent from the NPE unit to be toxic at effluent dilutions as low as 7.3 percent. The LPAE effluent was rendered completely non-toxic under these conditions. The results of this study also showed that high loadings of surfactants which biodegrade slowly may have an adverse impact on the activated sludge process by causing poor biosolids growth and settling, impaired BOD removal and loss in nitrification capability. These intermediates sorb strongly to sludge and may interfere in the proper functioning of activated sludge treatment.

Summary

As discussed previously, the biodegradation pathways using non-selected bacterial inocula for nonionic ethoxylates may be divided into two distinct areas, each dependent on the structure of the surfactant.

1. *Linear Alcohol Ethoxylates*
   These biodegrade via central fission initiation in which the hydrophobe is cleaved from the hydrophile to produce non-surface active, non-toxic intermediates which readily biodegrade to CO_2 and H_2O.

2. *Highly Branched Alkylphenol Ethoxylates and Highly Branched Alcohol Ethoxylates*
   Biodegradation of these nonionics is initiated by ω-hydrophile attack in which the POE chain is shortened. The evidence for this pathway is extensive for APE. Although mechanistic pathways for branched AE need further study, it has been reported that their branched alkyl groups direct initial attack to occur by an ω-hydrophile mechanism similar to that for APE. This attack, followed by shortening of the POE chain to approximately 2-3 EO units per mole of hydrophobe, produces intermediates which degrade more slowly than their intact surfactant precursors, have higher aquatic toxicity, and sorb strongly onto activated sludge thereby having the capability of upsetting the activated sludge process.

Currently, those surfactants with highly branched alkyl chains and those with aromaticity do not have an appreciable impact on domestic waste treatment from household sources since they are not used extensively in household laundry and dishwashing which are the largest volume end uses for surfactants. However, these highly branched surfactants are used in many industrial applications such as textiles and pulp and paper. When used in the industrial sector, they generally enter domestic waste treatment plants at much higher loadings than are found for surfactants from household sources. The fate of these surfactants in effluents, sludges, receiving waters and sediments will require further studies to ascertain their environmental impact.

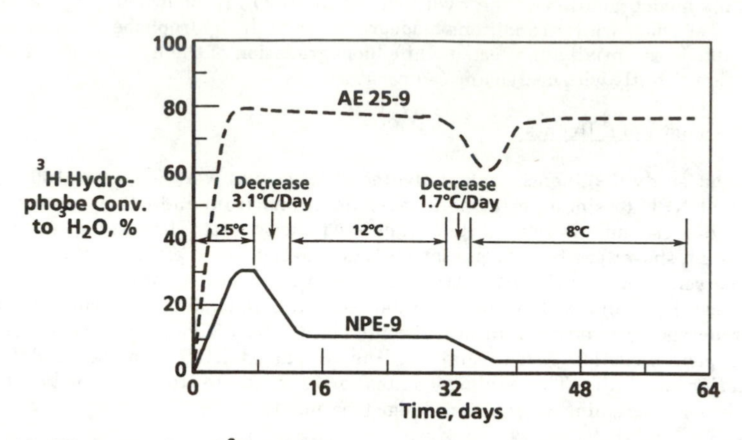

Figure 10. Biodegradation of ^{3}H-hydrophobe groups of C_{12-15}LPAE-9 and NPE-9 to ^3H_2O. (Reprinted with permission from Ref. 38. Copyright 1984 Tenside Detergents.)

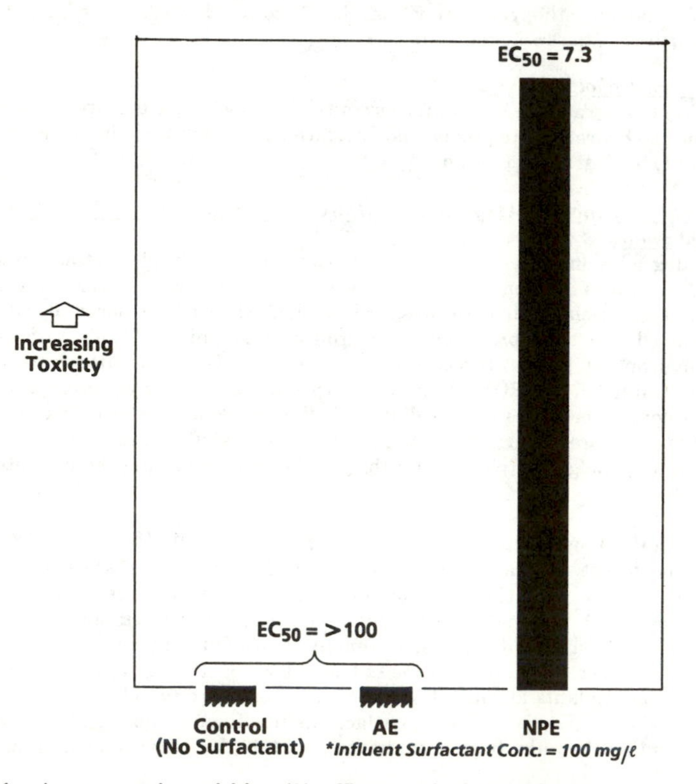

Figure 11. Acute aquatic toxicities (% effluent) of biotreated effluents* to fathead minnow.

Literature Cited

1. Haupt, D. E., Tenside 6, 332 (1983).
2. Swisher, R. D. Surfactant Biodegradation, Surfactant Science Series, Vol. 18, Marcel Dekker, Inc., New York (1987).
3. Mann, A. H.; Reid, V. W. J. Am. Oil Chem. Soc. 48, 794 (1971).
4. Huddleston, R. L.; Allred, R. C. J. Am. Oil Chem. Soc. 42, 983 (1965).
5. Abram, F. S. H.; Brown, V. M.; Painter, H. A.; Turner, A. H. Proceedings of IV Yugoslav Symposium on Surface Active Substances, Section D, Dubrovnik, October 1977.
6. Sykes, R. M.; Rubin, A. J.; Rath, S. A.; Chang, M. C. J. Water Poll. Control Fed. 51, 71 (1979).
7. Kravetz, L. J. Am. Oil Chem. Soc. 58, 58A (1981).
8. Birch, R. R. J. Am. Oil Chem. Soc. 61, 340 1984).
9. Steber, J.; Wierich, P. Tenside 20, 183 (1983).
10. Watson, G. K.; Jones, N. Gen. Microbiol. Quart., 6, 78 (1979).
11. Patterson, S. J.; Scott, C. C. ; Tucker, K. B. E. J. Am. Oil Chem. Soc. 44, 407 (1967).
12. Schöberl, P.; Kunkel, E.; Espeter, K. Tenside 18, 64 (1981).
13. Kravetz, L.; Chung, H.; Rapean, J. C.; Guin, K. F.; Shebs, W. T. Presented at American Oil Chemists' Society 69th Annual Meeting, St. Louis, May 1978.
14. Wickbold, R. Vom Wasser 33, 229 (1966).
15. Brown, D.; de Henau, H.; Garrigan, J. T.; Gerike, P.; Holt, M.; Kunkel, E.; Matthijs, E.; Waters, J.; Watkinson, R. J. Tenside 24 (1987).
16. Borstlap, C.; Kortland, C. FSA 69, 736 (1967).
17. Lashen, E. S.; Blankenship, F. A.; Booman, K. A.; Dupre, J. J. Am. Oil Chem. Soc. 43, 371 (1966).
18. Narkis, N.; Schneider-Rotel, M. Water Res. 14, 1225 (1980).
19. Gerike, P.; Jasiak, W. Surf. Cong. No. 8, 1, 195 (1984).
20. Pitter, P.; Fuka,T. Env. Protect. Eng. 5 47 (1979).
21. Yoshimura, K. J. Am. Oil Chem. Soc. 63, 1590 (1986).
22. Boyer, S. L.; Guin, K. F.; Kelley, R. M.; Mausner, M. L.; Robinson, H. F.; Schmitt, T. M.; Stahl, C. R.; Setzborn, E. A. Environ. Sci. Technol. 11, 1167 (1977).
23. Ahel, M.; Giger, W. Anal. Chem. 57, 1577 (1985).
24. Ahel, M.; Giger, W. Ibid. 57, 2584 (1985).
25. Giger, W.; Ahel, M.; Koch, M.; Laubscher, H. U.; Schaffner, S.; Schneider, J. Wat. Sci. Tech. 19, 449 (1987).
26. Holt, M. S.; McKerrell, E. H.; Perry, J.; Watkinson, R. J. J. Chrom. 362, 419 (1986).
27. Ahel, M.; Conrad, T.; Giger, W. Environ. Sci. Technol. 21, 697 (1987).
28. Gerike, P.; Fischer, W. K. Ecotoxicol. Env. Safety 5, 45 (1981).
29. Stumpf, P. K.; Barber, G. A. Comparative Biochemistry, Vol. 1, Academic Press, New York, 1960.
30. Gledhill, W. E. Appl. Microbiol. 30, 922 (1975).
31. Sturm, R. N. J. Am. Oil Chem. Soc. 50, 159 (1973).
32. Patterson, S. J.; Scott, C. C.; Tucker, K. B. E. J. Am. Oil Chem. Soc. 47, 37 (1970).
33. Kravetz, L.; Chung, H.; Guin, K. F.; Shebs, W. T.; Smith, L. S. Household Pers. Prod. Ind. 19, 46 and 72 (1982).
34. Rudling, L.; Solyom, P. Water Res. 8, 115 (1974).
35. Nooi, J. R.; Testa, M. C.; Willemse, S. Tenside 7, 61 (1970).
36. Unpublished Shell data.
37. Giger, W.; Brunner, P. H.; Ahel, M.; McEvoy, J.; Marcomini, A.; Schaffner, C. Gas, Wasser, Abwasser, 67, 111 (1987).
38. Kravetz, L.; Chung, H.; Guin, K. F.; Shebs W. T.; Smith, L. S. Tenside 21, 1 (1984).
39. Salanitro, J. P.; Langston, G. C.; Dorn, P. B.; Kravetz, L. Wat. Sci. Tech. 20, 125 (1988).

RECEIVED February 14, 1990

Chapter 10

Biodegradation of Polyethers

Fusako Kawai

Department of Biology, Kobe University of Commerce, Tarumi, Kobe 655, Japan

Polyethers are classified into biodegradable synthetic polymers. Among them, polyethylene glycols (PEGs) are most widely used as commodity chemicals in various industrial fields. Consequently, biodegradability studies have been primarily focused on PEGs. PEGs are aerobically or anaerobically metabolized by various bacteria. Biochemical routes have also been studied for aerobic degradation of the compound. On the other hand, bacterial degradation of polypropylene glycol and its metabolic route is suggested. Another polyether, polytetramethylene glycol, is also utilized by aerobic bacteria as the sole carbon and energy source.

Polyethers have a common structural formula: $HO[R-O]_nH$ ($R=CH_2CH_2$ for polyethylene glycol (PEG), CH_3CHCH_2 for polypropylene glycol (PPG) and $(CH_2)_4$ for polytetramethylene glycol (PTMG)). A number following an abbreviation of a polyether will indicate an average molecular weight (Mn). These materials are synthesized by ring-opening polymerization from alkylene oxides. Usually the reaction results in the formation of a mixture of homologous compounds differing in molecular weights. A mixture is fractionated into different ranges of molecular weights and assigned an Mn of every fraction. Among polyethers, PEGs are manufactured in large quantities and used as commodity chemicals in various industrial fields such as pharmaceuticals, cosmetics, lubricants, antifreezing agents, printing inks, etc. Most nonionic surfactants are made from alcohols and alkylphenols and ethylene oxide to form their polyethers. The ethylene oxide moiety of these nonionic polyethers is consumed faster than that of the polyethers made exclusively from ethylene oxide or propylene oxide. Ultimately, these will make a significant contribution to domestic and industrial waste water systems. Therefore, much concern has been paid to their biodegradability characteristics in the last 20 years.

On the other hand, PPGs are used in their original form for solvents for drugs and ingredients of paints, lubricants, inks, cos-

0097–6156/90/0433–0110$06.00/0

metics, etc., but most of them are transformed to polyurethanes or surface active agents. Another polyether, PTMG, is used exclusively as a constituent of polyurethane. Waste water containing these materials are emitted not only from synthetic processes, but also from domestic and industrial sources.

This report summarizes the biodegradability studies on polyethers and introduces future prospects (1).

Biodegradation of PEG

Analysis of PEG Samples. PEGs with molecular weights from 106 (dimer) to 20,000 are manufactured and used in many industrial fields. Although the physical properties of PEGs vary from viscous liquids to waxy solids according to their Mns. PEGs are completely water-soluble. PEGs are known as nontoxic materials and are often used as probe agents for the measurement of gastrointestinal permeability or absorption in man and other mammals. Meanwhile, more recently the potential toxicity resulting from repeated, topical applications of a PEG-based antimicrobial cream to burn patients and to open wounds in rabbits was indicated. The world-wide production of PEG amounts to approximately one million tons per year and its release into natural environments necessitates more information on the biodegradability of this material and the application of better microbial systems to waste water treatments.

Prior to beginning detailed degradation studies of this material, we have to know the ranges and contents of each species of PEG. We have assayed reagent grade PEG products for laboratory use by high-performance liquid chromatography (HPLC) and found that they included a respective range of species of PEGs, but their ranges differed from each other (Figure 1). It seems that we do not have to worry about the contamination of lower molecular weight species for certified reagent grade products when sufficient bacterial growth is introduced to PEGs. High temperature, an oxygen atmosphere and an initiator are supposed to cause autooxidation of PEG. In our experiments, we usually cultivated microorganisms in autoclaved PEG media at 30 °C with shaking, causing no apparent growth problems. However, in studies of PEG metabolism, we incubated intact cells or enzyme preparations with unsterilized PEG or tetraethylene glycol at pH 7.0 or 8.0 at 28-30°C with shaking. To check nonbiological degradation of PEGs, reaction mixtures without cells or enzyme preparations were incubated in parallel. Comparison of PEG solutions before and after incubation (for 10 days at most) revealed no reduction in the amount of PEG by the determination based on Stevenson's method, and no detectable amounts of oxidized products were formed based on an assay by thin layer chromatography, gas chromatography, and HPLC.

PEG-utilizing Bacteria. Since the first report of Fincher and Payne on a PEG-utilizing bacterium in 1962 (2), PEGs having an Mn higher than 1000 had long been considered bioresistant. In 1973, Pitter found that an activated sludge or a pure culture completely degraded PEG 1000 and 1500, but degraded PEG 3500 only slightly (3). Cox and Conway obtained adapted activated sludge cultures that can grow on PEGs with Mns up to 4000, but they did not degrade PEG 6000 (4). In 1975, we isolated various PEG-utilizing cultures by various enrichments on PEGs 400 to 20,000 and classified them into five groups

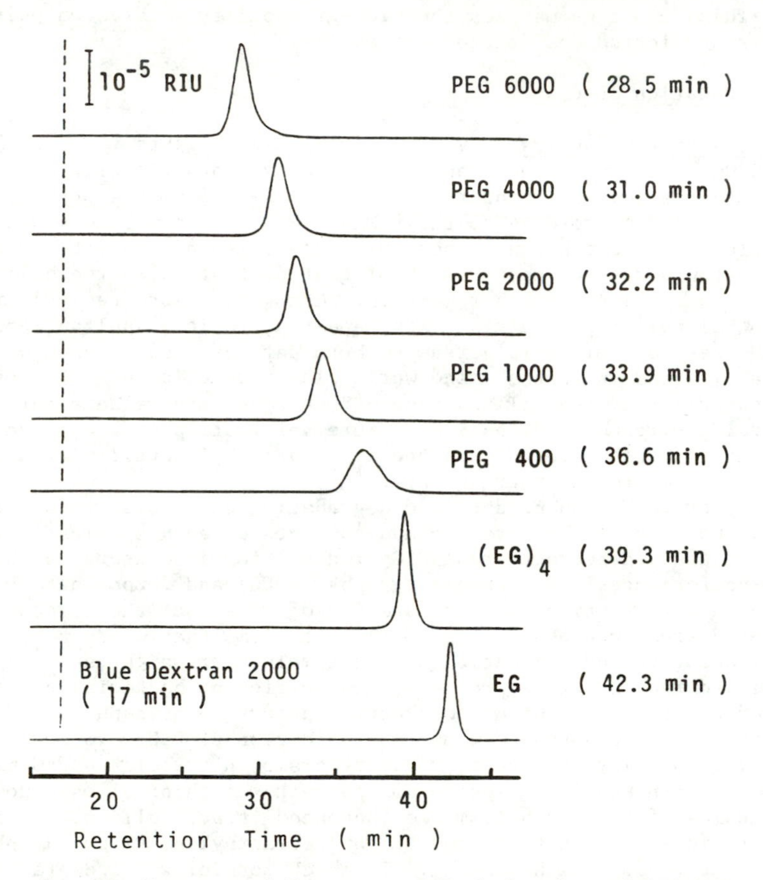

Figure 1. The Ranges and Contents of Each Species of PEG as Related to Mns. PEG materials used were products graded for laboratory use. They were analyzed by HPLC with a Hitachi Liquid Chromatograph 655 with a refractive index detector (Erma ERC-7510). The operational conditions were as follows: column, Toyo Soda TSK-GEL G3000PW 7.5 ϕ x 600 mm; sample size, 1%, 20 μl; eluent, distilled water; flow rate, 0.5 ml/min; pressure, 80 kg/cm^2; temperature, room temperature (16.5 ± 0.5°C). (Reproduced with permission from Ref. 1. Copyright 1987 CRC Press, Inc.)

according to their PEG-assimilating limits (5). PEG 6000 or 20,000 was assimilated only by symbiotic mixed cultures consisting of two bacterial strains. The representative culture E-1 consisted of *Flavobacterium* species and *Pseudomonas* species which individually could not assimilate PEGs (6). *Flavobacterium* species was dominant during the growth of the mixed culture. The mixed culture assimilated various PEGs with an Mn from 300 to 20,000. Oligomers could also be assimilated by the mixed culture in the presence of 0.01 % PEG 6000 as an inducer for PEG-metabolizing enzymes (7). As waste waters seem to contain PEGs with different Mns, a mixture of various PEGs (400, 600, 1000, 1540, 2000, 4000 and 6000) was used as the carbon source (each 0.1 % in a total 0.7 %) for biodegradation tests of the mixed culture (Figure 2). A mixture of PEGs were almost completely consumed during the 4-day culture. Observation of thin-layer chromatography of the culture showed that low molecular weight PEGs are decomposed more rapidly than high molecular weight PEGs.

On the other hand, Haines and Alexander reported that they isolated a soil bacterium identified as *Pseudomonas aeruginosa* which was capable of growing on and degrading PEG compounds ranging from monomer to PEG 20,000 (8). However, the original strain was lost. Hosoya et al. also found that two bacterial strains assimilated PEG 6000 (9). No further studies of this type have been noted by these authors to date.

In addition, anaerobic biodegradation of PEGs with Mns up to 20,000 was reported approximately at the same time in 1983 by three groups (10-12). Anaerobic digestion is also a useful process in sewage disposal systems and the studies on anaerobic metabolism of PEG are important to learning the fate of this compound in ecosystem. Research on this field is in its early stage, and further extensive studies remain to be done in the future.

Biochemical Routes for Aerobic Degradation of PEG. We also examined biochemical routes for aerobic degradation of PEG from the viewpoints of metabolizing enzymes and metabolites. Dehydrogenation of PEGs linked with electron acceptors such as NAD, falvins, ferricyanide, etc. has been observed with enzyme preparations obtained from PEG-grown bacteria by several researchers. We found that 2,6-dichloroindophenol (DCIP)-dependent membrane-bound dehydrogenases play a major role in the metabolism of PEG (13). We have purified membrane-bound PEG dehydrogenases from a PEG 6000-utilizing symbiotic mixed culture E-1 (14) and from PEG 4000-utilizing *Flavobacterium* sp. No. 203 (15). The former enzyme was inducibly formed and the latter was constitutively present in the cells. Both enzymes had wide substrate specificities toward various PEGs from diethylene glycol to PEG 20,000. Ethylene glycol was not oxidized. The prosthetic group of both enzymes was identified as PQQ which was discoverd in 1979 as the third coenzyme for oxidoreductases (25, 26). The enzyme was quite different from other alcohol dehydrogenases reported so far and the enzyme was considered to be a novel quinoprotein. According to our further studies, other PEG dehydrogenases of our various isolates could possibly be quinoproteins.

As described above, PEG seems to be initially oxidized by a dehydrogenase, indicating that terminal primary alcohol groups are oxidized. Several research groups obtained the acidic metabolites or carboxylated PEGs in culture filtrates of bacteria grown on PEGs.

Patterson et al. detected acidic metabolic intermediates from alcohol ethoxylate (16). The formation of mono- and diacid metabolites of the glycol has also been confirmed in mammals. Thus, oxidation of terminal alcoholic groups to carboxyl groups seemed to be the common pathway in PEG metabolism. We also found that tetraethylene glycol-monocarboxylic acid and dicarboxylic acid, triethylene glycol, diethylene glycol and ethylene glycol were formed in a reaction mixture of the symbiotic mixed culture with tetraethylene glycol (7). Further aldehydic compounds, the first metabolites of PEG compounds, were identified as the reaction products of PEG dehydrogenase with tetraethylene glycol and n-pentanol (17).

In conclusion, PEG was found to be aerobically metabolized via the oxidation of terminal alcohol groups, as shown in Figure 3. PEG is successively oxidized to an aldehyde and monocarboxylic acid, which is followed by the cleavage of the ether bond, resulting in PEG molecules that are reduced by one glycol unit. Simultaneous oxidation of two terminal alcohol groups of the molecule is also possible. Depolymerization might proceed via the same procedure as that in monocarboxylic acid. This sequence is repeated and eventually yields depolymerized PEG. The resultant glyoxylic acid may then be assimilated into central metabolic routes by known pathways: the oxidative dicarbonic acid cycle and the glycerate pathway.

On the other hand, Haines and Alexander reported that PEG 20,000 might be hydrolyzed by an extracellular enzyme of Pseudomonas aeruginosa (8). According to our experiments, PEG 6000 was not hydrolyzed by an extracellular enzyme (the supernatant fluid of a culture), but it was metabolized by washed cells and the culture could not utilize ethylene glycol or diethylene glycol which might be the hydrolysis products of PEG.

More recently, Pearce and Heydeman suggested non-oxidative removal of ethylene glycol units as acetaldehyde by a membrane-bound, oxygen-sensitive enzyme of a novel type, i.e., diethylene glycol lyase (18). Schöberl suggested that PEG was catabolized by C_1 step, liberating formate which was metabolized by a serine pathway (19).

We cannot deny the possibility of a metabolic route for PEGs other than the oxidative degradation pathway via carboxylated PEG, although further evidence on the point seems to be required. The metabolic route for PEGs supported by plural researchers is the oxidative route as shown in Figure 3.

Symbiotic Degradation of PEG. The symbiotic mixed culture composed of Flavobacterium species and Pseudomonas species was obtained as a PEG 20,000-utilizing microbial system, as described above. The symbiotic mechanism of the mixed culture was examined as follows.

As any nutrients or even the culture filtrate of the mixed culture on PEG 6000 did not support the growth of a single culture on PEGs, the symbiosis is not due to the nutritional requirement between the two strains. Although Pseudomonas species supplies no nutrient to Flavobacterium species, the latter grew well on a dialysis culture using a cellulose tube or a membrane filter, while poor growth of the former was observed. These results suggest that metabolites are exchanged between the two bacteria. However, the mixed culture was the optimal system for PEG degradation by metabolic cooperation of the two bacteria. The cellular contact between Flavobacterium species and Pseudomonas species seemed to be necessary for the efficient

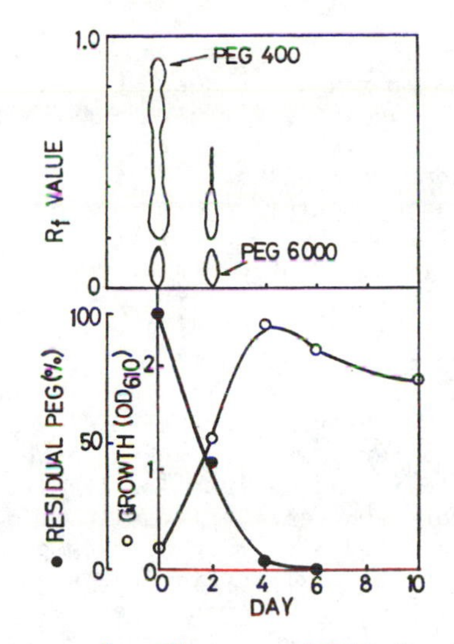

Figure 2. Degradation of a Mixture of PEG Samples with an Mn Ranging from 400 up to 6000. The medium was composed of PEGs (PEG 400, 600, 1000, 1540, 2000, 4000 and 6000), each 1 g (total 7 g), $(NH_4)_2HPO_4$ 3 g, K_2HPO_4 2 g, NaH_2PO_4 1 g, $MgSO_4 \cdot 7H_2O$ 0.5 g, and yeast extract 0.5 g in 1000 ml of tap water: the pH of the medium was adjusted to 7.2. The cultivation was carried out on 100 ml of the medium in a 500-ml shaking flask with an inoculum of 5 ml of a 7-day culture of a symbiotic mixed culture E-1. An aliquot of the culture was withdrawn and the pH of the rest of the culture was adjusted to 7.2 at the times indicated in the figure. The culture withdrawn was centrifuged to remove cells and the supernatant was subjected to thin layer chromatography on a Kiesel gel GF_{254} plate (E. Merck, 0.25 mm thick). The spots on a chromatogram corresponded to the residual PEG in the supernatant. Solvent system: ethylene glycol 80-methanol 20. Development at 5°C. Detection: Dragendorff reagent. (Reproduced with permission from Ref. 6. Copyright 1977 The Society of Fermentation Technology, Japan.)

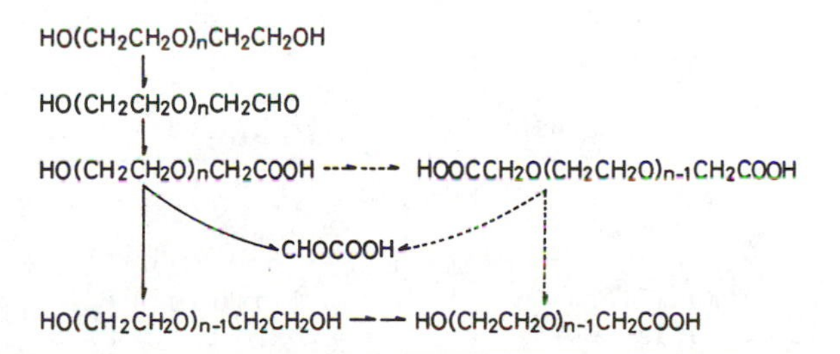

Figure 3. Proposed Oxidative Metabolic Pathway of PEG. (Reproduced with permission from Ref. 1. Copyright 1987 CRC Press, Inc.)

cooperation of the two bacteria. Three enzymes involved in the metabolism of PEG, PEG dehydrogenase, PEG-aldehyde dehydrogenase, and PEG-carboxylate dehydrogenase (ether-cleaving), were present in the cells of Flavobacteium species. The first two enzymes were not found in the cells of Pseudomonas species. PEG 6000 was neither degraded by intact cells of Flavobacterium species nor by those of Pseudomonas species, but it was degraded by their mixture. Glyoxylic acid, a metabolite liberated by the ether-cleaving enzyme, inhibited the growth of the mixed culture. The ether-cleaving enzyme was remarkably inhibited by glyoxylic acid: glyoxylic acid was metabolized faster by Pseudomonas species than by Flavobacterium species, and seemed to be a key material for the symbiosis. Therefore, PEG degradation by intact cells of Flavobacterium species seems to be stopped by glyoxylic acid at an early stage. On the other hand, Pseudomonas species is not involved in the degradation itself, but it is thought to remove the inhibitory substance, glyoxylic acid, quickly from the medium. In conclusion, Flavobacterium species obtains energy by the oxidation of PEG and utilizes glyoxylic acid as a carbon source. On the contrary, the growth of Pseudomonas species is supported by glyoxylic acid liberated by Flavobacterium species, resulting in the efficient removal of the excess acid from the medium and the smooth degradation of PEG (Figure 4) (20).

Biodegradation of PPG

PPG-utilizing Bacteria. PPG is the second synthetic polyether that resembles PEG in its chemical structure, but differs in its solubility in water. PPG is divided into two groups (diol and triol types), due to a straight or branched chain structure of the polymer. PPG with an Mn lower than approximately 700 (triol type) or 1000 (diol type) is readily water-soluble, but one with a higher Mn is oily and insoluble in water. Its use is expected to increase in the future. Nevertheless, the susceptibility of PPG to biological degradation was not reported until 1977, though the microbial assimilation of monomer, 1,2-propylene glycol, which is supplied by the petrochemical industry at a low price, had been known earlier. Fincher and Payne observed that a PEG-utilizing isolate could assimilate monomer and dimer as a sole carbon and energy source (2). Meanwhile, our PEG-utilizing isolates, or those isolated by Watson and Jones (21), did not grow on dimer or PPG.

PPG-utilizing bacteria were isolated by an enrichment culture containing PPG 2000 or 4000 from soils or activated sludges acclimated to PPG (22). Strain No. 7 was the most favorable strain and identified as Corynebacterium sp.

The strain grew on various PPGs with Mns 670-4000 (diol and triol types), monomer and dipropylene glycol, but did not assimilate PEGs. The strain also grew on a few PEG-PPG copolymers which contained a larger amount of PPG than PEG (Table I).

Biodegradation of PPG by Immobilized Cells. PPG-utilizing Corynebacterium sp. was immobilized in polyacrylamide gels. Entrapped cells in polyacrylamide gels could degrade approximately 80% of 0.1% PPG overnight. The cells were stable toward pH changes between pH 4 and 11 and were also stable against temperatures below 40°C. PPG was removed even by a continuous culture of immobilized cells. When the

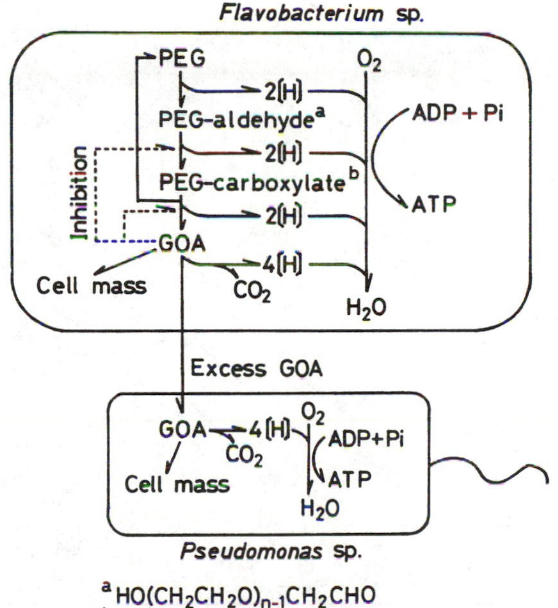

[a] $HO(CH_2CH_2O)_{n-1}CH_2CHO$
[b] $HO(CH_2CH_2O)_{n-1}CH_2COOH$

Figure 4. Symbiotic Degradation of PEG by Flavobacterium sp. and Pseudomonas sp. GOA: glyoxylic acid. (Reproduced with permission from Ref. 1. Copyright 1987 CRC Press, Inc.)

Table I. Growth Substrate Specificity of Corynebacterium sp. No. 7

Substrate	Growth (OD at 610 nm)
PPG diol type 670	2.23
1000	2.04
2000	2.45
triol type 1000	2.04
3000	2.48
4000	1.82
1,2-Propylene glycol	2.18
Dipropylene glycol	2.40
PEG 400	0.27
1000	0.30
2000	0.28
6000	0.30
Epan[a] 410	1.32
450	0.25
485	0.28
710	1.06
750	0.29
785	0.27
Ethylene glycol	0.30
Methanol	0.59
Ethanol	0.99
Propanol	1.94
2-Propanol	0.90
Glucose	2.51
Glycerol	1.75
None	0.30

[a] $HO(CH_2CH_2O)_\alpha(\overset{CH_3}{|}{CHCH_2O})_\beta(CH_2CH_2O)_\gamma H$

	Mn	PPG Content	PEG Content
Epan 410	1,330	1,200	130
450	2,400	1,200	1,200
485	8,000	1,200	6,800
710	2,220	2,000	220
750	4,000	2,000	2,000
785	13,000	2,000	11,330

cells were used repeatedly in batch reactions with everyday transfer to a new medium, the activity was maintained for eight days. On the contrary, the activity of suspended cells was lost in five days. Although the practical application of immobilized cells to treatment of PPG-containing water necessitates increasing the stability of the PPG-degrading activity, this technique seemed to be useful for the removal of PPG from waste water and other sources, etc.

Metabolic Route for Biodegradation of PPG. The aerobic metabolism of PPG by *Corynebacterium* sp. No. 7 was studied using dipropylene glycol (DPG) as a model substrate for biodegradation, since PPG contains molecules of different molecular weights (23). In studying PPG metabolism, unsterilized PPG or DPG was incubated with cells or enzyme preparations to avoid nonbiological degradation by autoclaving. PPG was not degraded at all by a culture filtrate or a cell-free extract of *Corynebacterium* sp., but was metabolized by washed, resting cells based on colorimetric determination by the cobaltothiocyanate method with slight modification. This result suggested that PPG was not metabolized by extracellular enzymes, but by intracellular enzymes including membrane enzymes. Commercially obtained DPG is a chemically synthesized compound which is expected theoretically to have several structural and optical isomers. These isomers were separated and identified by gas chromatography-mass spectrometry (GC-MS) analysis on a PEG 20M capillary column (0.25 mm i.d. by 25 m), as shown in Figure 5. Peak I was assigned to be structural isomer A, II and III to be B, and IV and V to be C: in peak I, optical isomers could not be separated; peaks II and III are diastereomers (the *R,R-S,S* and *R,S-S,R* complexes); peaks IV and V are also diastereomers (the *R,R-S,S* complex and the meso form). The ratio of structural isomers A, B and C was 36.3, 48.5 and 15.2%, respectively.

In nonshaken culture, DPG was scarcely degraded. With moderate shaking (60 to 70 rpm) at 30°C for 20 to 50 hr, metabolites were accumulated in a reaction mixture. Metabolites were characterized by GC-MS analysis as oxidized compounds of the terminal alcoholic groups [$OC(CH_3)CH_2OCH_2CH(CH_3)OH$, $OC(CH_3)CH_2OCH(CH_3)CH_2OH$ and $OC(CH_3)CH_2OCH_2\text{-}CO(CH_3)$] and 1,2-propylene glycol. Isomer A was degraded by 51.4%. The two diastereomers of isomer B (peaks II and III) were both degraded by 40.7% and those of isomer C (peaks IV and V) were both degraded by 18.5%. The results indicated that secondary alcohol groups were preferentially oxidized. On the other hand, dehydrogenation of PPG 2000 or DPG linked with DCIP and phenazine methosulfate was observed by the cell-free extract prepared from DPG-grown cells. Because of clouds caused by PPG attached to the cells, the activity of the cell-free extract prepared from PPG-grown cells was not measurable.

From these results, it seems likely that PPG might be metabolized and depolymerized via the same mechanism as that for PEG: oxidation of terminal alcohol groups leading to the cleavage of the ether linkage.

Biodegradation of PTMG

The third polyether PTMG is exclusively used for synthesizing polyurethanes which have superior properties to those synthesized from other polyethers, but its production is still limited because of the high cost of the polyether. As water-soluble oligomers are unsuitable for synthesizing polyurethanes, they are removed with water from a mixture of polymers and contained in the waste water of a synthetic chemical plant. Therefore, PTMG-utilizing microorganisms could be used for biological treatment of the waste water. Recently, PTMG-utilizing bacteria were isolated in our laboratory from soil and activated sludge samples by enrichment culture techniques (24). Although oxidation of 1,4-butanediol or bacterial utilization of this compound as a sole carbon source were reported, no report on microbial utilization of PTMG was found. Two strains that showed high growth on a PTMG medium were identified as *Alcaligenes denitrificans* subspecies *denitrificans* and *Pseudomonas maltophilia*, respectively.

The number and quantity of components contained in PTMG samples were analyzed by HPLC on a Toyo Soda TSK-GEL G1000PW column. Eight peaks were detected with PTMG 200 or 265 (Table II). Some oligomers up to about a polymerization degree of eight seemed to be water soluble. Contrary to the expected elution profile by gel filtration, the retention volume of the first peak coincided with that of 1,4-tetramethylene glycol. From the comparison of retention volumes of ethylene glycol, 1,2-propylene glycol, 1,4-tetramethylene glycol and their dimers, more hydrophilic compounds seemed to be eluted earlier than less hydrophilic compounds. This fact was confirmed by mass spectrometry analysis of each elution peak (monomer-octamer).

Table II. Elution Profile of PTMG by HPLC

Material	Retention Volume (ml)
PTMG 200 (n=2.5)	20.4, 21.2, 22.3, 23.8 25.6, 28.1, 30.9, 34.3
1,4-Tetramethylene glycol (monomer)	20.4
1,3-Butylene glycol	19.9
4-Hydroxybutyric acid	11.5
1,2-Propylene glycol	19.5
Ethylene glycol	19.4
Dipropylene glycol	18.6
Diethylene glycol	18.1
PEG 600	12.8
PEG 6000	11.1

Note: HPLC was carried out with a flow rate of 0.5 ml/min. The elution position of PEG 6000 shows the exclusion limit of the normal gel filtration by the column.

No detectable amount of decomposition and formation of oxidized material was observed by autoclaving a PTMG 265 medium at 120°C for 15 min even at pH 5.0. PTMG 265 or 200, ranging from 0.1 to 0.2%, completely disappeared in 7 days from the culture filtrates of both strains. PTMG 265 was also metabolized by their intact cells (Figure 6). The oligomers with a higher polymerization degree disappeared more rapidly than those with a lower polymerization degree. Sonic extracts of *Alcaligenes denitrificans* dehydrogenated PTMG coupling with an artificial electron acceptor and phenazine methosulfate. These results suggest that PTMG is degraded via the oxidation process which is linked to an electron transport system of the bacterium. On

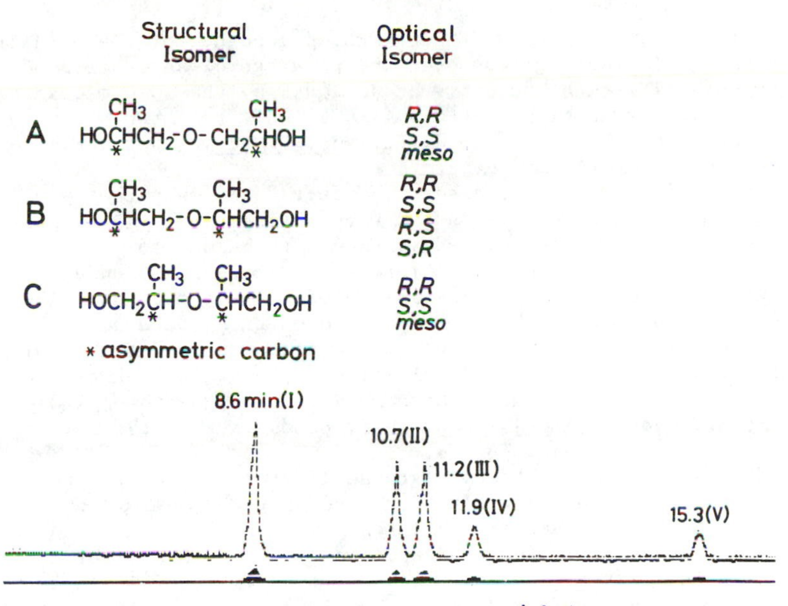

Figure 5. Presumed Structural and Optical Isomers Contained in DPG and Isolation of Isomers by Glass Capillary GC-MS (Total Ion Monitoring). (Reproduced with permission from Ref. 23. Copyright 1985 The Society of Fermentation Technology, Japan.)

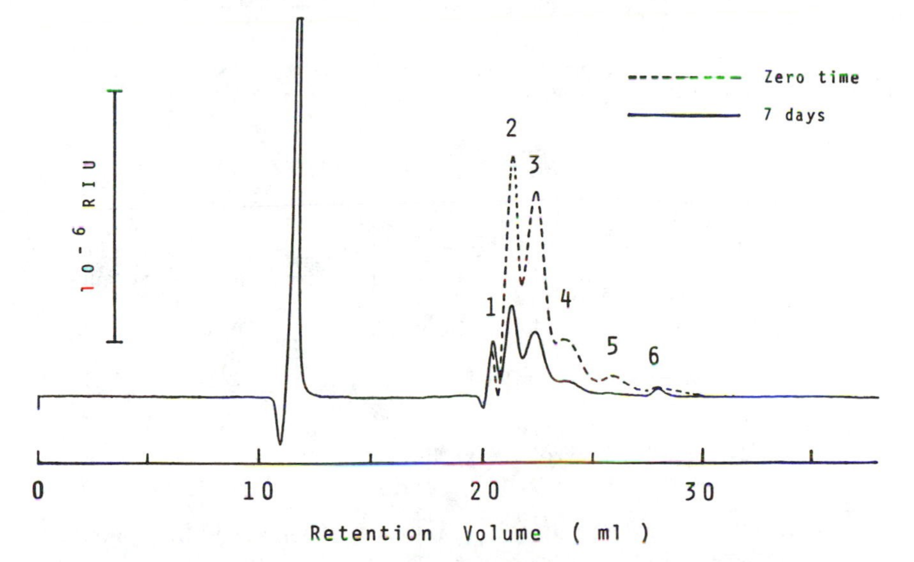

Figure 6. Degradation of PTMG 265 by Intact Cells of Alcaligenes denitrificans. HPLC was carried out with a flow rate of 1.0 ml/min. The numbers in the figure correspond to monomer to hexamer. The first peak corresponds to inorganic phosphate.

the bacterial oxidation of PEG and PPG, precedence of the oxidation of terminal alcoholic groups over the cleavage of an ether bond was suggested. PTMG might possibly be degraded by the same mechanism as those in the degradation of PEG and PPG. This research has been submitted for the publication to J. Ferment. Bioeng. (Kawai, F.; Moriya, F. Kobe University of Commerce.).

In conclusion, polyethers are biodegradable when they are liberated into natural environments and they are removed by biological treatment systems. The degradation seems to be due to the same oxidative mechanisms. However, different microorganisms and enzyme systems are involved in the metabolism of these compounds. The degradation of polyethers has been focused on PEGs, because of the quantities of them liberated into the environment and the resultant contribution to environmental pollution. The production of other polyethers is far less than that of PEGs, but it has been increasing every year. Hence, the issue of their biodegradabilities will be more salient in the near future.

The evaluation of the biodegradability of synthesized chemicals is important from the viewpoints of assessing the self-cleaning functions of nature, of removing pollutants in conventional sewage disposal systems by activated sludges, and of making new treatment systems using microorganisms which have specific activities to remove specific materials. In addition, the biodegradation studies of synthetic compounds bring us new information on microbial worlds, novel enzymes or reactions, novel genes, etc.

On the other hand, knowledge concerning biodegradability of synthetic polymers bring us new information on design of polymer molecules which can be biodegradable by ecosystems.

Acknowledgments

This research was supported in part by research grants from the Ministry of Education, Science and Culture, Japan.

Literature Cited:

1. Kawai, F. CRC Crit. Rev. Biotechnol. 1987, 6, 273.
2. Fincher, E. L.; Payne, W. J. Appl. Microbiol. 1962, 10, 542.
3. Pitter, P. Collect. Czech. Chem. Commun. 1973, 38, 2665.
4. Cox, D. P.; Conway, R. A. In Proc. 3rd Int. Biodegradation Symp.; Sharpley, J. M.; Kapalan, A. M., Eds.; Applied Science Publishers: London, 1976; p 835.
5. Ogata, K.; Kawai, F.; Fukaya, M; Tani, Y. J. Ferment. Technol. 1975, 53, 757.
6. Kawai, F.; Fukaya, M.; Tani, Y.; Ogata, K. J. Ferment. Technol. 1977, 55, 429.
7. Kawai, F.; Kimura, T.; Fukaya, M.; Tani, Y.; Ogata, K.; Ueno. T.; Fukami, H. Appl. Environ. Microbiol. 1978, 35, 679.
8. Haines, J. R.; Alexander, M. Appl. Environ. Microbiol. 1975, 29, 621.
9. Hosoya, H.; Miyazaki, N.; Sugisaki, Y.; Takanashi, E.; Tsurufuji, M.; Yamasaki, M.; Tamura, G. Agric. Biol. Chem. 1978, 42, 1545.
10. Schink, B.; Stieb, M. Appl. Environ. Microbiol. 1983, 45, 1905.
11. Dwyer, D. F.; Tiedje, J. M. Appl. Environ. Microbiol. 1983, 46, 185.
12. Grant, M. A.; Payne, W. J. Biotech. Bioeng. 1983, 25, 627.

13. Kawai, F.; Yamanaka, H. J. Ferment. Bioeng. 1989, 67, 300.
14. Kawai, F.; Yamanaka, H.; Ameyama, M.; Shinagawa, E.; Matsushita, K.; Adachi, O. Agric. Biol. Chem. 1985, 49, 1071.
15. Yamanaka, H.; Kawai, F. J. Ferment. Bioeng. 1989, 67, 324.
16. Patterson, S. J.; Scott, C. C.; Tucker, K. B. E. J. Am. Oil Chem. Soc. 1967, 44, 407.
17. Kawai, F.; Kimura, T.; Tani, Y.; Yamada, H.; Ueno, T.; Fukami, H. Agric. Biol. Chem. 1983, 47, 1669.
18. Pearce, B. A.; Heydeman, M. T. J. Gen. Microbiol. 1980, 118, 21.
19. Schöberl, P. Tenside Deterg. 1985, 22, 70.
20. Kawai, F.; Yamanaka, H. Arch. Microbiol. 1986, 146, 125.
21. Watson, G. K.; Jones, N. Water Res. 1977, 11, 95.
22. Kawai, F.; Hanada, K.; Tani, Y.; Ogata, K. J. Ferment. Technol. 1977, 55, 89.
23. Kawai, F.; Okamoto, T.; Suzuki, T. J. Ferment. Technol. 1985, 63, 239.
24. Kawai, F. Japanese Patent 208289, 1987.
25. Salisbury, S. A.; Forrest, H. S.; Cruse, W. B. T.; Kennnard, O. Nature, 1979, 280, 843.
26. Duine, J. A.; Frank, J. Jr.; Van Zeeland, J. K. FEBS Lett. 1979, 108, 443.

RECEIVED January 8, 1990

Chapter 11

Biodegradable Poly(carboxylic acid) Design

Shuichi Matsumura and Sadao Yoshikawa

Department of Applied Chemistry, Keio University, 3–14–1, Hiyoshi, Kohoku-ku, Yokohama 223, Japan

Poly(carboxylic acid)s containing hydroxyl, carbonyl, ether, ester or glycopyranosyl groups as biodegradable segments were prepared and their biodegradability and builder performances in detergents were compared. It was confirmed that these poly(carboxylic acid)s containing biodegradable segments in the polymer chain showed an improved biodegradablity. Polyvinyl-type poly(carboxylic acid)s, such as poly[(acrylic acid)-*co*-(vinyl alcohol)] and poly[(acrylic acid)-*co*-(2-cyclohexen-1-one)], were biodegraded, but the microbes utilizing this type of copolymers as a sole carbon source were relatively scarce in the environment. On the contrary, both poly(carboxylic acid) containing ester linkages on the backbone and partially oxidized polysaccharides containing unreacted glycopyranosyl groups were biodegraded well by activated sludge. These poly(carboxylic acid)s showed better detergency builder performances in detergents.

Polymeric polycarboxylates containing carbon-carbon backbones and a relating high charge density of carboxylate groups along the chain have been investigated for sodium tripolyphosphate (STPP) substitutes, and a number of poly(sodium carboxylate)s, such as poly(sodium acrylate), have been reported which give excellent builder performance when compared with STPP on an equal weight basis (1-4). These compounds; however, are extremely resistant to biodegradation, which is an important criterion for a compound acceptable as STPP replacements (4,5). Therefore, a design to develop a biodegradable poly(carboxylic acid) is needed. As biodegradable poly(sodium carboxylate)s, poly(sodium β-malate) (6,7), poly(sodium vinyloxyacetate) (8,9) and sodium polyglyoxylate (10) were recently reported.

In this report, poly(carboxylic acid) containing hydroxyl,

0097–6156/90/0433–0124$06.00/0

carbonyl, ether, ester or glycopyranosyl groups as biodegradable segments, which are susceptible to biodegradation, were prepared and their biodegradability and builder performances in detergents were compared.

Experimental

Materials. Poly[(sodium acrylate)-*co*-(vinyl alcohol)] P(SA-VA) (11) and poly[(sodium acrylate)-*co*-(2-cyclohexen-1-one)] P(SA-CHO) (12) were prepared by the radical polymerization of acrylic acid/ methyl acrylate with vinyl acetate and 2-cyclohexen-1-one, respectively, and subsequent hydrolysis. Poly(sodium glycidate) (PG) and poly(disodium epoxysuccinate) (PES) were prepared by the ring-opening polymerization of methyl glycidate and diethyl epoxysuccinate (13). Poly(β-DL-malic acid) sodium salt (PMLA) was prepared by the ring-opening polymerization of benzyl malolactonate according to the method of Vert et al.(14,7). Poly(DL-malic acid) sodium salt with both the α and β ester linkages of malic acid (PMLA-D) was prepared by the direct polymerization of DL-malic acid in dimethylsulfoxide under reduced pressure at 90 ℃(7). Partially oxidized polysaccharides containing unreacted glycopyranosyl groups as biodegradable parts in the polycarboxylate chain were prepared from cellulose (sodium dicarboxy cellulose : DCC), xylan (sodium dicarboxy xylan : DCXy) and starch (sodium dicarboxy starch : DCS) by the conversion of vicinal diols to dicarboxylates using periodic acid/chlorite oxidation (15). All polymers were dialyzed exhaustively against distilled water to remove any low-molecular-weight fractions prior to the test. Polymer structures and their codes are as follows :

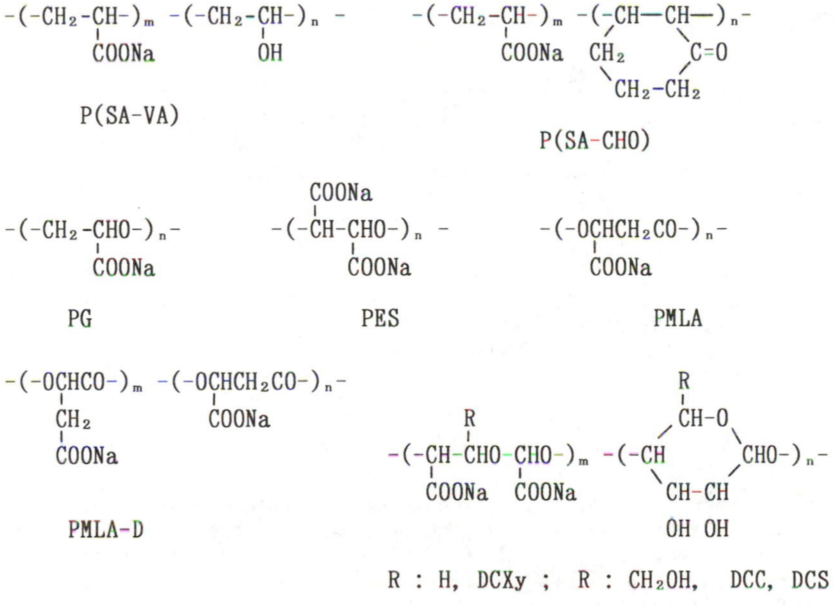

$-(-CH_2-CH(COONa)-)_m-(-CH_2-CH(OH)-)_n-$

P(SA-VA)

$-(-CH_2-CH(COONa)-)_m-(-CH—CH—)_n-$ (ring: CH—CH, CH_2, C=O, CH_2-CH_2)

P(SA-CHO)

$-(-CH_2-CH(COONa)O-)_n-$

PG

$-(-CH(COONa)-CH(COONa)O-)_n-$

PES

$-(-OCH(COONa)CH_2CO-)_n-$

PMLA

$-(-OCH(CH_2COONa)CO-)_m-(-OCH(COONa)CH_2CO-)_n-$

PMLA-D

$-(-CH(R)(COONa)-CHO-CH(COONa)O-)_m-(-CH(CH(R)-O)CHO-)_n-$ (ring: CH-O, CH-CH with OH OH)

R : H, DCXy ; R : CH_2OH, DCC, DCS

Measurements. Molecular weight and molecular weight distributions were measured on a gel permeation chromatographic system (GPC) with commercial GPC columns (TSK-GEL G5000PW + G2500PW, TOSOH Co. Ltd., 0.1M phosphate buffer containing 0.3M NaCl as an eluent) calibrated with a poly(ethylene glycol) standard. Residual polymers in culture broth were directly analyzed by GPC after ultrasonification with a small amount of nonionic surfactant, if necessary. Total organic carbon (TOC) concentration was measured by a commercial TOC analyzer. Bacterial growth was followed by the turbidity of the culture broth at 660 nm (OD_{660}). The five-day biochemical oxygen demand (BOD_5) was determined by the oxygen consumption method using activated sludge in the closed bottle according to the Japanese Industrial Standard (JIS K 0102). In this procedure, the concentration of test substrates was 8 to 25 mg/L, and the basal medium was BOD diluent water containing 21.75 g/L K_2HPO_4, 8.5 g/L KH_2PO_4, 44.6 g/L $Na_2HPO_4\ 10H_2O$, 1.7 g/L NH_4Cl, 22.5 g/L $MgSO_4\ 7H_2O$, 27.5g/L $CaCl_2$ and 0.25g/L $FeCl_3$. The activated sludge used in this experiment was freshly obtained from a municipal sewage plant in Yokohama city.

Isolation of Polymer-Degrading Bacteria. The microbes capable of degrading poly(carboxylic acid) were isolated by an enrichment culture technique from soil and activated sludge (8). The enrichment culture medium consisted of 0.1-0.3% polymers in the inorganic medium (0.2 % NH_4Cl, 0.02 % KH_2PO_4, 0.02 % $MgSO_4 \cdot 7H_2O$, 0.0002 % $CaCl_2$, 0.0001 % $FeSO_4 \cdot 7H_2O$, 0.0002 % $MnSO_4 \cdot 4H_2O$, 0.0007 % $ZnSO_4 \cdot 7H_2O$, and a trace amount of $CuSO_4$ and Vitamin $B_1 \cdot HCl$). The pH of the medium was adjusted to 7.2. Independent, discrete colonies on the same 0.5% polymer medium containing agar were picked up and purified again on the same agar medium. Identification of the isolated strains was made by the taxonomic examinations.

Detergency. The detergency test (12) was first conducted with a standard heavy duty detergenct formulation containing 20% sodium dodecylbenzene sulfonate, 25% STPP/ODA, 5% sodium silicate, 3% sodium carbonate, 0.5% carboxymethyl cellulose (CMC) and 46.5% sodium sulfate to determine the detergency of the STPP/ODA formulation as a basis for comparison with the poly(sodium carboxylate)s. In the experimental formulas the STPP was replaced with an equal weight of the polymers. Reflectance was measured after washing the artificially soiled cotton cloth test pieces prepared from an aqueous dispersion method (16) in a Terg-O-Tometer at 25 ℃. All tests were conducted in water with 3° DH of hardness. Detergency was expressed as a value relative to 10 for STPP and 0 for disodium 3-oxapentanedioate (ODA).

Results and Discussion

Table I shows the five-day biochemical oxygen demand (BOD_5) of some poly(sodium carboxylate)s and poly(vinyl alcohol)s (PVA). The BOD_5 test according to JIS K 0102 is only an indication of susceptibility of products to biodegradation, but the test is often an useful tool for rapid screening. As shown in Table I, PMLA and

PMLA-D containing ester linkages showed the best biodegradability of the polymers tested in this study. Also, poly(sodium carboxylate) containing more than 50% unreacted glycopyranosyl group (DCXy and DCS) and poly(sodium carboxylate) containing an ether linkage (PG and PES) showed a good biodegradability. The BOD_5 values of polyvinyl-type copolymers [P(SA-VA) and P(SA-CHO)] were lower than those of poly(sodium carboxylate)s containing ester, ether or the glycopyranosyl group, but were higher than that of the corresponding poly(sodium acrylate) (PSA). It seems that the introduction of a hydroxyl or carbonyl group into the polymer chain improves their biodegradability. But this improved biodegradability doesn't necessarily mean satisfactory or acceptable biodegradation. Also the JIS method is only a screening tool and polymers which give a low biodegradation in this test may be more rapidly and extensively biodegraded in systems such as soil or activated sludge where more organisms exist and more time can be provided for acclimation. This fact is supported by the results with P(SA-VA) and P(SA-CHO) where only about 10% degradation was noted in the JIS test, but more extensive degradation occurred via the pure cultures as described in the following section.

Table I. Five-day biochemical oxygen demand (BOD_5) by oxygen consumption method (Japanese Industrial Standard, JIS K 0102)

Compound	ThOD (mg O/g)	BOD_5 (mg O/g)	BOD_5/ThOD (%)
PVA-9510	1818	129	7.1
PVA-12300	1818	121	6.7
DSA	1011	282	27.9
TSA	986	297	30.1
PSA-520	936	75	8.0
PSA-1070	936	13	1.3
PSA-2050	936	23	2.5
PSA-8730	936	0	0
PSA-4250	936	0	0
P(SA-VA)-8300[90]*1	980	122	12.4
P(SA-VA)-11710[61]*1	1139	121	10.6
P(SA-CHO)-17060[73]*1	1365	120	8.8
P(SA-CHO)-7310[75]*1	1333	176	13.2
PG-1140	655	158	24.1
PES-960	364	145	39.8
PMLA-6210	637	277	43.5
PMLA-D-3700	637	357	56.1
DCC-6500[76]*2	657	10	1.5
DCXy-32200[39]*2	855	140	16.4
DCS-36300[48]*2	817	83	10.2

ThOD : Theoretical oxygen demand. DSA : Dimeric sodium acrylate, TSA : trimeric sodium acrylate. Polymer codes indicate number-average molecular weights($\overline{M}n$). *1 Figures in the brackets indicate the mol% of the first monomer (M1) in the copolymer. *2 Figures in the brackets indicate the degree of dicarboxylation in mol%.

It will be useful to measure the molecular-weight distributions of the polymers by GPC before and after the biodegradation, and whether the main chain of the polymer will be cleaved by the microbes. Figure 1 shows the GPC profiles of the cell-free filtrate of the biodegradation media of poly(sodium carboxylate)s before and after the biodegradation test (BOD_5) according to JIS K 0102 as a typical example. It was found that the amount of polymer containing a biodegradable unit decreased after 5 days incubation and the biodegradation of the polymer undoubtedly occurred even if the BOD_5 value was low. On the other hand, GPC profiles of PSA with a molecular weight of 8730 showed no significant changes after 5 days incubation.

Among these polymers, the polyvinyl-type poly(sodium carboxylate)s, which are capable of large scale industrial production, and partially oxidized polysaccharides, which are available from renewable raw materials, have been attractive to the industrial field. Therefore, their biodegradation behavior was further analyzed.

Biodegradation of Polyvinyl-Type Poly(sodium carboxylate). PVA is the only substance which is known to be biodegradable in the class of polyvinyl-type synthetic polymer. It may be biodegraded by oxidizing hydroxyl group of PVA to the corresponding carbonyl group and subsequent hydrolysis as shown below (17, 18).

$$-CH_2-\underset{OH}{CH}-CH_2-\underset{OH}{CH}- \longrightarrow -CH_2-\underset{O}{\overset{\|}{C}}-CH_2-\underset{O}{\overset{\|}{C}}- \longrightarrow -CH_2COONa + CH_3\underset{O}{\overset{\|}{C}}-$$

As a design to develop a polyvinyl-type poly(sodium carboxylate), acrylate copolymers containing hydroxyl or carbonyl groups which are susceptible to the enzymatic reaction, were prepared. It is presumed that the copolymer is first cleaved at a hydroxyl or carbonyl group as in the case of PVA, then the resultant acrylate oligomer is further assimilated by the microbes. The biodegradation of oligomeric acrylic acid (11), in fact, occurs as shown in Table I.

$$-(-CH_2-\underset{OH}{CH}-)_a-(-CH_2-\underset{COONa}{CH}-)_b-(-CH_2-\underset{OH}{CH}-)_c-(-CH_2-\underset{COONa}{CH}-)_d-(-CH_2-\underset{OH}{CH}-)_e-$$

$$\Big\downarrow \text{ Microbes}$$

$$-(-CH_2-\underset{COONa}{CH}-)_b- \quad + \quad -(-CH_2-\underset{COONa}{CH}-)_d-$$

Proposed biodegradation mechanism of P(SA-VA).

From dimeric to oligomeric sodium acrylate having a molecular weight of 500 was easily biodegraded. However, it seems that the specific microbes utilizing this polyvinyl-type copolymers as a sole carbon source are relatively scarce in the environment, and the biodegradability of these types of copolymers differ according to

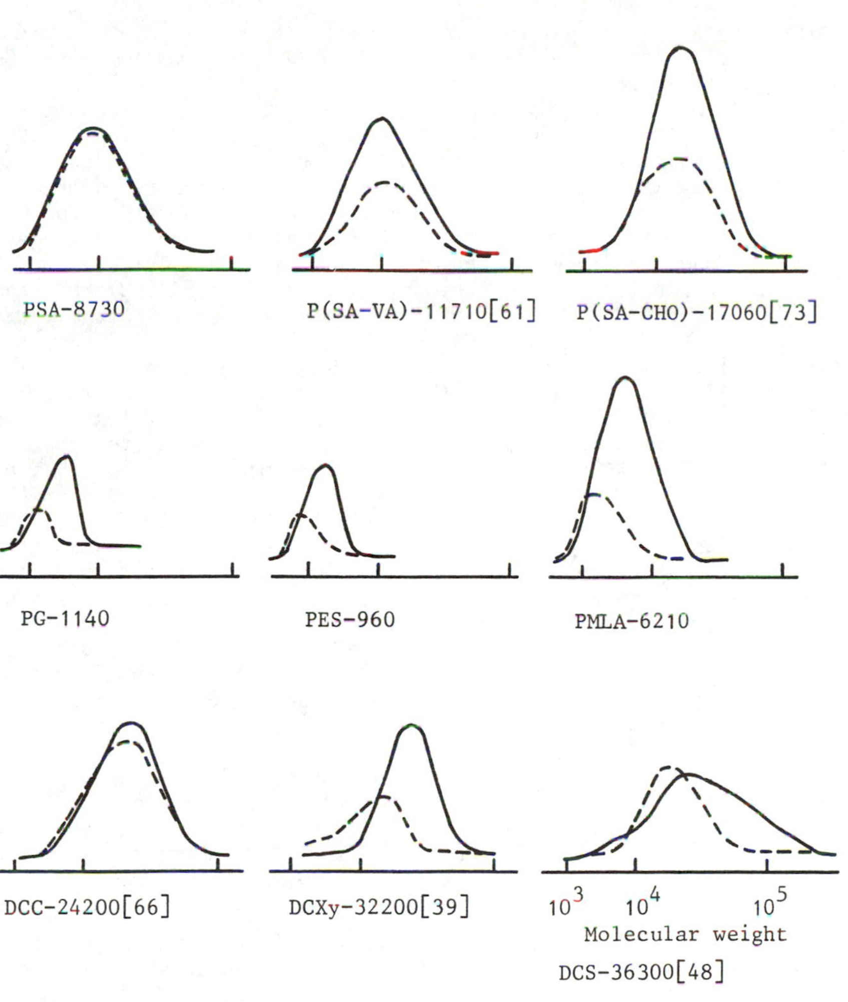

Figure 1. GPC profiles of polymers before and after the biodegradation by BOD_5 test in the closed bottle (JIS K 0102).

——, 0 day ; - - -, 5 days.

All molecular weight ranges (X axis scale) are the same as specified for DCS-36300[48].

the seasons of the year. Under these circumstances, it was necessary to isolate the polymer-degrading microbes to prove the biodegradability of the polymers. P(SA-VA)-degrading symbiotic bacterial strains of *Pseudomonas* sp. C1 and C2, and P(SA-CHO)-degrading fungal strain of *Trichoderma* sp. KM802 were isolated by an enrichment culture technique from soil and activated sludge. Biodegradation (11) of P(SA-VA)-8160 by *Pseudomonas* sp. C1 and C2 is shown in Figure 2. Figure 3 shows the growth curve of *Trichoderma* sp. KM802, pH and the residual polymer in the culturing broth of P(SA-CHO)-4620. It was confirmed that the P(SA-CHO)-degrading strain, *Trichoderma* sp. KM802, can grow on a P(SA-CHO) as a sole carbon source, and the polymers decreased gradually. The GPC profiles of P(SA-CHO)-4620 before and after the biodegradation by *Trichoderma* sp. KM802 is also shown in Figure 4. It was found that the amount of polymer in the culture broth decreased gradually and the biodegradation of P(SA-CHO) by the microbe occurred. GPC profiles also showed that the high-molecular-weight fractions degraded equally as well as the low-molecular-weight fractions of the polymer. Substrate specificity towards molecular weight in the biodegradation by *Trichoderma* sp.KM802 was not clearly recognizable. From the GPC profiles, it was indicated that when an enzyme of microbes attacks a polymer chain, further cleavage of the same chain is more probable than that of another chain. From GPC analysis, it was found that about 65% of the polymer of P(SA-CHO) was degraded after 20 days incubation. TOC values of the cell-free culture filtrate after 20 days incubation showed that about 63% of the organic carbon of P(SA-CHO) was mineralized by the microbes. These data agreed approximately with that obtained by GPC, indicating that the oligomeric fractions produced by the biodegradative cleavage of the P(SA-CHO) were not accumulated in the culture. Small differences between the TOC values and GPC values will be ascribed to the biodegradation intermediates in the culture broth. Table II shows the substrate specificity for growth of the P(SA-CHO)-degrading strain, *Trichoderma* sp. KM802 at 30 ℃ for 10 days incubation. *Trichoderma* sp. KM802 grows well on dimeric to oligomeric sodium acrylate and cyclohexanone as well as P(SA-CHO). P(SA-CHO) is first cleaved at the 2-cyclohexen-1-one moiety which is susceptible to the enzymatic reaction, then the resultant acrylate oligomer is further assimilated by the microbes.

Table II. Substrate specificity for growth of P(SA-CHO)-degrading strain, *Trichoderma* sp. KM802 at 30 ℃ for 10 days in shaking tube

Substrate	OD_{660}
P(SA-CHO)-4620[62]	0.280
P(SA-CHO)-9890[72]	0.172
Cyclohexanone	0.936
DSA	0.496
TSA	0.242
PSA-520	0.097
PSA-1070	0.068
PSA-9020	0.004

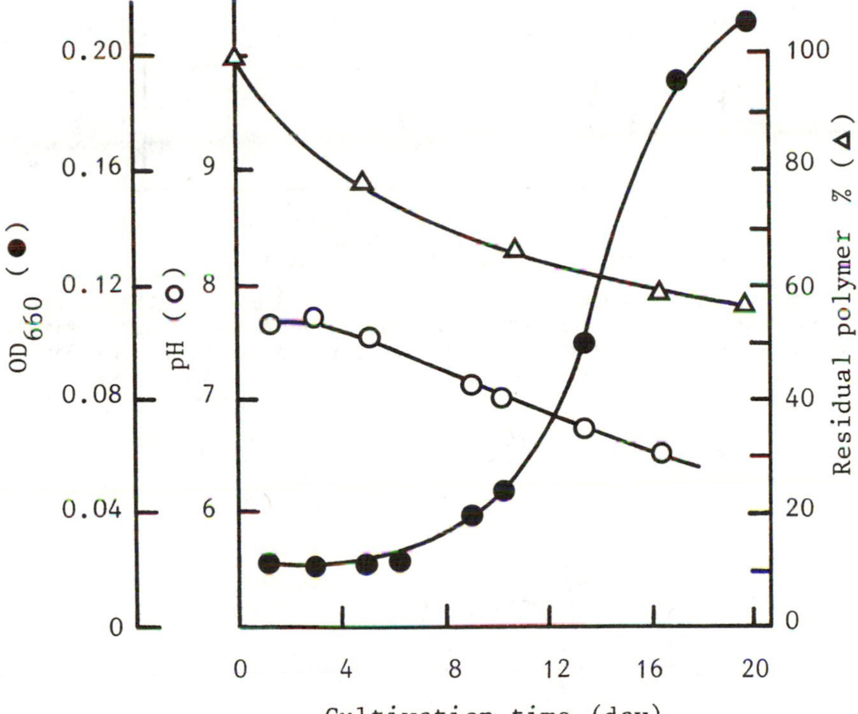

Figure 2. Biodegradation of P(SA-VA)-8160[61] by P(SA-VA)-degrading strains, *Pseudomonas* sp. C1 and C2. Double inoculum of C1 and C2 was cultivated in an inorganic medium containing 0.2% polymer as the sole carbon source in a shaking flask at 30 ℃. (Reproduced with permission from Ref. 11. Copyright 1988 The Society of Polymer Science.)

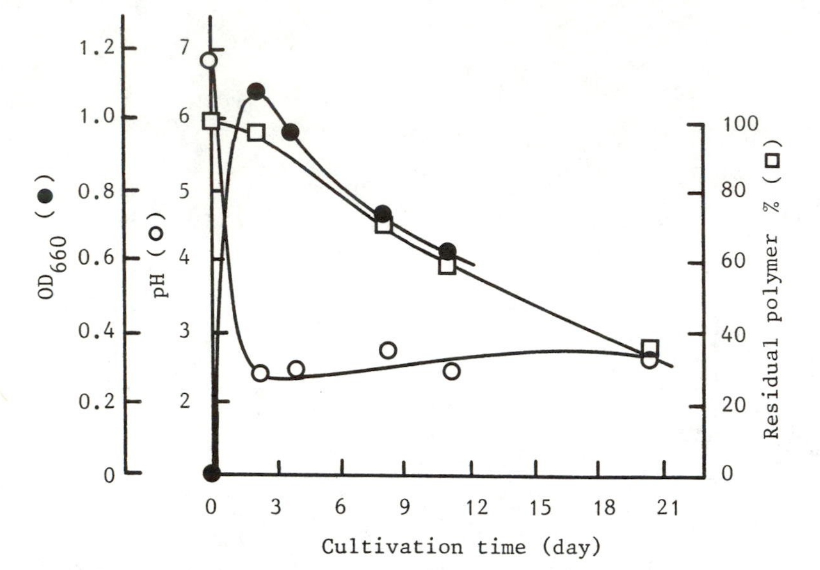

Figure 3. Biodegradation of P(SA-CHO)-4620[62] by P(SA-CHO)-degrading strain, *Trichoderma* sp. KM802. KM802 was cultlivated in an inorganic medium containing 0.1% polymer as the sole carbon source in a shaking flask at 30 ℃.

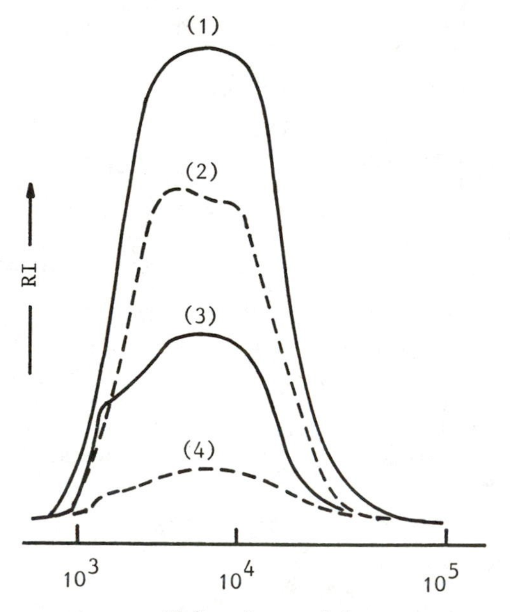

Figure 4. GPC profiles of P(SA-CHO)-4620[62] before and after the biodegradation by *Trichoderma* sp. KM802. Cultivation conditions, same as those of Figure 3.
(1) 0 day ; (2) 8 days ; (3) 20 days ; (4) 30 days.

Biodegradation of poly(sodium carboxylate) containing a glycopyranosyl group. Poly(sodium carboxylate)s containing unreacted glycopyranosyl groups as the biodegradable segments was obtained by the partial oxidation of polysaccharides. From the BOD test and GPC analysis, it was confirmed that the partially oxidized polysaccharides from cellulose, xylan and starch showed biodegradability. Figurue 5 and Figure 6 show the biodegradation of DCS-2100[44] by activated sludge from a municipal sewage plant. It was found that DCS-2100[44] was easily biodegraded. About 75% of the polymer fractions of DCS was biodegraded by activated sludge. Two bacterial strains, *Ochrobactrum anthropi* KM807 and *Pseudomonas* sp. KM806, were isolated from the activated sludge as DCS-degrading strains. Among the partially oxidized polysaccharides, xylan derivatives showed a better biodegradability. Two bacterial strains, *Arthrobacter* sp. KM810 and *Flavobacterium* sp. KM811, were isolated from the activated sludge as DCXy-degrading strains.

Detergency Building Performance. Detergency building performances of poly(sodium carboxylate)s were examined using a heavy duty detergent formulation on standard soiled cotton cloths. The detergency expressed as a value relative to 10 for STPP and 0 for disodium 3-oxapentanedioate(ODA) was shown below.

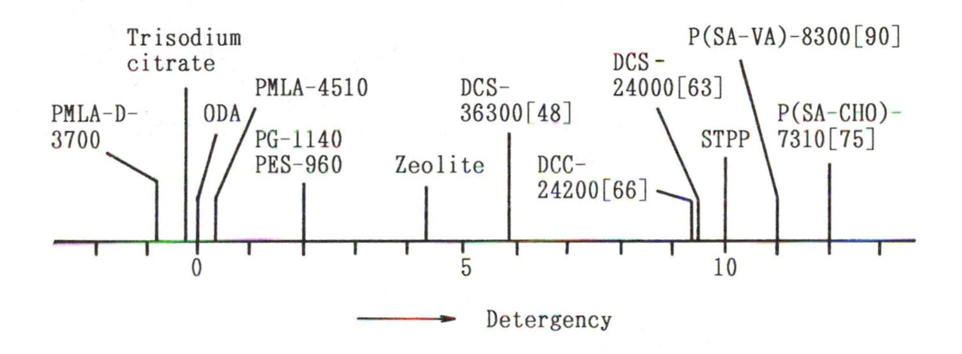

Poly(sodium acrylate) and their copolymers, and partially oxidized polysaccharides showed better detergency building performance than those of ODA and trisodium citrate with some of them superior to those of STPP. It will be anticipated that the 2-cyclohexen-1-one moiety of P(SA-CHO) will act as a booster of detergency by the hydrophobic effect of this group as well as a biodegrading segment in the polymer. Detergency building performance of PMLA and PMLA-D containing ester linkages and PG and PES containing ether linkages were less effective than those of STPP, but similar to those of ODA and trisodium citrate.

Conclusion

Poly(sodium carboxylate)s containing hydroxyl, ester, carbonyl or glycopyranosyl groups in the polymer chain showed an improved biodegradability, suggesting these functional groups to be useful as

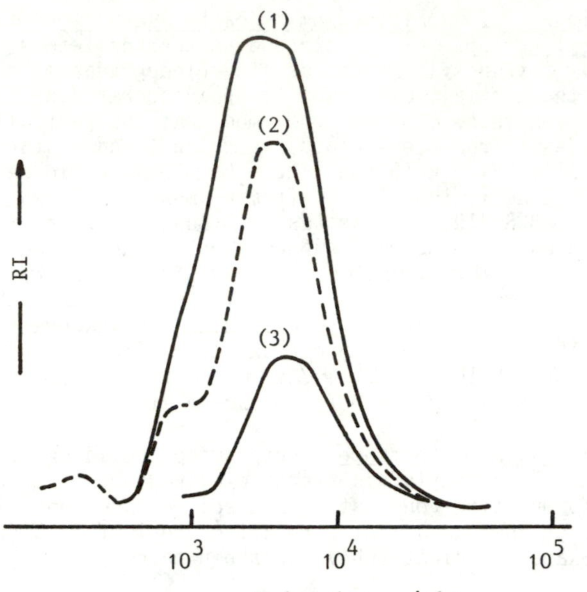

Figure 5. GPC profiles of DCS-2100[44] before and after the biodegradation by activated sludge (initial MLSS 10 ppm). Activated sludge was cultivated in an inorganic medium containing 0.2% DCS-2100[44] in a shaking flask at 30 ℃. (1) 0 day ; (2) 7 days ; (3) 13 days.

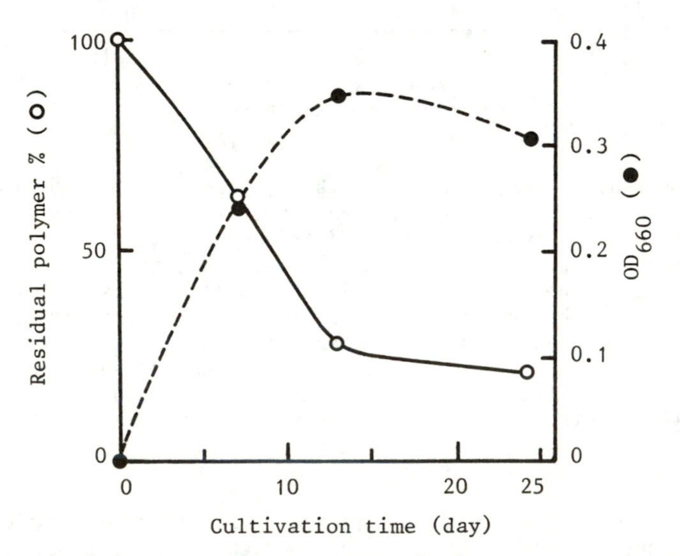

Figure 6. Biodegradation of DCS-2100[44] by activated sludge. Cultivation conditions, same as those of Figure 5.

biodegrading units in the polymer. Among these groups, the ester and glycopyranosyl groups were the most effective as biodegrading segments in the polymer. Introduction of hydroxyl or carbonyl groups to the polyvinyl-type carboxylate copolymer improved their biodegradability, but it seems that the probability of appearance of the microbes utilizing these types of copolymers as their sole carbon source is not so high in the environment, compared with those of poly(sodium carboxylate)s containing ester or glycopyranosyl groups. Polyvinyl-type poly(sodium carboxylate)s and poly(sodium carbosylate)s containing glycopyranosyl groups showed a better builder performance in detergents, compared to those containing ester or ether linkages.

Literature Cited

1. Abe, Y.; Matsumura, S. et al. J. Jpn. Oil Chem. Soc. (Yukagaku) 1981, 30, 757 ; 1982, 31, 586 ; 1984, 33, 211, 219 ; 1985, 34, 202, 456 ; 1986, 35, 167 ; 1987, 36, 874.
2. Crutchfield, M.M. J. Am. Oil Chem. Soc. 1978, 55, 58.
3. Matzner, E.A. ; Crutchfield, M.M. ; Langguth, R.P. ; Swisher, R.D. Tenside 1973, 10, 239.
4. Kemper, H.C. ; Martens, R.J. ; Nooi, J.R. ; Stubbs, C.E. Tenside 1975, 12, 47.
5. Abe, Y. ; Matsumura, S. ; Yajima, H. ; Suzuki, R. ; Masago, Y. J. Jpn. Oil Chem. Soc. (Yukagaku) 1984, 33, 228.
6. Braud, C. ; Bunel, C. ; Vert, M. Polym. Bull. (Berlin) 1985, 13, 293.
7. Abe, Y. ; Matsumura, S. ; Imai, K. J. Jpn. Oil Chem Soc. (Yukagaku) 1986, 35, 937.
8. Matsumura, S. ; Takahashi, J. ; Maeda, S. ; Yoshikawa, S. Makromol. Chem., Rapid commun. 1988, 9, 1.
9. Matsumura, S. ; Takahashi, J. ; Maeda, S.; Yoshikawa, S. Kobunshi Ronbunshu 1988, 45, 325.
10. Gledhill, W.E. ; Saeger, V.W. J. Ind. Microbiol. 1987, 2, 97.
11. Matsumura, S. ; Maeda, S. ; Takahashi, J. ; Yoshikawa, S. Kobunshi Ronbunshu 1988, 45, 317.
12. Matsumura, S. ; Maeda, S. ; Yoshikawa, S. ; Chikazumi, N ; Senda, T. J. Jpn. Oil Chem. Soc. (Yukagaku) 1989, 38, 612.
13. Matsumura, S. ; Hashimoto, K. ; Yoshikawa, S. J. Jpn. Oil Chem. Soc. (Yukagaku) 1987, 36, 874.
14. Vert, M. ; Lenz, R.W. Polym. Prepr. 1978, 20, 608.
15. Nieuwenhuizen, M.S. ; Kieboon, A.P.G. ; H. van Bekken Starch 1985, 37, 192.
16. Okumura, O. ; T. Tokuyama, ; Sakatani, T. ; Tsuruta, Y. J. Jpn. Oil Chem. Soc. (Yukagaku) 1981, 30, 432.
17. Sakai, K. ; Hamada, N. ; Watanabe, Y. Agric. Biol. Chem. 1985, 49, 1901.
18. Suzuki, T. ; Tsuchii, A. Process Biochem. 1983, 13.

RECEIVED January 8, 1990

Chapter 12

Biodegradation of Synthetic Polymers Containing Ester Bonds

Yutaka Tokiwa, Tadanao Ando, Tomoo Suzuki, and Kiyoshi Takeda

Fermentation Research Institute, 1-1-3, Higashi, Tsukuba, Ibaraki, Japan 305

Oligomers of synthetic polymers are biodegradable, whereas only few polymers are biodegradable. Polyethylene adipate with $\overline{Mn}$ 3000 and polycaprolactone with $\overline{Mn}$ 25,000 were completely degraded by *Penicillium* spp. Aliphatic and alicyclic polyesters, ester type polyurethanes (I), copolyamide-esters (II) and copolyesters (III) of aliphatic and aromatic polyesters were hydrolyzed by lipases from various microorganisms and hog pancreas, and hog liver esterase. When aliphatic polyesters were used as enzyme substrates, it was found that their melting points (Tm) had a effect, in addition to their chemical structure, on biodegradability. It was assumed that the hydrogen bonds in the I, II chains and aromatic ring in the III chains, which were related to high Tm of I, II and III, influenced their biodegradation by lipases.

A variety of low-molecular-weight artificial synthetic compounds are biodegradable, whereas only few synthetic polymers are biodegradable. Among synthetic polymers, aliphatic polyesters are generally known to be susceptible to biological attack (1-5).

We report here that polyethylene adipate (PEA) and polycaprolactone (PCL) were degraded by *Penicillium* spp., and aliphatic and alicyclic polyesters,ester type polyurethanes, copolyesters composed of aliphatic and aromatic polyester (CPE) and copolyamide-esters (CPAE) were hydrolyzed by several lipases and an esterase. Concerning these water-insoluble condensation polymers, we noted that the melting points (Tm) had a effect on biodegradability.

Materials and Methods

Materials. PCL and polypropiolactone (PPL) were prepared by ring opening polymerization of ε-caprolactone (6) and β-propiolactone respectively in benzene in a nitrogen atmosphere at 60 °C with a di-

0097–6156/90/0433–0136$06.00/0

ethylzinc-water catalyst system. Poly-DL- β -methylpropiolactone ($\overline{Mn}$: 8190; Tm:167-171 °C; poly-DL- β -hydroxybutyrate) was made from DL-β- methylpropiolactone by the method of Yamashita et al.(7) with a triethylaluminum-water catalyst system. Poly-D- β- methylpropiolactone (PHB) produced by Alcaligenes eutrophus was kindly supplied by ICI. PCL-diols ($\overline{Mn}$: 530, 1250, 2000, 3000) were purchased from Aldrich Chemical. Other saturated aliphatic polyesters were synthesized by a melt polycondensation technique (8), and unsaturated polyesters were synthesized by high temperature solution polycondensation (9). All alicyclic and aromatic polyesters were from Nihon Chromato except polyethylene terephthalate (PET) from Asahikasei, polytetramethylene terephthalate (PBT) from Union Carbide, copolyester of PET and polycyclohexylenedimethyl succinate (PETG) from Eastman Chemical Products. All diisocyanates were from Tokyo Chemical. Polyurethanes were synthesized by a solution polycondensation method (10). CPE and CPAE were synthesized by the transesterification reaction (11) and the amide-ester interchange reaction (12) respectively. All nylons were from Aldrich Chemical except nylon 6 ([η]=1.30 dl/g in m-cresol at 25 °C) from Toyokasei and nylon 11 and nylon 12 from Nihon Lilsan.

Achromobacter sp. and Candida cylindracea lipases (Meito Sangyo) were purified by gel filtration on Sephadex G-100 (2.6 x 79 cm) from crude preparations. Ultracentrifugally homogeneous preparations of Geotrichum candidum and Rhizopus delemar lipases were from Seikagaku Kogyo and partially purified preparations of R. arrhizus lipase and hog liver esterase were from Boehringer Mannheim Yamanouchi. A partially purified preparation of hog pancreas lipase was obtained from Worthington Biochemicals. One unit of enzyme liberated one μ mole of fatty acid from olive oil per min at pH 7.0 and 37 °C.

Culture. Fungi were cultured on a rotary shaker at 180 rpm at 30 °C. The composition of cuture medium was as follows; PEA (or PCL), 1.0 g; $(NH_4)_2SO_4$, 1.0 g; KH_2PO_4, 0.2 g; K_2HPO_4, 1.6 g; $MgSO_4 \cdot 7H_2O$, 0.2 g; NaCl, 0.1 g; $CaCl_2 \cdot 2H_2O$, 0.02 g; $FeSO_4 \cdot 7H_2O$, 0.01 g; $Na_2MoO_4 \cdot 2H_2O$, 0.5 mg; $Na_2WO_4 \cdot 2H_2O$, 0.5 mg; $MnSO_4$, 0.5 mg in one liter of distilled water. pH was adjusted to 7.2. After sterilization, PEA (or PCL) in the culture medium was dispersed by shaking. The particle size of PEA and PCL in the medium was about 0.1-3 mm, 1-5 mm respectively.

Assay of Enzymatic Hydrolysis of Synthetic Solid Polymers. Hydrolysis of solid polymers was measured by the rate of their solubilization, and the measurement process does not necessarily involve complete hydrolysis into the constituent parts. The rate was determined by measuring the water-soluble total organic carbon (TOC) concentration at 30 °C in the reaction mixture using a Beckman TOC analyzer (Model 915-B). In the substrate and enzyme controls, enzyme or substrate was omitted from the reaction mixture.

Determination of Molecular Weight. The number average molecular weight ($\overline{Mn}$) was measured by the vapor pressure equilibrium method with a Hitachi Molecular Weight apparatus (Model 117).

Measurement of Melting Point. Melting points of polyesters (Tm) were measured using a Yanaco Micro Melting Point apparatus (Model MP-S3).

Determination of Carboxylic Group Terminals. The carboxylic group terminals were determined by alkali titration (13). After 200 mg of the sample was dissolved in 7.5 ml of benzyl alcohol in a nitrogen atmosphere at 110 °C, the solution was titrated with 0.05N potassium hydroxide using phenolphthalein as an indicator. A blank test was carried out by the same method but omitting the sample.

Estimation of Molecular Weight Distribution of Polyamide Blocks. The molecular weight distribution of the polyamide blocks was estimated by gel permeation chromatography (GPC) using two instruments, model HLC-802R (Toyo Soda Industry Co., Ltd.) and model GPC-244 (Waters Associates, Inc.). Polystyrene and nylon oligomer were used as standards.

Results and Discussion

Degradation of Aliphatic Polyesters by Fungi. Polyethylene adipate (PEA) with $\overline{Mn}$ 3000 was almost completely degraded by Penicillium sp. strain 14-3 isolated from soil of a factory producing polyurethane (Figure 1). As strain 14-3 grew, PEA was rapidly degraded and disappeared in 120 hours. As shown in Figure 1, water-soluble TOC were found in the culture medium. Adipic acid and ethylene gylcol were detected in the culture fluid as degradation products. Furthermore the strain assimilated some other aliphatic polyesters.

Figure 2 illustrates the time course of polycaprolactone (PCL; $\overline{Mn}$ 25.000) degradation by Penicillium sp. strain 26-1 isolated from soil. The isolate almost completely degraded PCL in 12 days. Among degradation products of PCL, ε- hydroxycaproic acid was detected. The fungus assimilated various polyesters. In general, assimilation of aliphatic polyesters by the fungus was better the greater the number of carbon atoms between the ester bonds. Polyesters with side chains were generally less assimilated than without side chains. The fungus also assimilated unsaturated aliphatic polyesters, but hardly assimilated alicyclic and aromatic polyesters.

Polyester-Degrading Enzyme. A polyester-degrading enzyme from Penicillium sp. strain 14-3 was purified into a homogeneous state ultracentrifugally and electrophoretically (14). In substrate specificity, this enzyme degraded various kinds of saturated and unsaturated aliphatic polyesters and polycyclohexylenedimethyl adipate as alicyclic polyester but not aromatic polyesters. A kind of terminal groups, such as hydroxy, hexahydrophthalic acid and hexahydrophthalic acid glycidil ester terminal did not affect so much on the enzyme activity. The enzyme further hydrolyzed various plant oils, triglycerides and methyl esters of fatty acid. Therefore the polyester-degrading enzyme was assumed to be a kind of lipase.

Hydrolysis of Polyester by Lipase. Aliphatic polyester, PEA and PCL were hydrolyzed by lipases from Achromobacter sp., C. cylindracea, G. candidum, R. arrhizus, R. delemar, hog pancreas and hog liver esterase (15). Especially R. arrhizus and R. delemar lipases were found capable of hydrolyzing various kinds of polyesters (Table I).

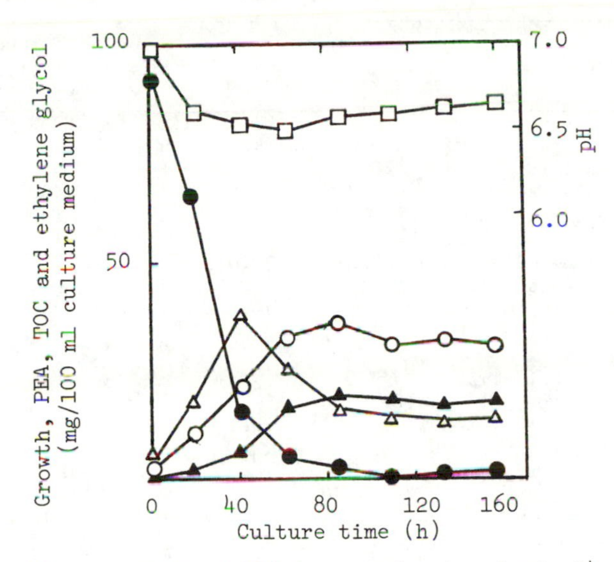

Figure 1. Time courese of PEA degradation by strain 14-3. The strain was cultured at 30 °C. Growth (○), PEA (●), water-soluble TOC (△), ethylene glycol (▲) and pH (□) are shown.

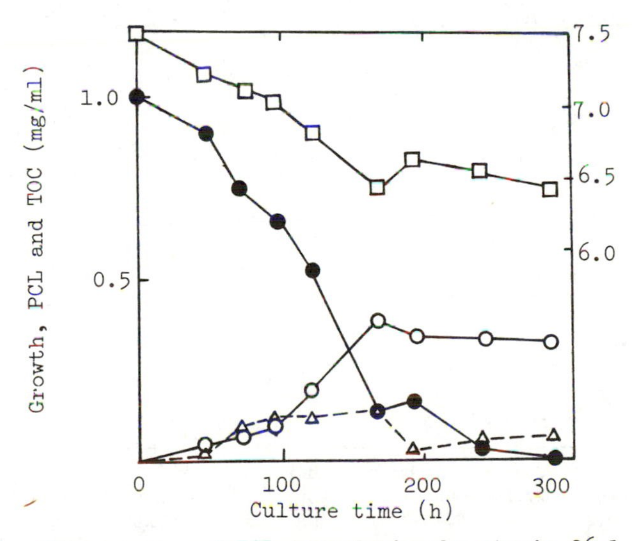

Figure 2. Time course of PCL degradation by strain 26-1. The strain was cultured at 30 °C. Growth (○), PCL (●), water-soluble TOC (△) and pH (□) are shown.

Table I. Hydrolysis of Polyesters by R. delemar and R. arrhizus Lipases

Polyester	$\overline{Mn}$	Tm (°C)	Powder size	TOC formed by lipases (ppm) R. delemar	R. arrhizus
Polyethylene adipate	2720	48.5	C	8360	9290
Polyethylene suberate	4050	64.5	B	1020	1620
Polyethylene azelate	4510	52.1	B	3080	3770
Polyethylene sebacate	1570	74.5	A	550	980
Polyethylene decamethylate	1610	86.0	B	180	240
Polytetramethylene succinate	4240	117	A	150	210
Polytetramethylene adipate	1790	72.0	B	3360	2900
Polytetramethylene sebacate	2440	65.8	A	980	3300
Polyhexamethylene sebacate	5820	74.0	B	380	1160
Poly-2,2-dimethyltrimethylene succinate	2370	76.5	A	240	50
Poly-2,2-dimethyltrimethylene adipate	2020	36.5	E		150
Polyglycolide		226-234	E		0
Copolyester of glycolide and lactide(molar ratio, 92:8)		200-210	E		0
Polypropiolactone	4270	95.0	A	2240	1600
Poly-DL-β-methylpropiolactone	8190	167-171	E	10	60
Poly-D-β-methylpropiolactone (PHB)	25000	175-181	A	0	0
Polycaprolactone (PCL)	6740	59.0	A	310	3610
Poly-cis-2-butene adipate	2700	56.9-59.8	C	580	550
Poly-cis-2-butene sebacate	6190	60.8-62.5	B	300	3430
Poly-trans-2-butene sebacate	3560	57.0-59.0	A	340	1190
Poly-2-butyne sebacate	4930	61.9-63.0	C	670	910
Polyhexamethylene fumarate		113-117	A	35	0
Poly-cis-2-butene fumarate		300	B	30	40
Polytetramethylcyclobutane succinate	3440	63.5-84.0	E	0	0
Polycyclohexylenedimethyl succinate	3910	123-130	B	130	120
Polycyclohexylenedimethyl adipate	3250	108-114	B	200	160
Polytetramethylene terephthalate		230-240	C	0	0
Polyethylene tetrachlorophthalate	1670	78.0-84.2	A	0	0
Poly-2,2-dimethyltrimethylene isophthalate		66.5-75.0	A	0	0
Poly-p-hydroxybenzoate		300	A		0
Poly-3,5-dimethyl-p-hydroxy-benzoate		300	A		0
Poly-4-ethoxy-3,5-dimethyl-benzoate		259-267	A		0

The particle size of each polyester powder was ranked A, B, C, D, or E, corresponding respectively to roughly less than 0.25 mm, less than 0.50 mm, less than 1.0 mm, 0.25-1.5 mm, 0.25-3 mm. Each reaction mixture contained 400 μ mol of phosphate buffer (pH 7.0), 1 mg of surfactant Plysurf A210G, 300 mg of the polyester powder and 60 μ g of R. arrhizus lipase (or 300 μg of R. delemar lipase) in a total volume of 10.0 ml. In the case of R. delemar lipase, surfactant was omitted and pH of phosphate buffer was 6.0. In the substrate and enzyme controls, enzyme or substrate was omitted from the reaction mixture. The reaction mixtures were incubated at 30 °C for 16 hours.

Effect of the Particle Size of PEA Powders on the Hydrolysis by *R. delemar* Lipase. In the case of PEA, small particles were hydrolyzed better than large ones as shown in Figure 3. So it was assumed that the enzymatic hydrolysis depends on the amounts of surface area of polyester powders.

Effect of Molecular Weight of Polyester on the Hydrolysis by Rhizopus lipase. Using three kinds of polyesters, PCL-diol (I), polyhexamethylene adipate (II), and a copolyester (III) made from 1,6-hexamethylenediol and a 70:30 molar ratio mixture of ε- caprolactone and adipic acid, the effects of the $\overline{Mn}$ of polyester on the hydrolysis by lipase were examined (Figure 4). $\overline{Mn}$ did not affect the rates of hydrolysis by *R. arrhizus* and *R. delemar* lipases when $\overline{Mn}$ was more than about 4000. This would indicate these lipases randomly spilts ester bonds in polmer chains. In contrast, when $\overline{Mn}$ was less than about 4000, the rates of the enzymatic hydrolysis were faster with the smaller $\overline{Mn}$ of polyesters. This corresponded to the fact that Tm was lower with the smaller $\overline{Mn}$ of polyesters.

The rates of hydrolysis of copolyester III by both lipases were much higher than those of homopolymers I and II. III was crystalline, but showed lower Tm than the homopolymers. This would show that III have less order and more amorphous regions than the homopolymers do.

Effect of ε- Caprolactone and Adipic Acid Molar Ratio for Copolyester III on the Hydrolysis by *R. delemar* Lipase. The hydrolysis of various copolymers by *R. delemar* lipase was examined to see whether there was an optimum chemical structure or not. $\overline{Mn}$ of those copolyesters was selected from 1740 to 2220, to diminish the effect of molecular weight. Optimum molar ratio of ε- caprolactone and adipic acid was about from 90:10 to 70:30 (Figure 5). The Tm at the optimum molar ratio was the lowest of all. So it seemed that the existence of optimum molar ratio came from the lowest Tm, which would show the most amorphous material, rather than the optimum chemical structure.

Relationship Between Tm and the Biodegradability of Polyester by Lipases. The relationship between Tm and the biodegradability of saturated aliphatic polyester is shown in Figure 6. For the same series polyesters, the biodegradabilities decreased with increasing Tm.

In general, Tm was represented by the following formula:

$$Tm = \Delta H / \Delta S$$

where ΔH is the change of enthalpy in melting and ΔS is the change of entropy in melting. It is known that the interactions among polymer chains mainly affect the ΔH value and that the internal rotation energies corresponding to the rigidity (flexibility) of the polymer molecule remarkably affect the ΔS value. The high Tm of aliphatic polyamide (nylon) is caused by the large ΔH value based on the hydrogen bonds among polymers chains. Nylon is not biodegradable though nylon oligomer is biodegradable. On the other hand, the high Tm of aromatic polyester is caused by the small ΔS value with increase in rigidity of the polymer molecule based on an aromatic ring. Aromatic polyester is not biodegradable.

Hydrolysis of Polyurethanes by Lipase. Effects of $\overline{Mn}$ of PCL-diol moiety on the hydrolysis of polyurethanes, which were composed of

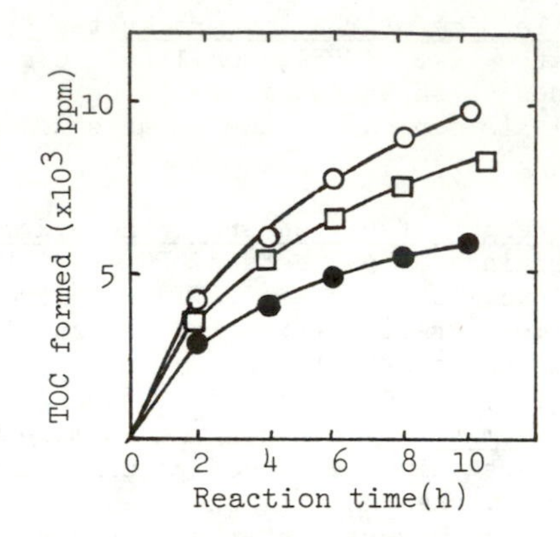

Figure 3. Effect of the particle size of PEA powders on the hydrolysis by R. delemar lipase. Reaction mixtures were incubated at 30 °C. Particle size: -○-, 0-0.25 mm; -□-, 0-1.00 mm, -●-, 0.25-1.00 mm.

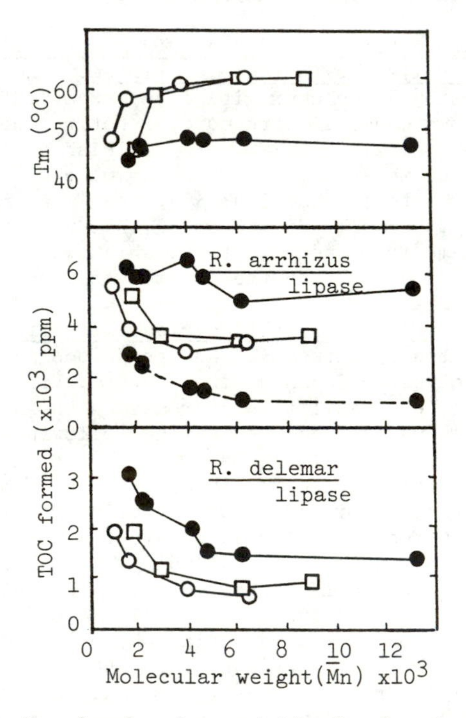

Figure 4. Effects of molecular weight of polyester on the hydrolysis by lipases. Three kinds of polyesters were used: PCL-diol (□), polyhexamethylene adipate (○) and their copolyester (●). The dashed line shows the result when one tenth enzyme concentration was used.

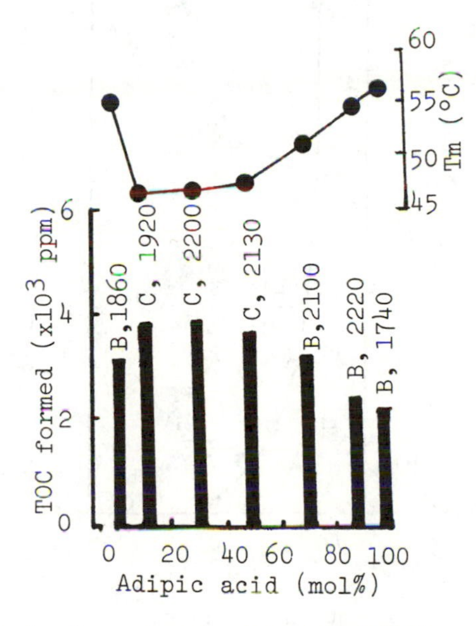

Figure 5. Effect of ε-caprolactone and adipic acid molar ratio for copolymers made from 1,6-hexamethylenediol and a mixture of ε-caprolactone and adipic acid on the hydrolysis by R. delemar lipase. Four orders numbers in this figure showed Mn of each polyester. B and C in this figure showed the particle size of each polyester as same as Table I.

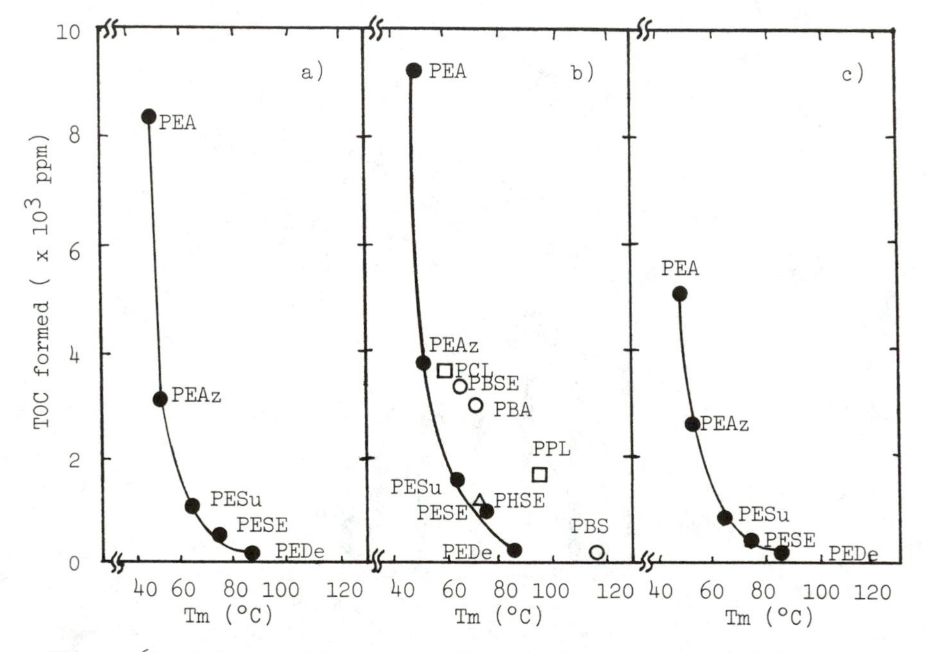

Figure 6. Relationship between Tm and the biodegradability of polyesters by R. delemar (a) and R. arrhizus (b) lipases, and PEA-degrading enzyme from Penicillium sp. strain 14-3 (c). PESu: polyethylene suberate; PEAz: polyethylene azelate; PESE: polyethylene sebacate; PEDe: polyethylene decamethylate; PBS: polytetramethylene succinate; PBA: polytetramethylene adipate; PBSE: polytetramethylene sebacate; PHSE: polyhexamethylene sebacate; PPL: polypropiolactone.

PCL-diol and diphenylmethane-4,4'-diisocyanate (MDI), by R. delemar lipase were examined. These polyurethanes have both the hydrogen bonds among polymer chains and aromatic rings in the polymer molecules. R. delemar lipase could hydrolyze the polyurethanes though the rate of hydrolysis toward polyurethanes decreased as compared to that toward PCL-diol. The rate of hydrolysis decreased with decreasing the $\overline{Mn}$ of PCL-moiety of polyurethanes (Figure 7).

The effects of chemical structure of diisocyanate component on the hydrolysis of polyurethanes by R. delemar lipase were examined (Figure 8). The rates of hydrolysis of the polyurethanes containing MDI or tolylene-2,4-diisocyanate (TDI) were smaller than that of the polyurethane containing 1,6-hexamethylene-diisocyanate (HDI).

Thus it was assumed that the rigidity of the polyurethane molecules based on the aromatic rings, rather than the hydrogen bonds among the polyurethane chains, would influenced their biodegradability by R. delemar lipase.

Hydrolysis of Copolyesters (CPEs) Containing Aromatic and Aliphatic Ester Blocks by Lipase (16). CPEs were synthesized by the transesterification reaction between aromatic polyesters (PET, PBT, PETG, PEIP) and aliphatic polyester (PCL). The susceptibility of CPEs to hydrolysis by R. delemar lipase dropped off rapidly during the initial stage of the transesterification reaction and increase gradually as the reaction proceeded. The susceptibility to hydrolysis decreased with increase in aromatic polyester content (Figure 9). The susceptibility to hydrolysis by the lipase of CPEs composed of PCL and polyethylene isophthalate (PEIP), the latter being used as a low Tm (103 °C) aromatic polyester, were greater than those of other CPEs as shown in Figure 9. It was assumed that the rigidity of the aromatic ring in the CPE chains influenced their biodegradability by this lipase.

Hydrolysis of Copolyamide-esters (CPAEs) by Lipase (17). CPAEs were synthesized by the amide-ester interchange reaction between polyamide and polyester. The length of the polyamide blocks was measured after hydrolysis of ester bonds in CPAE by alkali at 30 °C. The infrared spectra after hydrolyzing ester bonds on CPAEs showed that the ester bonds were almost completely removed. The molecular weight distribution of polyamide blocks was examined by GPC (Table II). The following samples were used: CPAE-1 (reaction time for synthesis, 1 hr) and CPAE-2 (reaction time, 4 hr) composed of nylon 6 and PCL at a 50/50 molar ratio, CPAE-3 (reaction time, 1 hr) and CPAE-4 (reaction time, 4 hr) composed of nylon 12 and PCL at a 50/50 molar ratio, CPAE-5 (reaction time, 4 hr) composed of nylon 6 and PCL at a 20/80 molar ratio, and CPAE-6 (reaction time, 4 hr) composed of nylon 12 and PCL at a 20/80 molar ratio. In addition, the $\overline{Mn}$ of polyamide blocks of

Table II. Polymerization Degree of the Main Component of the Polyamide Blocks of CPAEs 1-6

Standard for GPC	CPAE-1	CPAE-2	CPAE-3	CPAE-4	CPAE-5	CPAE-6
Polystyrene	94	32	53	16		
Each nylon oligomer		9-10		2-6	7-8	2-4

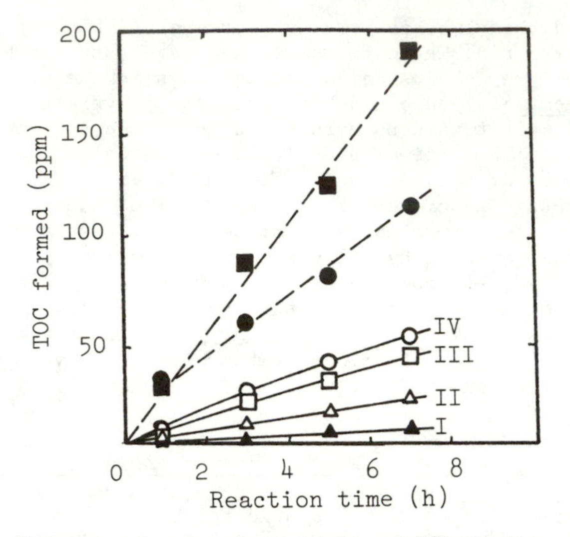

Figure 7. Effects of molecular weight of PCL-diol parts on the hydrolysis of polyurethanes by R. delemar lipase. Each reaction mixture for biodegradability assay contained 15.6-37.2 mg of polyurethane film (14.4-27.1 mg as polyester moiety) on the cover glass (3.2 cm^2) in a total volume of 10 ml. In this condition, no effect of amount of polyurethane was observed. ($\overline{Mn}$)s of PCL-diols parts of polyurethanes I, II, III and IV were 530, 1250, 2000, 3000 respectively. The dashed lines show PCL-diol ($\overline{Mn}$ 2000) (■) and PCL-diol ($\overline{Mn}$ 3000) (●).

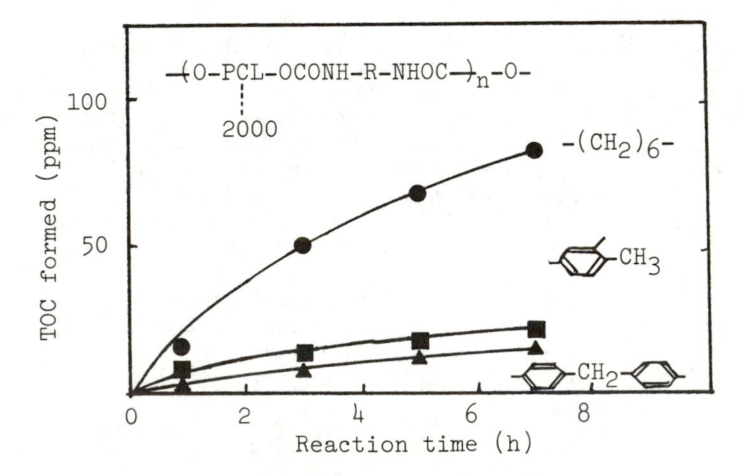

Figure 8. Effects of chemical structure of diisocyanate component on the hydrolysis of polyurethanes by R. delemar lipase. HDI (●), TDI (■) or MDI (▲) was used as a diisocyanate of the polyurethane containing PCL-diol ($\overline{Mn}$ 2000). Assay conditions are the same as in the case of Figure 7.

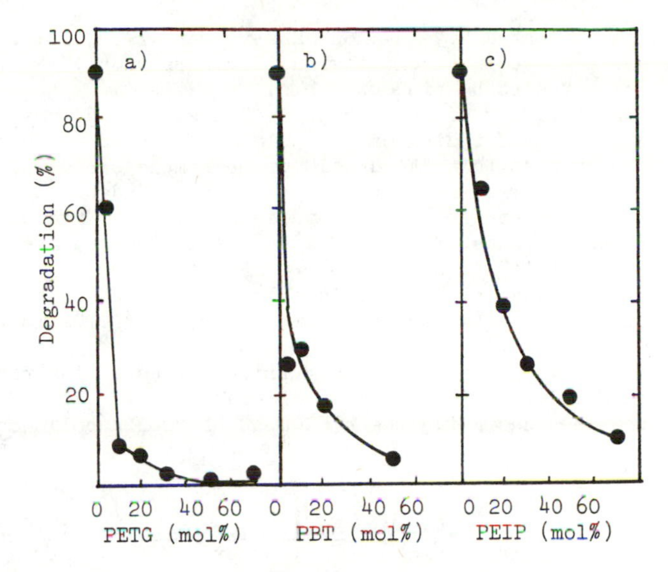

Figure 9. Effect of molar ratio of PCL and aromatic polyester on the biodegradability of CPE by R. dememar lipase. (a), (b), and (c) indicate PCL-PETG, PCL-PBT, and PCL-PEIP systems, respectively. Each reaction mixture for biodegradability assay contained CPE powder or its films (20 mg as polyester moiety) in a total volume of 1.0 ml. Reaction mixtures were incubated at 37 °C for 16 hours. Formation of the water-soluble TOC was in proportion to substrate amounts (up to 50 mg as PCL moiety) in this reaction system. (Reproduced from Reference 16. Copyright 1981 John Wiley.)

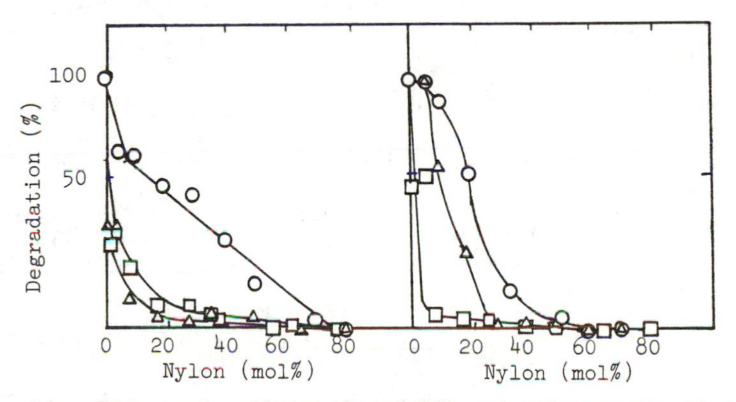

Figure 10. Effect of molar ratio of PCL and nylon on the biodegradability of CPAE by R. delemar lipase. The reaction time for each CPAE synthesis was 4 hours. The basic structures of nylon were of two types. One was $\{NH(CH_2)_nCO\}_m$ (left); the other was $\{NH(CH_2)_6NHCO(CH_2)_nCO\}_m$ (right). Left: nylon 6 (○); nylon 11(△); nylon 12 (□); right : nylon 6,6 (○); nylon 6,9 (△); nylon 6,12 (□). Assay conditions are the same as in the case of Figure 9 except CPAE was used in stead of CPE. (Reproduced from Reference 17. Copyright 1979 John Wiley & Sons, Inc.)

CPAEs 1-6, as estimated by end-group assay, were 3750, 1030, 2870, 850, 830, and 480, respectively. Nylon oligomers with low molecular weight are expected to be biodegradable.

The susceptibility of CPAEs to hydrolysis by R. delemar lipase decreased with the shortening of the polyamide blocks and with increasing polyamide content (Figure 10). The simple blends of nylon and PCL at 270 °C for 10 min retained high biodegradability of PCL. So it was assumed that the amount and distribution of hydrogen bonds, based on the amide bonds, in the CPAE chains influenced their biodegradability by this lipase.

The new biodegradable synthetic polymer, CPAE can be formed into any desirable shape. A transparent thin film (about 0.02 mm thickness) was made from CPAE.

It would be very important that various types of interactions among macromolecular chains, which are related to Tm, are taken into consideration when designing the biodegradable solid polymers.

Literature Cited

1. Darby, R. T.; Kaplan, A. M. Appl. Microbiol. 1968, 16, 900.
2. Potts, J. E.; Clendinning, R. A.; Ackart, W. B.; Niegisch, W.D. Am. Chem. Soc., Polymer Preprints 1972, 13, 629.
3. Fields, R. D.; Rodriguez, F.; Finn, R. K. J. Appl. Polym. Sci. 1974, 18, 3571.
4. Diamond, M. J.; Freedman, B.; Garibaldi, J. A. Int. Biodetn. Bull. 1975, 11, 127.
5. Huang, S. J.; Bell, J. P.; Knox, J. R.; Atwood, H.; Bansleben, D.; Bitritto, M.; Borghard, W.; Chapin, T.; Leong, K. W.; Natarjan, K.; Nepumuceno, J.; Roby, M.; Soboslai, J.; Shoemaker, N. Proc. 3rd Int. Biodegradation Symp., 1976. p 731.
6. Lunberg, R. D.; Koleske, J. V.; Wischmann, K. B. J. Polym. Sci. 1969, A-1, 7, 2915.
7. Yamashita, Y.; Tuda, T.; Ishikawa, Y.; Miura, S. J. Chem. Soc. Jpn, Ind. Chem. Sec. 1963, 66, 110.
8. Carothers, W. H.; Arvin, J. A. J. Am. Chem. Soc. 1929, 51, 2560.
9. Batzer, H.; Holtschmidt, H.; Wiloth, F.; Mohr, B. Makromol. Chem. 1951, 7, 82.
10. Marvel, C. S.; Johnson, J. H. J. Am. Chem. Soc. 1950, 72, 1674.
11. Charch, W. H.; Shivers, J. C. Text. Res. J. 1959, 29, 538.
12. Kiyotsukuri, T.; Takada, K.; Imamura, R. Chem. High Polym. Jpn. 1970, 27, 410.
13. Waltz, J. E.; Tayler, G. B. Anal. Chem. 1947, 19, 448.
14. Tokiwa, Y.; Suzuki, T. Agric. Biol. Chem. 1977, 41, 265.
15. Tokiwa, Y.; Suzuki, T. Nature 1977, 270, 76.
16. Tokiwa, Y.; Suzuki, T. J. Appl. Polym. Sci. 1981, 26, 441.
17. Tokiwa, Y.; Suzuki, T.; Ando, T. J. Appl. Polym. Sci. 1979, 24, 701.

RECEIVED January 8, 1990

Chapter 13

Biodegradable Polymers Produced by Free-Radical Ring-Opening Polymerization

William J. Bailey[1], Vijaya K. Kuruganti[1], and Jay S. Angle[2]

[1]Department of Chemistry and [2]Department of Agronomy, University of Maryland, College Park, MD 20742

Traditionally, synthetic polymers have always been designed to be resistant to microbial attack and are usually stabilized with antioxidants, antiozonants, heat stabilizers and UV stabilizers in order to protect them from environmental degradation. The environmental stability of some polymeric materials, such as construction materials and electrical insulations, is very important for their long term performance. However, the use of disposable plastics is growing rapidly and these materials are used for a short term and then discarded, as in the food and packaging industry. It is highly desirable to have these disposable plastics maintain the required properties during their short term service but degrade in a safe manner when exposed to soil micro-organisms for a long duration. Biodegradable polymers also find specific applications in medicine and agriculture.

Most synthetic polymers are not biodegradable because they have not been on the earth long enough for the micro-organisms to evolve to utilize them as food. On the other hand, all naturally occurring polymers, such as starch, cellulose, proteins, nucleic acids, and lignin, are biodegradable since micro-organisms have evolved which contain the necessary enzymes for the degradation of these natural polymers. These enzymes are a result of centuries of biological adaptations by the micro-organisms to these natural polymers.

Microbial degradation of natural polymers involves initial hydrolysis of these polymers by water-soluble extracellular enzymes to yield low molecular weight oligomers, which are then taken inside the cell. These low molecular weight oligomers are further hydrolyzed and oxidized by various intracellular enzymes. However, since most natural polymers evolved in an aqueous system and contain very polar groups, they decompose on heating before they melt. This limits their fabrication to only solution processes and not by relatively inexpensive commercial processes, such as injection molding, melt spinning and melt extrusion. Moreover, most natural polymers are very sensitive to moisture and do not have the desirable moisture barrier characteristics of synthetic polymers, such as polyethylene.

0097–6156/90/0433–0149$06.00/0

On the other hand, most addition polymers are not biodegradable because the common micro-organisms do not contain the required extracellular enzymes capable of hydrolyzing or cleaving polymers containing only carbon to carbon bonds in the backbone[1,2] to low molecular weight oligomers. Recently, Albertsson, et al.,[3] determined the degradation of polyethylene (M_n = 18,000; M_w = 84,000) to be less than 0.2% by weight in a 10- year period. Studies on the biodegradability of common plastics, such as polystyrene, polyethylene and polyesters, were performed by Potts, et al.,[4] who found that of all the common plastics, only low melting, low molecular weight polyesters showed any reasonable rate of biodegradation. It is not surprising that some synthetic polyesters are biodegradable, since poly(β-hydroxybutyric acid) is a naturally occurring material that many bacteria and fungi use for energy storage in the same way that animals use fat.[5]

This indicates the possibility of making addition polymers biodegradable by the introduction of ester linkages in to the backbone. Since the free radical ring-opening polymerization of cyclic ketene acetals, such as 2-methylene-1,3-dioxepane (I, Scheme I), made possible the introduction of ester groups into the backbone of addition polymers, this appeared to be an attractive method for the synthesis of biodegradable addition polymers.

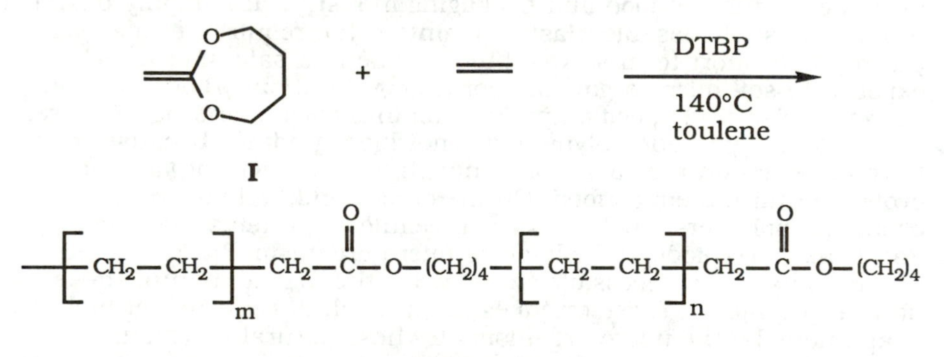

Scheme I

In our laboratory, it has been shown that copolymers of ethylene and 2-methylene-1,3-dioxepane with 10% ester-containing units are highly degradable, while copolymers with 6% ester-containing units are slowly biodegradable.[6] The current research was undertaken to investigate the biodegradability of copolymers of styrene and cyclic ketene acetals I and II (Scheme II). A typical use of biodegradable polystyrene would be the production of biodegradable styrofoam cups and boxes, which are used only once and discarded, often to despoil our streets, lakes, and streams. In addition, as part of an effort designed to produce more reactive cyclic monomers, monomer III was synthesized and polymerized to yield polymer VI (Scheme III) with little or no ring opening. It was hoped that polymer VI would represent a new class of biodegradable polymers.

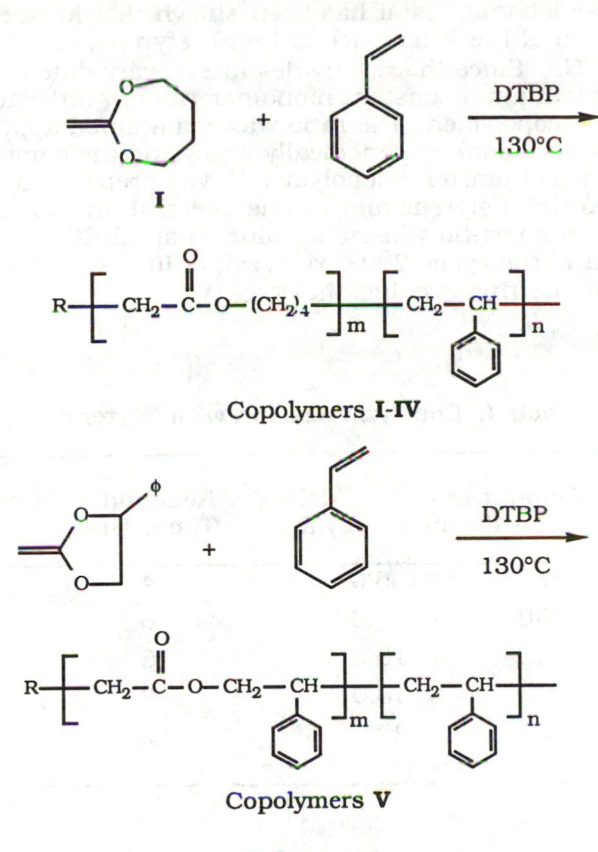

Scheme II

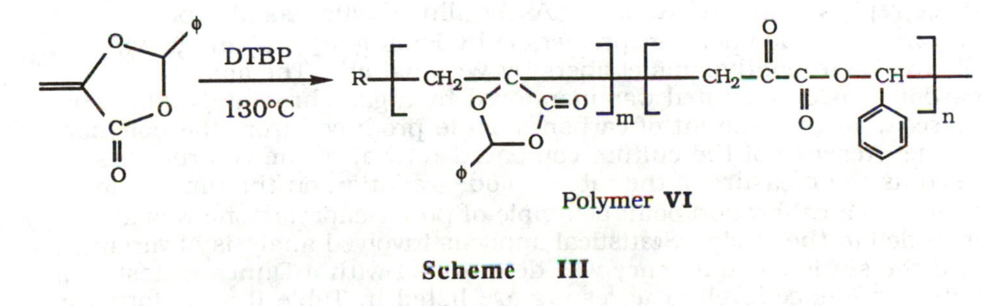

Scheme III

RESULTS AND DISCUSSION

In these laboratories, it has been shown that for the copolymerization of the ketene acetal I with styrene, $r_1 = 0.021$ and $r_2 = 23.6$ at 120°C.[7] Since the reactivities are so very different, it is necessary to maintain a constant monomer ratio in order to synthesize a homogeneous copolymer. The ratio was maintained approximately constant (10% variation) by periodically analyzing the samples in a 60 MHz ^1H NMR spectrometer. Copolymer V was prepared by taking an equimolar mixture of styrene and ketene acetal II and carrying out the polymerzation in a tert-butylbenzene solution at 120°C. Solution polymerization of the cyclic 2-alkoxy acrylate III resulted in a polymer with essentially no ring opening. (See Table I.)

Table I. Copolymerization with Styrene

Copolymer	Amount of I or II, Mol% In Feed	Amount of I or II, Mol% In Copolymer	Reaction Time, Hrs.	Conversion of I or II, Wt%
Copolymer I	75	5.0	4	7.0
Copolymer II	80	7.6	6	11.6
Copolymer III	85	10.8	6	13.0
Copolymer IV	90	18.0	8	17.6
Copolymer V	50	23.0	12	23.0

The rate of biodegradability was determined by a modification of the method developed by Ennis and Kramer.[8] A pure culture of aspergillus flavus (ATCC9643) was used as the innoculum instead of mixed microflora in order to achieve better reproducibility among the three replicates of each sample. Aspergillus flavus has also been shown to degrade poly-ε-caprolactone by Huang, et al.,[9] and poly(*p*-phenylene terephthalamide) fibers by Watanabe.[10] The amount of carbon dioxide liberated was monitored by a gas chromatograph. The increase in the amount of carbon dioxide produced from the polymer in the presence of the culture compared to that of the control, was used as the measure of the rate of biodegradation on the time scale used. As a calibration point a sample of poly-ε-caprolactone was also included in the study. Statistical analysis involved analysis of variance, and the statistical difference was determined with a Duncan's test at a 95% confidence level. The results are listed in Table II and plotted in Figure I.

Table II. Biodegradation Studies

Sample	Cummulative amount of CO_2[a] released after 14 days	35 days
Copolymer I	97	103
Copolymer II	98	115[b]
Copolymer III	109	118[b]
Copolymer IV	109	122[b]
Copolymer V	112	125[b]
Poly-ε-caprolactone	119	152[b]
Poly-ε-caprolactone[c]	124	168[b]
Polymer VI	129	187[b]

a Expressed as percent of control.
b Significantly different from the control.
c Sample had a larger surface area as a film compared with commercially available (Aldrich) pellets.

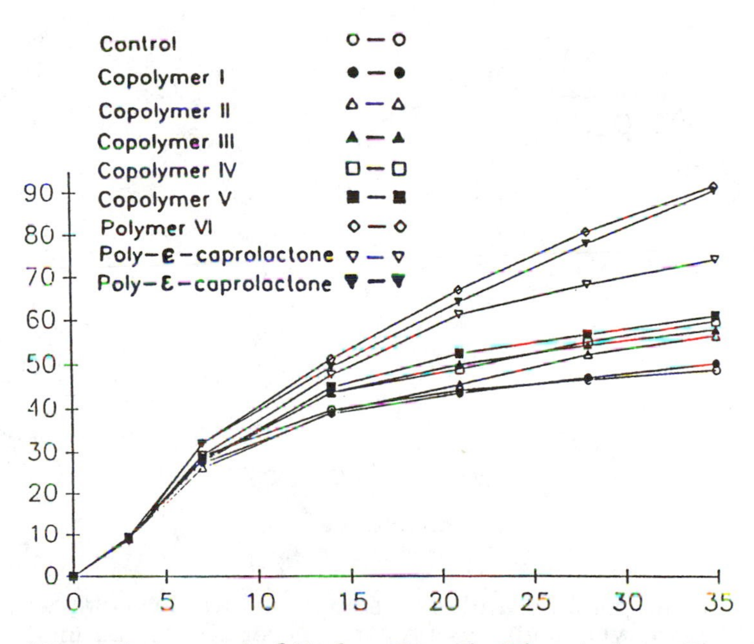

Figure I. Amount of Carbon Dioxide Liberated vs. Time

Table III. Copolymerization of Styrene and Cyclic Ketene Acetal I

Copolymer	Amount of DTBP*, Mol%	Reaction Time, Hrs.	Amount of Styrene Added, mL	$[\eta]_{int}$	M_v[11]	Elemental Analysis C	H	T_g, ^0C
I	0.84	4	11.7	0.185	25,000	90.70	7.76	87
II	1.26	6	25.5	0.180	24,000	89.95	7.82	-
III	1.26	6	20.0	0.160	21,000	88.83	7.84	-
IV	1.68	8	15.0	0.115	13,000	86.65	7.90	53

* DTBP (0.42 mol%) was added at the end of every 2-hr. interval.

From these results it can be concluded that the introduction of ester linkages in the backbone of polystyrene indeed (See Table III.) makes it biodegradable. The higher the number of ester units, the faster the degradation of the copolymers. The biodegradation of polymer VI may involve initial hydrolysis to yield low molecular weight poly(2-hydroxyacrylic acid), which may then be consumed by the fungus (Scheme IV).

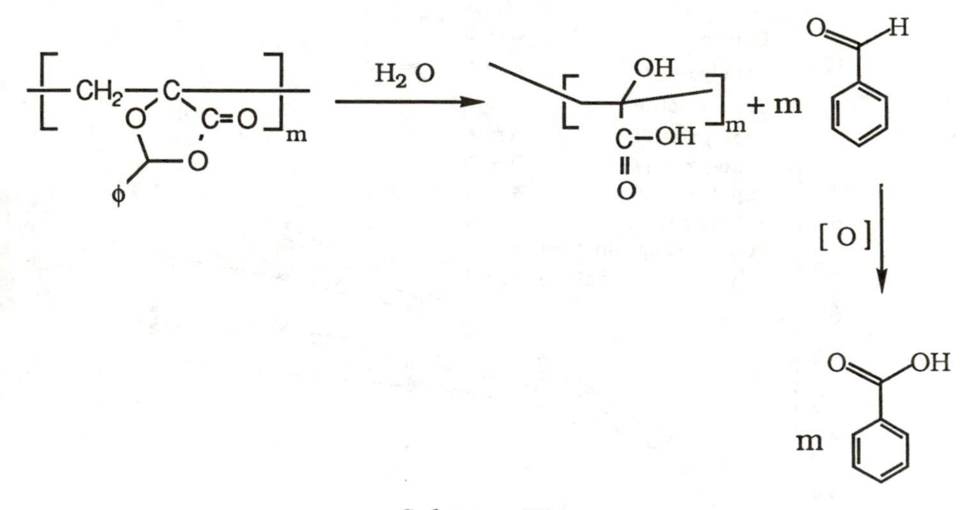

Scheme IV

To establish this point an authentic sample of low molecular weight poly(2-hydroxyacrylic acid) was prepared according to the method of Mulders and Gilain[12] (Scheme V).

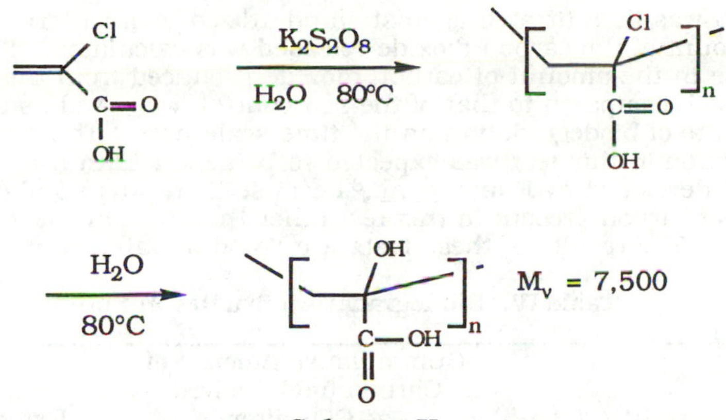

Scheme V

All these polymers were tested for biodegradation in the presence of soil micro-organisms. A conventional sandy loam was employed. Schematic representation of the degradation apparatus was shown in Figure II.

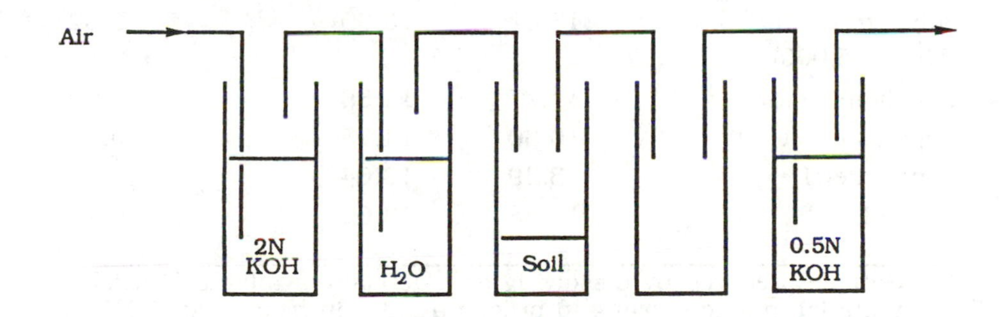

Figure II. Degradation apparatus.

Air was drawn through a 2 N KOH solution to remove any carbon dioxide present in the air which was then remoisturized. The carbon dioxide released from the sample was trapped by a 0.5 N KOH solution as potassium carbonate, which was precipitated as barium carbonate by the addition of barium chloride. The unreacted KOH in solution was then titrated against standardized hydrochloric acid and the amount of the carbon dioxide released was calculated. The increase in the amount of carbon dioxide produced from the polymer in the soil, compared to that of the soil control, was used as a measure of the rate of biodegradation on the time scale used. The rate of degradation in this test was expected to be slower than that of the method developed by Ennis, et al.,[8] because there was no alternate source of carbon present in this test other than that in the soil and polymer. The results of these tests are listed in Table IV.

Table IV. Biodegradation Studies in Soil

Sample	Cummulative Amount of Carbon (mg) Evolved as CO_2 after 41 Days	102 Days	Extent of Degradation,%
Poly-ε-caprolactone[a]	128.08	194.88[b]	51.40
Polymer VI	18.57	30.54[b]	4.96
Poly(2-hydroxyacrylicacid)[c]	18.17	21.60[b]	3.90
Copolymer II	15.54	26.03[b]	2.76
Copolymer III	14.85	23.29[b]	2.18
Polystyrene (Mv = 35,000)	13.37	15.99	0.52
Copolymer V	11.77	14.86	0.29
Copolymer VI	10.30	14.75	0.26
Copolymer I	8.39	12.04	---
Control	10.88	13.62	---

a Sample prepared by dissolving commercially available (Aldrich) material in chloroform and precipitating it in hexane at -70°C.
b Significantly different from control at a 95% confidence level.
c Prepared according to Ref. 15. Mv = 7,500.

From these results it can be concluded that low molecular weight poly(2-hydroxyacrylic acid) is biodegradable and the rate of degradation is similar to that of polymer VI. This supports the possibility that polymer VI is hydrolyzed initially to give low molecular weight poly(2-hydroxyacrylic acid), which was then consumed by soil micro-organisms. Additional studies are required to determine the mechanism of degradation of poly(2-hydroxyacrylic acid). The amount of carbon dioxide released by the degradation of the polystyrene sample (M_v = 35,000) was not significantly different from that of soil controls. Of all the copolymers only copolymer II containing 7.6 mol% ester groups and copolymer III containing 10.8 mol% ester groups showed any significant degradation over that of control. This result may be due to different types of microorganisms being present in the different soil samples. In addition, since the rate of degradation was very slow, longer term testing would be required before any conclusions could be made.

EXPERIMENTAL

Typical copolymerization of styrene with cyclic ketene acetal I (20 g, 0.174 mol) was carried out in a 250-mL, three-necked flask fitted with a 10-cm glass addition funnel and containing an appropriate initial feed ratio in 25 g of tert-butylbenzene at 130°C and initiated by di-tert-butyl peroxide, under a nitrogen atmosphere. The monomer ratio was maintained approximately constant (±10% variation) by the gradual addition of the more reactive styrene. The polymerization was followed by analyzing samples of the reaction mixture in a 60 MHz ^{1}H NMR. The areas corresponding to the vinylic protons in stryene (2H, dd δ 5.0-5.8) and ketene acetal I (2H, s δ 3.35) were chosen for the calculation of the ratio of these two monomers. Samples were analyzed more frequently (every 15 min.) for the first hour after the reaction started and less frequently (every 25 min.) thereafter. All copolymers were white powders : IR($CHCl_3$) 3050, 3020, 2920, 1725(s), 1600, 1495, 1450, 1155, 695 cm^{-1}; ^{1}H NMR (200 MHz, $CDCl_3$) δ 1.00-2.05 (broad,—CO_2—CH_2—$C\underline{H}_2$—$C\underline{H}_2$—$C\underline{H}_2$—) and (ϕC$\underline{H}$—$C\underline{H}_2$—), 2.30 [t (broad —$C\underline{H}_2$—CO_2—)], 4.03 [t (broad, —CO_2—$C\underline{H}_2$—)], 6.30-7.35 (broad, aromatic H's); ^{13}C NMR (200 MHz, $CDCl_3$) δ 34.07, 28.29, 25.49, 24.53(—$\underline{C}H_2$—CO_2CH_2—$\underline{C}H_2$—$\underline{C}H_2$—$\underline{C}H_2$—), 64.03 (—$CO_2\underline{C}H_2$—), 43.56, 40.33 (ϕ$\underline{C}$H—$\underline{C}H_2$—), 125.47, 127.78, 145.07 (aromatic C's), 173.28 (—$\underline{C}O_2$—). Glass transition temperatures (T_g's) of copolymers I and IV were determined by the use of a differential scanning calorimeter. The samples were heated at 10°C/minute in air, to 200°C, then cooled and reheated. Glass transition data was taken from the "second heating".

Copolymer V was prepared by taking 10 g (0.06 mol) of monomer II and 6.2 g (0.06 mol) of styrene in 16.08 g (0.12 mol) of tert-butylbenzene in a 100-mL flask fitted with 10-cm glass dropping funnel. The copolymerization was initiated by the addition of 0.35 g (2 mol%) of di-tert-butyl peroxide under a nitrogen atmosphere at 120°C. After a six hour period, another 0.35 g (2 mol%) of di-tert-butyl peroxide was added and the copolymerization was allowed to proceed for a total of 12 hrs. to yield 7.79 g of yellowish white powdery copolymer: IR ($CHCl_3$) 3060, 3020, 2930, 1730 (S), 1600, 1455, 1498, 1160, 755, 700 cm^{-1}; ^{1}H NMR (200 MHz, $CDCl_3$): δ 1.30-1.08 (broad, —$C\underline{H}_2$—C$\underline{H}$—C_6H_5—), 2.30-2.45 (broad doublet, —$C\underline{H}_2$—CO_2—),

3.26 (broad quintet, —C<u>H</u>—C_6H_5—), 4.03 (broad doublet, —CO_2—C<u>H_2</u>—), 7.16 (broad aromatic H's); ^{13}C NMR ($CDCl_3$): δ 37.13, 40.46, 67.40 (—<u>C</u>H_2CO_2—<u>C</u>H_2—<u>C</u>HC_6H_5—), 125.62, 127.2, 127.9, 128.6, 139.9, 145.23 (aromatic C's), 171.15 (—CO_2—). <u>Anal</u>. Calcd. for $(C_8H_8)_{0.77}$ $(C_{10}H_{10}O_2)_{0.23}$: C, 86.52; H, 7.21. Found: C, 86.64; H, 7.52; $[\eta]_{int}$ = 0.133 (at 30°C in benzene). Monomer II was prepared as previously reported.[13] Poly-ε-caprolactone [$[\eta]_{int}$= 0.373 (in benzene at 30°C], M_v = 22,800][14] was purchased from Aldrich Chemical Company. Polymer VI [$[\eta]_{int}$ = 0.283 (in toluene at 30°C)] was prepared in tert-butylbenzene according to Ref. 15.

Biodegradability tests were carried out in a 25-mL Erlenmeyer flask containing 20-mg polymer samples and 10-mL of medium. All the polymers were fine powders except poly-ε-caprolactone which was either used as a commercially available pellet or it was dissolved in 10-mL of chloroform and the chloroform was removed by evaporation under vaccuum with a Rotovapor to form a thin coating on the inside wall of the flask. The medium consisted of 5 g of hydrolyzed casien (BBL Trypticase soy agar) and 1 g of dextrose in 1 L of ASTM[16] nutrient salt solution. A pure culture of Aspergillus flavus (ATCC 9643) was obtained from the American type culture collection. The fungus was stored as a spore suspension washed from Sabouraud's dextrose agar (Difco) with 0.1% (v/v) Triton X-100 in sterile glass-distilled water. The spore suspension was prepared according to ASTM[16] and 0.1 mL of the suspension was used as innoculum. All the procedures were carried out under a sterile hood and after innoculation, the flasks were flushed with oxygen at a gauge pressure of 30 psi for at least 2 minutes. Flushing was achieved by allowing oxygen gas into the flasks through sterile entrance syringe needle connected by Tygon tubing to a gas cylinder and by allowing the exit gas to escape through another sterile syringe needle connected by a tube to a flask containing 95% ethanol. The flasks were incubated at 25°C for 35 days. Periodically 0.3 mL of the head-space gas was withdrawn and was injected into a gas chromatograph which was fitted with Supelco, chromosorb-packed (60/80 mesh) column. Certified standard carbon dioxide (37.1 mol%) cylinder was purchased from Air Products, Inc. The retention time of CO_2 was t_r = 1.70±0.02 at a 100°C column temperature and a flow rate of 20 mL/min. After each determination, the flasks were flushed with oxygen. All samples had three replicates. Control flasks did not contain polymer test material. Statistical analysis involved multiple analysis of variance by using a Duncan's test at a 95% confidence level.

Soil biodegradation tests were done by taking a sandy loam as the source of the soil micro-organisms.[17] All polymer samples were sieved through ASTM sieves and all the samples except poly-ε-caprolactone, had a particle diameter between 90 and 53 μm. Poly-ε-caprolactone sample was prepared by dissolving a commercially available (Aldrich) sample in chloroform and reprecipitating it at -70°C in hexane in order to get a nice fluff. Into a 250-mL Erlenmeyer flask 100- g portions of air dry sieved soil was placed and to this, 0.5- g of polymer samples were added and thoroughly mixed. The moisture in the flasks was adjusted to 20% of water holding capacity and the flasks were connected to the CO_2 collection apparatus. Into the absorption tube 25-mL portions of 0.5 N KOH were pipetted and the absorption tubes were connected to the CO_2 collection apparatus by rubber stoppers. The flow rate of air was adjusted to 1 bubble every 2 seconds. In order to remove any CO_2 from the air, it was drawn through a 25-mL tube containing 2 N KOH and 2-3 drops of Tropolein indicator. The indicator imparts a light brown color to 2 N KOH solution and would turn colorless when all the KOH was converted into K_2CO_3. The carbon dioxide-free air was remoisturized and passed through the sample flask. The carbon dioxide released by the sample was trapped in an absorption tube. The samples were incubated at room temperature and the absorption tubes were changed after regular intervals. The amount of CO_2 evolved was determined by pipetting out 10-mL of the KOH solution from the absorption tube into a 250-mL beaker containing 25-mL of 2 N $BaCl_2$ and three drops of phenolphthalein indicator, and titrating with 0.4559 N HCl to the end point (disappearance of pink color). Each sample had three replicates. Soil control flasks consisted of only soil and blank flasks contained neither polymer nor soil and polymer control flasks contained only polymer. The weight of carbon evolved as CO_2 gas was calculated by mgC = (B - V)NED, where:

V = volume (mL) of acid to titrate samples,
B = volume (mL) of acid to titrate blank,
N = normality of acid,
E = equivalent weight of carbon (6), and
D = dilution factor (25/10).

Statistical analysis involved multiple analysis of variance using a Duncan's test at a 95% confidence level.

ACKNOWLEDGMENTS

The authors are grateful to the Interx Research Corp., a subsidiary of Merck and Co., Inc., and the Polymer Program of the National Science Foundation for partial support of this research. The authors are also thankful to Dr. J. N. Hansen, and his group members (Department of Chemistry & Biochemistry, UMCP), for providing sterile hood facilities.

REFERENCES

1. M. Alexander, Microb. Ecol., **2**, 17-27 (**1975**).
2. S. J. Huang, in Encyclopedia of Polymer Science and Engineering, Vol. 2, H. F. Mark, C. G. Overberger, N. M. Bikales, G. Menges and J. I. Kroschmitz, eds., John Wiley and Sons, New York, 1985, p. 220.
3. A. C. Albertsson and S. Karlsson, Polym. Mats. Sci. Eng., **58**, 65-69 (**1988**).
4. J. E. Potts, R. A. Clendinning, W. B. Ackart, and W. D. Niegisch, Polym. Prepr., Am. Chem. Soc., Div. Polym. Chem., **13**, 629-634 (**1972**).
5. J. R. Shelton, D. E. Agostini and J. B. Lando, Polym. Prepr., Am. Chem. Soc., Div. Polym. Chem., **12**(2), 483 (**1971**).
6. W. J. Bailey and B. Gapud, Polym. Prepr., Am. Chem. Soc., Div. Polym. Chem., **25**(1), 111 (**1984**).
7. W. J. Bailey, in Ring-Opening Polymerization, E. Goethals and T. Segusa, eds., ACS Sym. Series No. 286, American Chemical Society, Washington, D. C., 1982, p. 947.
8. D. Ennis and A. Kramer, J. Food Science, **40**, 181 (**1975**).
9. C. V. Benedict, W. J. Cook, P. Jarrett, J. A. Cameron, S. J. Huang and J. P. Bell, J. Appl. Polym. Sci., **28**, 327-334 (**1983**).
10. T. Watanabe, Sen-I. Gakkaishi, **43**(4), 192-197 (**1987**).
11. W. R. Krigbaum and P. J. Flory, J. Polym. Sci., **11**, 37 (**1953**).
12. J. Mulders and J. Gilain (Solvay & cie., Brussels, Belgium); Chem. abstr. 82:171706g, U. S. Pat. 4,107,411 (Aug. 15, 1978).
13. W. J. Bailey, S. R. Wu and Z. Ni, Makromol. Chem., **183**, 1913-1920 (**1982**).
14. J. V. Koleske and R. D. Lundberg, J. Polym. Sci., Part A2, **7**, 897 (**1969**).
15. W. J. Bailey and P. Z. Feng, Polym. Prepr., Am. Chem. Soc., Div. Polym. Chem., **28**(1), 154-155 (**1987**).
16. ASTM DG 21-70, 1976 Annual Book of Standards, American Society for Testing Materials, Philadelphia, PA., 1976, pp. 781-785.
17. G. Stotzky, in Methods of Soil Analysis, Agronomy 9, C. A. Black, ed., Amer. Soc. of Agron., Madison, Wisconsin, 1965, p. 1550.

RECEIVED February 16, 1990

Chapter 14

In Vitro and In Vivo Degradation of Poly(L-lactide) Braided Multifilament Yarns

Brian C. Benicewicz[1,3], S. W. Shalaby[1], Alastair J. T. Clemow[2], and Zale Oser[2]

[2]Ethicon, Inc., Somerville, NJ 08876–0151
[3]Johnson and Johnson Orthopedics, New Brunswick, NJ 08901

The first results on the potential of poly(L-lactide) for bioabsorbable surgical devices were reported by Kulkarni et al (2, 3). It was reasoned that PLA would be a useful material for bioabsorbable implants since its hydrolytic degradation product, lactic acid, is a normal intermediate of carbohydrate metabolism (glycolysis). These authors determined that poly(D, L-lactide) degraded at a faster rate than poly(L-lactide) and that poly(L-lactide) could be generally characterized as a slowly degrading polymer. More recently, Vert (4-7) and Tunc (5, 9) have conducted extensive studies on the use of polylactide for internal fixation devices such as plates, screws, and rods. A slow degradation rate is a desirable property for these applications. In another series of papers by Pennings et al (10-12) the fiber properties of both solution- and melt-spun poly(L-lactide) were reported. High tensile strengths were obtained by hot drawing fibers spun from good solvents such as dichloromethane and chloroform. In vitro degradation studies were performed on the solution-spun poly(L-lactide) fibers. In spite of the interest in polylactide as a bioabsorbable polymer for more than 20 years, there is very little data on the in vivo and in vitro properties of melt-spun poly(L-lactide) fibers. Jamshidi et al. (13, 14) have recently reported on the in vitro and in vivo degradation of loosely bundled melt-spun PLA fibers. We wish to report the results of our studies on the in vitro and in vivo degradation of poly(L-lactide) braided multifilament yarns produced by melt spinning as models of synthetic bioabsorbable ligaments.

Experimental

The polymers were made according to procedures described in Ref. 15. A representative example is given here. A 1-liter round-bottom flask equipped with an overhead mechanical stirrer was flame dried under a vacuum to remove moisture from the interior surfaces of the vessel. When cool, the vacuum was released by the introduction of dry nitrogen gas. To the flask was added, in a dry nitrogen glovebox, 500.0 grams (3.469 moles) of pure L-lactide. Then, 0.35 ml of a 0.33 molar catalyst solution (stannous octoate in toluene) containing 1.155×10^{-4} moles of catalyst was

[1]Current address: Los Alamos National Laboratory, Los Alamos, NM 87544

0097–6156/90/0433–0161$06.00/0

added using a dry glass syringe. The molar ratio of monomer to catalyst was 30.000:1. Subsequently, 1.06 ml (4.66 x 10^{-3} mole) of warm 1-dodecanol was added to provide a molar ratio of monomer to 1-dodecanol of about 750:1. The reaction vessel was closed and a high vacuum was applied to remove the toluene. After a few hours, the vacuum was released to dry nitrogen gas and the contents of the flask were kept under a slight positive pressure of nitrogen. The flask was placed into an oil bath preheated to a temperature of 120°C and stirring was started as the monomers melted and continued until the viscosity of the reactants became too high to stir. The stirrer was raised above the reaction mixture and the reaction was continued for 3 days at 120°C. The polymer was removed from the flask, cut into pieces, and ground in a mill. The ground polymer was placed in a vacuum oven to dry and devolatilized at 110°C for 18 hours under vacuum. The inherent viscosity of these polymers measured at a concentration of 0.1 g/dL in chloroform at 25°C was typically 2.1 $\pm$ 0.2 dL/g. The yields of polymer were approximately 98%. Poly(L-lactide-co-glycolide) 95/5 was also produced in a similar manner.

The polymer was extruded at 240°C to produce 28-strand multifilament yarn. The yarn was hot-stretched and fiber tensile strengths in the range 3-6 g/den (1-2 dpf) could be obtained. The yarn was placed on braider bobbins on a 12-carrier machine with 7-ply core. The braid was made with 51 picks per inch and hot-stretched 25% at 215°F.

The in vitro absorption and breaking strength retention (BSR) were determined by placing the braided yarn in a phosphate buffer of pH 7.27 at 50°C for the required period of time. At the end of each period sufficient samples were removed from the bath, filtered, rinsed with water and dried to constant weight at 25°C under reduced pressure. The in vitro absorption of braided samples was determined in terms of the percent mass loss. The change in molecular weight of the dry samples and a control (not subjected to the phosphate buffer) was measured in terms of inherent viscosity (I.V.) in chloroform (16). Similarly, the BSR was determined through measurement of the breaking strength of the samples at the specified time period and expressed as the percentage of the original unimplanted strength.

The in vivo breaking strength retention was determined by implanting two strands of the ethylene oxide sterilized braided samples in the dorsal subcutis of each of a number of Long-Evans rats. The number of rats used was a function of the number of implantation periods, employing 4 rats per period giving a total of eight (8) examples for each of the periods. The periods of in vivo residence were 7, 31, 92, 184, 276 or 365 days. The ratio of the mean value of 8 determinations of the breaking strength (determined with an Instron tensile tester employing the following settings: a gauge length of 1 inch, a chart speed of 1 inch/minute, and a crosshead speed of 1 inch/minute) at each period to the mean value (of 8 determinations) obtained for the fiber prior to implantation constituted its breaking strength retention for that period.

Results and Discussion

The properties of the braided yarns are given in Table 1. Annealing of the braids was conducted at 100°C for 24 hours in a nitrogen-atmosphere oven under tension. The annealed braids were used for all of the in vitro and in vivo studies reported herein.

The in vitro BSR and absorption studies were conducted in pH 7.27 phosphate buffer at 50°C. Our initial studies indicated a slow absorption rate and a temperature of 50°C was used instead of 37° to obtain results at a faster rate. These data indicate a fairly steady loss in BSR (Table 2) for both the homopolymer and copolymer. The molecular weight changes paralleled the BSR changes throughout most of the testing period. Hydrolysis of the polymers began immediately as evidenced by the I.V. measurements. The in vitro absorption results given in Table 3 indicate that very little weight loss occurred while the braids retained any of their original strength. However, after the braids lost all of their strength, the rate of weight loss increased substantially. This behavior is similar to that of the polyglycolide-based sutures (17). The addition of 5 mole percent glycolide had a noticeable effect on the degradation of the PLA-based braids. This effect was seen in both the in vitro BSR and absorption data. The data in Table 2 show a 56% retention of original strength after 12 weeks for the homopolymer and a 57% retention of original strength after 8 weeks for the copolymer. The in vitro absorption data in Table 3 express similar results. The PLA homopolymer retained 58.5% of its original weight after 52 weeks, whereas the copolymer registered virtually the same weight retention after 26 weeks. Thus, it appears from the in vitro results of this study that even small amounts of glycolide in poly(L-lactide) will accelerate the hydrolytic processes.

The same annealed braids were also subjected to in vivo BSR testing. The results are shown in Figure 1 for both the poly(L-lactide) and 95/5 poly(L-lactide-co-glycolide). These data showed that during the first 4-5 months, both sutures retained

Table 1. Tensile Properties of Braided Poly(L-lactide) and 95/5 Poly(L-lacticc-*co*-glycolide) Yarns

	PLA	PL/G 95/5
Load at break (lb) (unannealed)	6.10 ± 0.29	7.05 ± 0.23
diameter (mils)	10.8 ± 0.38	15.4 ± 0.3
Load at break (lb) (annealed)[a]	5.17 ± 2.24	7.89 ± 0.1
diameter (mils)	8.7 ± 0.2	14.4 ± 0.25

[a]Annealed under tension, 100°C, 24 hours, nitrogen atmosphere.

Table 2. In Vitro BSR (pH 7.27 buffer, 50°C) of Poly(L-lactide) and 95/5 Poly(L-lactide-*co*-glycolide) Braids

	PLA		PL/G 95/5	
Time, weeks	I.V., dL/g	% BSR	I.V., dL/g	% BSR
0	1.56	100	1.47	100
1	1.47	95.7	-	-
3	1.40	91.3	-	-
4	1.24	88.4	1.11	84.9
6	1.19	83.0	-	-
8	1.17	76.8	0.83	57.0
12	0.94	56.1	0.38	14.7
16	0.80	28.6	0.36	0
26	0.29	0	-	-

approximately the same percentage of their original strength. After this period of time, the curves of Figure 1 diverge, and the copolymer-based braids were observed to lose strength faster than those made of the homopolymer. It appears that the incorporation of 5 mole % glycolide in polylactide also accelerates the in vivo hydrolytic processes. (Table 4).

Table 3. In Vitro Absorption (pH 7.27 buffer, 50°C) of Poly(L-lactide) and 95/5 Poly(L-lactide-*co*-glycolide) Braids

	% Weight Remaining	
Time, weeks	PLA	PL/G 95/5
0	100	100
4	100	99.8
12	99.5	95.3
26	95.9	58.3
39	91.9	20.5
52	58.5	2.1
65	23.3	1.2

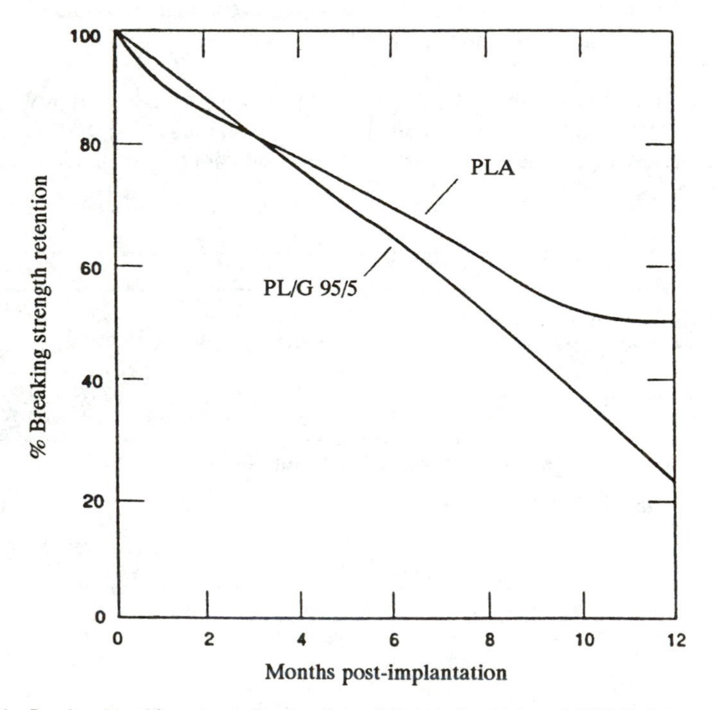

Figure 1. In vivo breaking strength retention of Poly(L-lactide) and 95/5 Poly(L-lactide-*co*-glycolide).

Table 4. In Vivo BSR (pH 7.27 buffer, 50°C) of Poly(L-lactide) and 95/5 Poly(L-lactide-*co*-glycolide) Braids

	% BSR	
Time. days	PLA	PL/G 95/5
0	100	100
7	98	101
31	90	93
92	82	81
184	70	65
276	56	–
365	52	26[1]

[1]Time period was 368 days.

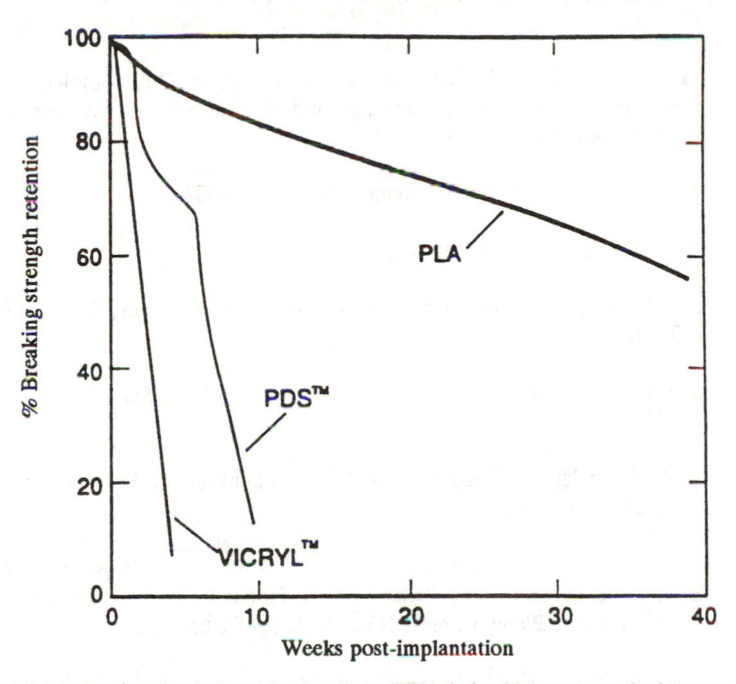

Figure 2. In vivo breaking strength retention of Poly(L-lactide) compared with commercial absorbable surgical sutures.

The in vivo BSR data for poly(L-lactide) generated in this study are plotted in Figure 2 with in vivo BSR data for two commercial sutures. VICRYL™ and PDS™. It is clearly seen that the melt-spun, braided PLA braids retain their strength much longer than those currently used as absorbable surgical sutures. In fact, their longer-term presence would probably be undesirable in many common surgical procedures. However, their good tensile strength properties and extended BSR and absorption profiles will be beneficial in orthopedic applications, such as tendon and ligament repairs.

References

1. R. K. Kulkarni, K. C. Pani, C. Neuman and F. Leonard. Arch. Surg., **93,** 839 (1966).

2. R. K. Kulkarni, E. G. Moore, A. F. Hegyelli and F. Leonard. J. Biomed. Mater. Res. **5,** 169 (1971).

3. P. Christel, F. Chabot, J. L. Leray, C. Morin and M. Vert. in "Biomaterials," Ed. G. D. Winter, D. F. Gibbons and H. Plenk, Jr., John Wiley & Sons. Inc., NY, 1982 pp 271-80.

4. M. Vert and F. Chabot, Makromol. Chem., Suppl. **5,** 30 (1981).

5. P. Christel, F. Chabot, and M. Vert. Trans. Soc. Biomaterials. **7,** 279 (1984).

6. M. Vert, P. Christel, F. Chabot and J. Leray, in "Macromolecular Biomaterials," Ed. G. W. Hastings and P. Ducheyne, CRC Press. Boca Raton, FL. 1984, pp 119-42.

7. D. C. Tunc, Trans. Soc. Biomaterials. **6,** 47 (1983).

8. D. C. Tunc, Polymer Preprints, **27(1),** 431 (1986).

9. B. Eling, S. Gogolewski and A. J. Pennings. Polymer, **23,** 1587 (1982).

10. S. Gogolewski and A. J. Pennings. J. Appl. Polym. Sci., **28,** 1045 (1983).

11. J. W. Leenslag, S. Gogolewski and A. J. Pennings. J. Appl. Polym. Sci., **29,** 2829 (1984).

12. S. H. Hyon, K. Jamshidi, and Y. Ikada, in "Polymers as Biomaterials," Ed. S. W. Shalaby, A. S. Hoffman, B. D. Ratner and T. A. Horbett, Plenum Press, NY. 1984, pp 51-65.

13. K. Jamshidi, S. H. Hyon, T. Nakamura, Y. Ikada, Y. Shimizu, and T. Teramatsu. in "Adv. Biomater., Biol. Biomech. Perf. Biomater.", Ed. P. Christel, A. Meunier, and A. J. C. Lee. Elsevier Science Pub., Amsterdam. 1986, pp 227-32.

14. B. C. Benicewicz. A. J. T. Clemow, Z. Oser, and S. W. Shalaby. Eur. Pat. Appl. 241-252, 1987.

15. A. Schindler and D. Harper. J. Polym. Sci.: Polym. Chem. Ed., **17,** 2593 (1979).

16. R. J. Fredericks, A. J. Melveger and L. J. Dolegiewitz. J. Polym. Sci.: Polym. Physics. Ed., **22,** 57 (1984).

RECEIVED February 16, 1990

Chapter 15

Bioabsorbable Fibers of *p*-Dioxanone Copolymers

R. S. Bezwada, S. W. Shalaby, and H. D. Newman, Jr.

Ethicon, Inc., Somerville, NJ 08876–0151

Growing interest in synthetic, absorbable copolymers and their use in the production of fibers, including sutures, and drug delivery systems has been addressed in the patent and technical literature over the past fifteen years[1-11]. However, most investigators have focused their attention on copolymers which are derived primarily from α-hydroxy acids [3-6], anhydrides[7,8] or oxalate esters[9-11]. The preparation of high molecular weight poly(p-dioxanone)[12], a polyether-ester, and its conversion to a useful absorbable suture with unique properties has now led to the interest in exploring the potential of poly(p-dioxanone) containing other comonomeric units in the chain.[13,14].

Hence, the studies which form the basis of this report were conducted to determine the effect of incorporating different quantities of glycolyl moieties into the poly(p-dioxanone) chains on the physical and biological properties of drawn monofilaments.

EXPERIMENTAL

Materials and Major Characterization Methods.

Polymerization grade p-dioxanone and glycolide monomers were obtained as described in early reports [3,12]. Varian XL-300 ^{1}H and ^{13}C NMR, DuPont DSC, Waters-Model 150C GPC, and X-ray diffractometers were used in the characterization of the copolymers prior to processing. Inherent viscosities (I.V.) were obtained in hexafluoroisopropanol (HFIP) at 25°C and 0.1 dl/g concentration using a Ubbehlohde viscometer. Melt rheological studies and polymer extrusion were conducted using an Instron Model 3211 A stock solution of stannous octoate of known concentration (e.g., 0.33 M) in toluene was prepared under anhydrous conditions. The tensile properties of the fibers were determined using an Instron Tensile Tester, Model 1130 (strain rate = 2"/min. and gauge length = 2").

0097–6156/90/0433–0167$06.00/0

Polymerization Scheme and Typical Examples.
General Polymerization Scheme and Preparation of Copolymer I.
A copolymer of p-dioxanone and glycolide at 95/5 by weight was prepared by charging 95g (0.9306 mole) of p-dioxanone, 0.266 ml of 1-dodecanol, and 0.0984 ml of a toluene solution of stannous octoate (0.33 molar) into a 250 ml, round bottom, two-neck flask. The contents of the reaction flask were held under high vacuum at room temperature for about 16 hours. The flask was fitted with a mechanical stirrer and an adaptor with a hose connection. The reactor was purged with nitrogen before being vented with nitrogen. The reaction mixture was heated to 110°C, and maintained there for 5 hours while stirring. A small sample of this polymer was removed for analysis (Inherent Viscosity = 1.21 dl/g), and 5.0g (0.04308 mole) of glycolide was added to the reaction mixture. The temperature was raised to 140°C, and maintained for 1 hour. The temperature was then lowered to 90°C, and maintained for 65 hours at 90°C. The copolymer was isolated, ground, and dried about 48 hours/80°C/0.1 mm Hg to remove any unreacted monomer. A weight loss of 13.7% was observed.

Preparation of Copolymers II and III. Using the above general copolymerization scheme, two more copolymers of p-dioxanone/glycolide at 90/10 and 80/20 initial weight composition were prepared. The following amounts of reactants and catalysts were used.

	Copolymer II	Copolymer III
PDO/glycolide, initial weight ratio	90/10	80/20
p-dioxanone (PDO)	90 g(0.8816 mole)	80g(0.7836 mole)
1-dodecanol	0.26 ml	0.217 ml
$Sn(oct)_2$(0.33 molar)	0.098 ml	0.048 ml
glycolide	10 g(0.0862 mole)	20g(0.1723 mole)

Conversion of copolymers to fibers and pertinent tensile date. Copolymers I-III described in Table I were melt extruded, and the extrudates were oriented by drawing and then annealed. The tensile properties of the unannealed and annealed fibers are summarized in Table II and III, respectively.

Effect of initiator type on copolymer properties. Copolymer IV of p-dioxanone/glycolide at 90/10 weight composition was prepared using diethylene glycol (DEG) as the initiator. Fiber properties of the resulting copolymer were determined and compared with those of Copolymer II, which was made using 1-dodecanol as an initiator (as shown in Table IV).

In-Vivo Absorption and breaking strength retention studies. For comparing PDS monofilament *in-vivo* properties with copolymeric fibers, a second 90/10 PDO/glycolide copolymer (II-A) was prepared and processed following similar schemes to those used for copolymer II.

COPOLYMER CHARACTERIZATION

Analytical data of copolymers I to III are summarized in Table I.

TABLE I
COPOLYMERIZATION CONDITIONS AND SOME PROPERTIES OF RESULTING COPOLYMERS

Copolymer	I	II	III	PDS*
p-oxanone/glycolide-initial weight ratio	95/5	90/10	80/20	100/0
Final comp. PDS/PGA, mole %[(a)]	--	87.3/12.7	82.7/17.3	100/0
I.V. dl/g	1.63	1.44	1.64	1.8
M.P. (hot-stage)	96-100°C	100-105°C	95-100°C	--
% Conversion	86.3	86.2	93.6	95
Crystallinity, % [(b)]	--	45%	40%	50%
DSC Data[(c)], Tg, °C	--	-8	-2	-10 to -15
Tm, °C	--	97	125	110-115

(a) using NMR
(b) using X-ray diffraction
(c) second heat, using a heating rate of 20°C/min. in N_2
* Typical properties of poly(p-dioxanone) made according to reference #12

SOURCE: Ref. 14.

TABLE II
EFFECT OF DRAWING SCHEMES ON TENSILE PROPERTIES OF MONOFILAMENT FIBERS

		Tensile Data				
Copolymer Example No.	Drawing Conditions	Diameter (mils)	Straight Kpsi	Knot Kpsi	Elong. %	Young's Modulus Kpsi
I	4 x at 58°C followed by 1.562 x at 75°	7.1	88	53	49	211
II	5 x at 52°C followed by 1.3 x at 72°C	7.3	87	49	61	143
III	5 x at 50°C followed by 1.2 x at 71°C	7.5	65	43	94	81

SOURCE: Ref. 14.

Ethylene oxide sterilized monofilaments of copolymer II-A and a poly(p-dioxanone) were compared in terms of their in-vivo absorption and BSR (breaking strength retention) profiles. The BSR and absorption studies were conducted in rats according to the procedures described elsewhere[1,15]. The comparative data of these studies are given in Table V.

DISCUSSION OF RESULTS

During the melt polymerization of PDO, about 75% conversion is attained after 5-6 hrs/110°C. Addition of glycolide to the melt polymerized PDS facilitates the copolymerization of the unreacted PDO with glycolide, thereby, creating a copolymer soft segment at the end of the poly(p-dioxanone) chain. Incorporation of such segments provided more compliant fibers as compared with those made from PDS. The copolymers of PDO/glycolide provided monofilaments with higher strength and improved absorbability and decreased propensity to retain their in-vivo breaking strength.

The properties of the copolymers of p-dioxanone/glycolide at 95/5, 90/10 and 80/20 by initial weight are summarized in Table I. It was observed that the increase of glycolide content from 0 to 20% was associated with a decrease in crystallinity from 50% to 40%. Tg of the copolymers increased with the increase of glycolide ratio but only a slight increase in the value of Tm could be observed. Copolymer III based on 20% glycolide displayed a Tm @ 125°C.

Copolymers I-III were extruded into monofilaments and the tensile properties after drawing and annealing are shown in Tables II and III. The reported data indicate that copolymeric monofilaments can be made to have equivalent or higher tensile strength as compared with those made from PDS.

TABLE III
TENSILE PROPERTIES OF DRAWN, ANNEALED[(a)] MONOFILAMENTS MADE OF COPOLYMERS I TO III AND PDS

Copolymer	I	II	III	PDS*
Fiber Properties				
Diameter (mils)	7.5	7.3	7.5	8.65
Straight Tensile Strength Kpsi	79	85	61	75
Knot Strength, Kpsi	50	62	51	45
Elongation, %	34	39	55	28
Young's Modulus, Kpsi	281	283	201	278

SOURCE: Ref. 14.
(a) All fibers were annealed at constant strain for 12 hrs. @ 60°C
* Typical properties of PDS made according to reference #12

Using diethylene glycol (DEG) as an initiator, a copolymer (IV) of PDO/glycolide at 90/10 by weight was preapred and its properties are compared with those of a 90/10 copolymer (II) made in the presence of 1-dodecanol (Table IV). Although the tensile properties of IV are comparable those of II, the percent BSR of the latter is higher. This may be associated with a difference in the copolymeric chain microstructure and/or molecular weight. However, a lower modulus of IV suggest a major contribution of the microstructure to the overall properties of the fibers. A second 90/10 PDO/glycolide copolymer (II-A) made similarly to copolymer II was converted to drawn annealed monofilaments which displayed higher tensile properties and compliance than those of PDS. The monofilaments of the copolymer II-A exhibited faster in vivo absorption. At four weeks the in-vivo BSR of PDS is 69%, whereas the BSR of the copolymer is only 12%. At 119 days, monofilaments of the copolymer absorbed (in-vivo) completely, whereas PDS monofilaments absorb in about 180-210 days.

TABLE IV
EFFECT OF INITIATOR TYPE ON FIBER PROPERTIES

Copolymer	II	IV
Composition of PDO/glycolide initial weight ratio	90/10	90/10
Initiator Type	1-dodecanol	diethylene glycol
I.V., dl/g.	1.44	1.88
Fiber Properties (annealed for 12 hr/60°C		
Diameter, (mils)	7.3	7.4
Straight Tensile, Kpsi	85	88
Knot Strength, Kpsi	62	55
Elongation, %	39	30
Y.M., Kpsi	283	204
In-Vitro BSR, % remaining at 4 days/50°C	49	34

SOURCE: Ref. 14.

CONCLUSION

Melt polymerization of p-dioxanone, proceeds to about 75% conversion at 110°C. Addition of glycolide at this stage allows for the copolymerization of the unreacted p-dioxanone with glycolide, to produce amorphous glycolide/p-dioxanone segments at the end of the poly(p-dioxanone) chains. Incorporation of such segments provides much more compliant monofilaments as compared with those made from poly(p-dioxanone). The copolymers of p-dioxanone/glycolide provided monofilaments with higher strength and improved absorbability and more rapid loss of in-vivo breaking strength than the homopolymer of p-dioxanone.

TABLE V
COMPARATIVE PROPERTIES OF PDS AND COPOLYMER II-A FIBERS

	PDS*	Copolymer II-A
Polymer I.V.	1.80	1.82
Tm, °C (Hot stage microscope)	110-115°C	98-102°C
Fiber Properties		
Diameter (mils)	8.65	7.8
Straight Tensile strength, Kpsi	75	86
Knot strength, Kpsi	45	53
Elongation, %	28	48
Y.M. Kpsi	278	221
In-Vivo BSR, % at		
3 weeks	--	30
4 weeks	69	12
In-Vivo Absorption, % remaining at		
91 days	96	23%
119 days	--	0
180-210 days	0	--

SOURCE: Ref. 14.
*Typical properties of PDS made according to reference #12

REFERENCES

1. S. W. Shalaby, Chap. 3 in "High Technology Fibers: Part A (M. Lewin & J. Preston, Eds.) Marcel Dekker, New York, 1985.
2. S. W. Shalaby, Vol. 1, in "Encyclopedia of Pharmaceutical Technology" (J. C. Boylan & J. Swarbrick, Eds.) p. 465, Wiley, New York (1988).
3. S. W. Shalaby & D. D. Jamiolkowski, U.S. Pat. (to ETHICON, Inc.) 4,605,730 (1986).
4. M. N. Rosensaft and R. L. Webb, U.S. Pat. (to American Cyanamid) 4,243,775 (1981).
5. S. W. Shalaby & D. D. Jamiolkowski, Polym. Preprints, 26 (2), 200 (1985)
6. A. Kafrawy, D. D. Jamiolkowski & S. W. Shalaby, J. Bioact. Biocomp. Polym., 2, 305 (1987).
7. A. J. Domb, E. Ron, & R. Langer, Macromolecules, 21, 1925-1929 (1988).
8. A. J. Domb, & R. S. Langer, U.S. Pat. (to M.I.T.) 4,757,128 (1988).
9. S. W. Shalaby & D. D. Jamiolkowski, U.S. Pat. (to ETHICON, Inc.) 4,205,399 (1980).

10. S. W. Shalaby & D. D. Jamiolkowski, U.S. Pat. (to ETHICON, Inc.) 4,209,607 (1980).
11. S. W. Shalaby & D. D. Jamiolkowski, U.S. Pat. (to ETHICON, Inc.) 4,141,087 (1979).
12. N. Doddi, C. C. Versfelt, and D. Wasserman, U.S. Pat. (to ETHICON, Inc.) 4,052,988 (1977).
13. R. S. Bezwada, S. W. Shalaby, H. D. Newman, and A. Kafrawy, U.S. Pat. (to ETHICON, Inc.) 4,643,191 (1987).
14. R. S. Bezwada, S. W. Shalaby and H. D. Newman, U.S. Pat. (to ETHICON, Inc.) 4,653,497 (1987); J. Appl. Biomater. (in press).
15. T. N. Salthouse, Chap. 2 in "Biocompability in Clinical Practice, Vol. I. (D. F. Williams, Ed.) C.R.C. Press, Boca Raton, Fl., 1982.

RECEIVED February 14, 1990

AGRICULTURAL POLYMER UTILIZATION
Monomer Source

Chapter 16

Monomers and Polymers Based on Mono- and Disaccharides

Stoil K. Dirlikov

Coatings Research Institute, Eastern Michigan University, Ypsilanti, MI 48197

Polymers containing mono- or disaccharides in the main chain or as pendant groups are receiving increasing attention. They possess a number of attractive, useful, and unique properties such as high hydrophilicity, optical activity (chirality), etc. In addition, some of these polymers have biological activity. Others are anticipated to be tractable and combine useful properties of both natural polysaccharides and synthetic polymers. Finally, polymers based on saccharide monomers are expected to undergo easier biodegradation than the synthetic polymers. Consequently, this class of biodegradable polymers has potential utility in reducing "plastic" pollution which is a major problem in our modern world. Many "saccharide monomers" such as glucose, sucrose, etc. or their derivatives, sorbitol, gluconic acid, and others, are industrially produced in large scale from sugar waste, starch, etc. and are available at a low price normally less than one dollar per pound. Other mono- and disaccharides are potentially available from different renewable resources (biomass), again at low cost. Glucosamine is available from chitin and cellobiose, which is the cellulose dimer, can be obtained from cellulose by enzymatic degradation.

0097–6156/90/0433–0176$06.00/0

Our results on the preparation, characterization, and potential applications of different mono- and disaccharides as monomers in polymer chemistry are discussed in the present report.

ISOSORBIDE

Sorbitol (I) is an attractive and inexpensive raw material available from renewable resources. It is industrially produced by hydrogenation of glucose (II) from sugar waste and is available at about 50 cents/pound. Alternative sorbitol can be produced form cellulose.

Sorbitol easily dehydrates into its 1:4-3:6-dianhydro derivative, isosorbide (III), in the presence of an acid catalyst such as sulfuric, hydrochloric, or toluenesulfonic acid or a cation exchange resin in a one-step preparation in high yield (75-85%)(1):

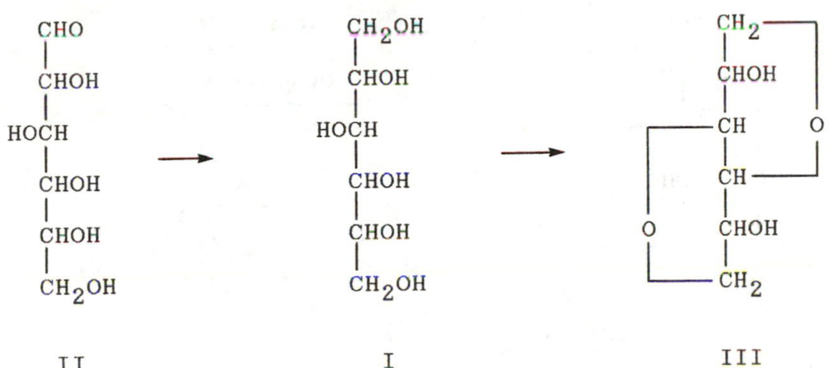

Isosorbide is a rigid diol with two non-equivalent hydroxyl groups (endo-5 and exo-2) and two tetrahydrofuran rings in its structure:

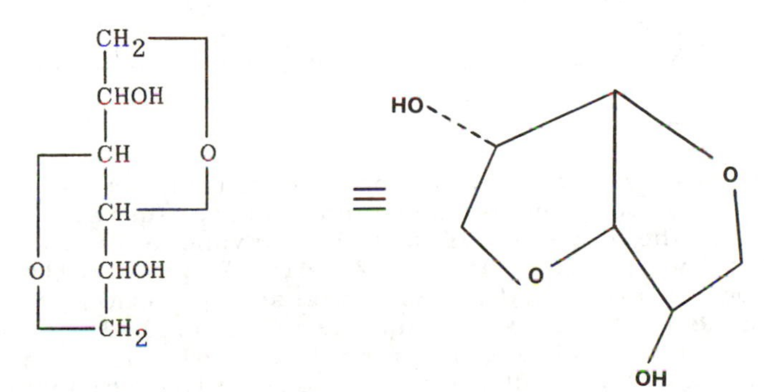

Its polymers, therefore might have good complexation ability.

A literature search shows that isosorbide has been used in the preparation of polyesters (2) and polycarbonates (3,4).

We have prepared and characterized three linear isosorbide containing polyurethanes with toluene diisocyanate (TDI), 4,4'-diphenylmethane diisocyanate (MDI), and 1,6-hexamethylene diisocyanate (HMDI): P(I-TDI), P(I-MDI), and P(I-HMDI). These polyurethanes have been synthesized as described in the experimental section by solution polymerization of isosorbide with the corresponding diisocyanate in dimethylacetamide using dibutyltin dilaurate as the catalyst at 75°C for 24 hours. All polymers have been isolated in quantitative yield by precipitation in methanol or water (5).

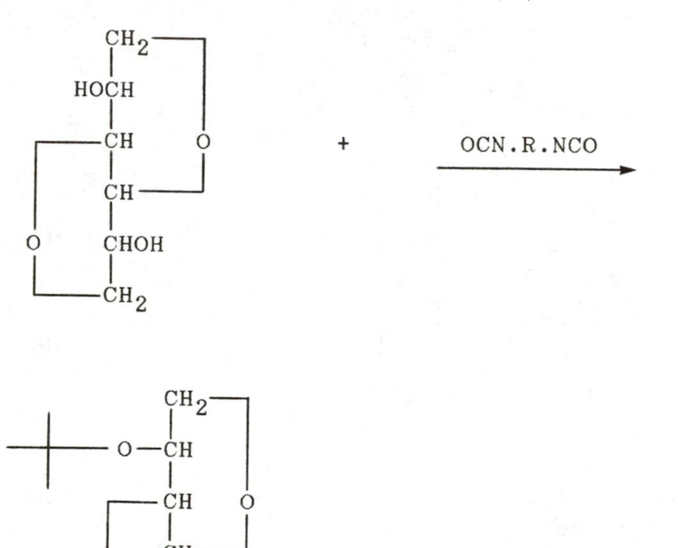

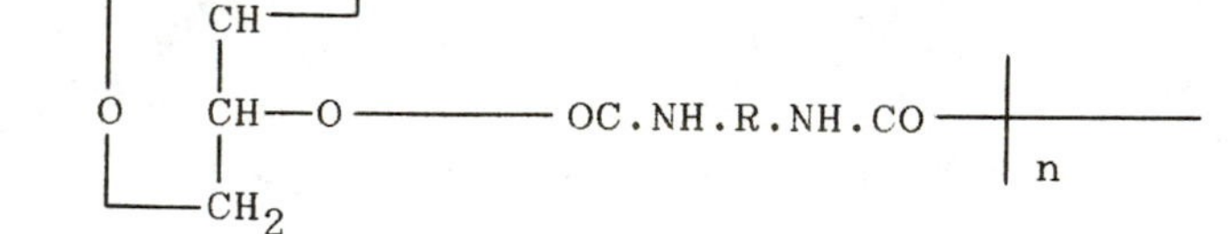

All polyurethanes have been obtained by reacting 1:1 molar ratios of isosorbide and diisocyanate. A slight excess of about 5-10% of the diisocyanate or of the diol resulted in the generation of soluble polyurethanes with lower molecular weight and possessing either functional isocyanate or hydroxyl groups, respectively.

All three polymers possess high molecular weights in the range of 20-25,000 by gel permeation chromatography. They are soluble in polar solvents such as dimethylacetamide, dimethyl sulfoxide, and dimethylformamide, and insoluble in non-polar solvents such as chloroform and carbon tetrachloride. The other properties of these polymers depend on the diisocyanate used in their preparation.

The isosorbide polyurethane based on the aliphatic diisocyanate P(I-HMDI) is flexible. It is a thermoplastic with a glass transition temperature of 110°C and softening temperature of 190°C. Both transitions are well below its degradation onset which occurs at approximately 260°C. It forms good films by evaporation from its solutions and colorless transparent compression moldings.

The isosorbide polyurethanes based on the aromatic diisocyanates P(I-TDI) and P(I-MDI), possess more rigid structures with both polymers forming brittle films and brittle compression moldings. Their glass transition temperatures are above their decomposition temperature of 260°C. The thermostability of isosorbide polyurethanes correspond to that of conventional polyurethanes with similar structure based on 1,4-cyclohexanedimethanol for which degradation temperature of 260°C has been determined.

The limiting oxygen index (LOI) corresponds to the minimum oxygen content in air which supports combustion in the polymer. LOI of P(I-MDI) is 25 and it is not flammabe since atmospheric air has only 21% oxygen. In contrast, the conventional polyurethane with a similar structure based on MDI and 1,4-cyclohexanedimethanol has LOI of 20 and it is flammable in air. It indicates that isorsorbide polyurethanes require higher oxygen content to support combustion and they have lower flammability than the corresponding "synthetic" polymers.

Isosorbide polyurethanes, especially those based on aliphatic isocyanates, may be useful in the same applications as conventional polyurethanes i.e. thermoplastics, coatings, and foams. In fact, excellent rigid foams have been obtained from P(I-MDI)(5). Isosorbide has a low melting point of 61°C and it is suitable for use in reactive injection molding processes alone or in the form of a mixture with other conventional diols. In addition, its polymers may also find specific applications due to the anticipated high complexation ability of the two tetrahydrofuran rings in their isosorbide units.

ISOMANNIDE

Isomannide or 1,4:3,6-dianhydromannitol (IV) is an isomer of isosorbide. It has been prepared by dehydration of mannitol (V) according to a procedure similar to that for the preparation of isosorbide (6) which has been described previously.

Isomannide is a rigid diol with a low melting point of 85°C. It has two adjacent tetrahydrofuran rings in a "clam" configuration and two equivalent hydroxyl groups, both in the endo-configuration, suitably situated for complexation via their oxygen atoms. One could speculate that isomannide will have better complexation ability than isosorbide since its four oxygen atoms are oriented in the same direction.

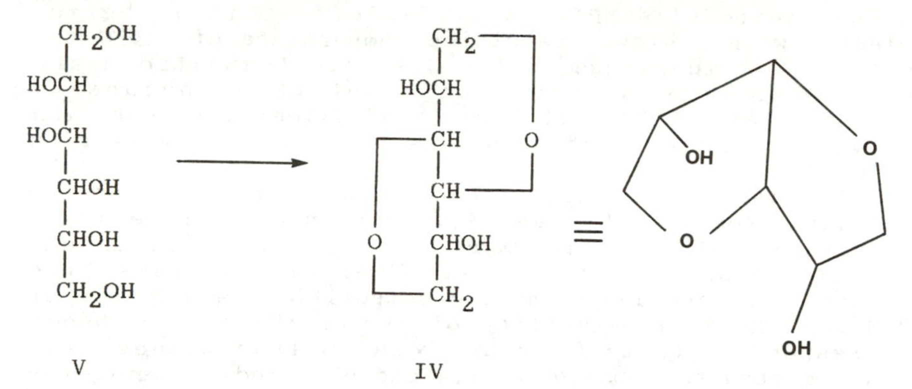

There is a growing interest in polymers which contain tetrahydrofuran rings in their main chains because of their good/excellent ability to complex with different cations. The applications, however, are limited by difficult multi-step preparations. Isomannide polymers appear to be attractive candidates and easy alternatives in such applications. Isomannide polyurethanes have been prepared by the same procedure described for isosorbide polyurethanes in the experimental section. Their properties are under investigation.

DIMETHYL ISOSORBIDE

Other molecules of the same type, which are expected to have good complexation ability, are the methyl or ethyl ethers of isosorbide, isomannide, and isoidide. They appear to have the potential as solvents with high solvation power and low reactivity. All these molecules possess rigid structures with two adjacent tetrahydrofuran rings in "clam-like" configuration and four oxygen atoms with glyme distribution, $-CH_2OCH_2CH_2OCH_2CH_2OCH_2$. They are expected, therefore, to exhibit stronger solvating power and complexation ability in comparison to that of tetrahydrofuran and other glymes. Dimethyl isosorbide (VI) and dimethyl isomannide (VII) illustrate the structures of these materials.

Dimethyl isosorbide (VI) has been prepared in quantitative yield from isosorbide and dimethyl sulphate according to known procedure (1). It is a liquid possessing a low vapor pressure at room temperature and boiling point of 95°C at 0.1 mm Hg. In addition, it also possesses optical activity. It is expected to be a relatively inexpensive solvent (for an optical active solvent with 100% purity of the optical isomer) in comparison to other "solvating" solvents such as tetrahydrofuran (one dollar per liter), dioxane (one dollar per liter), and pyridine (two dollars and 50 cents per liter).

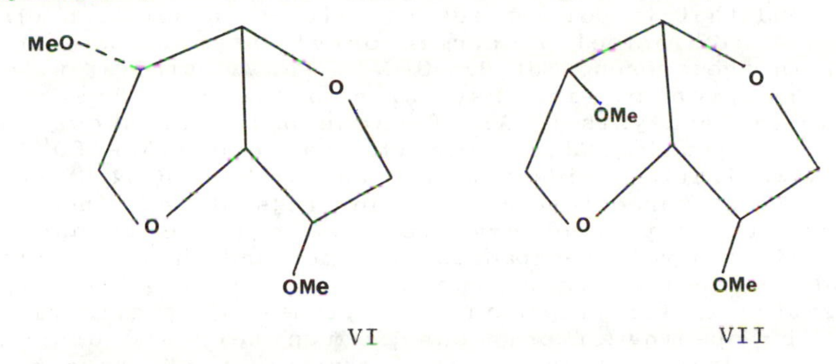

Dimethyl isomannide (VII) has both methoxy groups in endo-position and might even have better complexation than dimethyl isosorbide (as discussed above for isomannide). However, it does not have optical activity.

Dimethyl isoidide is the third possible isomer with two exo groups.

1,4:2,5:3,6-TRIANHYDROMANNITOL

1,4:2,5:3,6-trianhydromannitol (VIII) is another attractive monomer. It is obtained from isosorbide in a two-step synthesis with about 50% overall yield as outlined below (7,8).

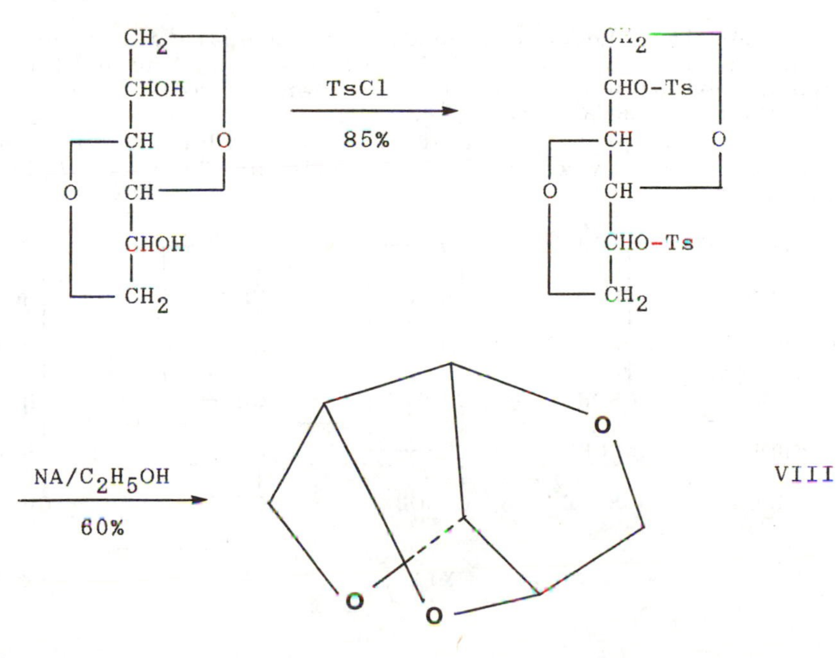

1,4:2,5:3,6-trianhydromannitol has a low melting point of 64°C. It is a highly strained molecule with three adjacent tetrahydrofuran rings. In our studies, we have found that it polymerizes in melt or in solution (in methylene dichloride) at normal or elevated temperature (70°C) in the presence of $BF_3.O(C_2H)_2$ as a catalyst with the formation of an insoluble polymer. Thermogravimetric analysis (TGA) of these polymers shows an onset of thermodegradation in the range of 135-150°C. The polymerization evidently proceeds with opening of one or two of the three tetrahydrofuran rings of the monomer. As a result of this observation, we believe it has a potential as a volume-expansion monomer and this property is under investigation. Another possibility for its utilization is for preparation of glyme-like cyclic oligomers by opening of only one of its tetrahydrofuran rings by polymerization or copolymerization of the monomer in solution at milder condition and lower temperatures.

1,4-LACTONE OF 3,6-ANHYDROGLUCONIC ACID (LAGA)

LAGA (IX) is available from gluconic acid (X) according to the following scheme. Gluconic acid (X) is an inexpensive (70 cents/lb.), commercially available raw material produced by oxidation of glucose (II) from sugar waste. Cellulose could be used as an alternative resource. Gluconic acid exists in its "acidic" form only in aqueous solutions. Upon evaporation, it forms its crystalline 1,4- (XI) or 1,5-lactones (XII) depending on reaction conditions.

LAGA is a known compound. Its preparation from glucose involves a long, multi-step blocking and deblocking procedure which proceeds in low overall yield (9-13).

We have been able to develop a one-step procedure (14) for the preparation of LAGA by acid catalyzed dehydration directly from gluconic acid or from its 1,4- or 1,5-lactones:

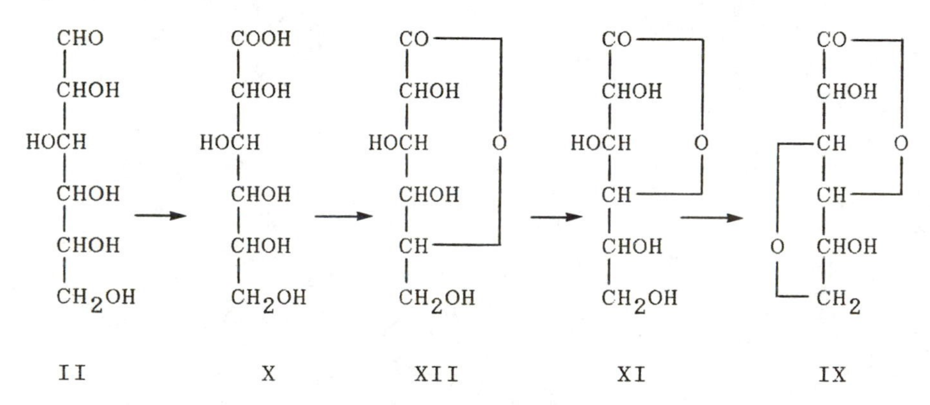

It was found that the 1,5-lactone of gluconic acid rearranges into 1,4-lactone of LAGA in the course of the reaction. The formation of the 3,6-anhydro ring proceeds easily in the presence of DOWEX 50WX4 ion exchange resin as a catalyst at 110°C with the removal of water as an water/toluene azeotrope. The process was carried out as described for the preparation of isosorbide from sorbitol. The procedure, however, has not been optimized and the yield of LAGA was only 37%. The optimization of the procedure will certainly increase the yield of LAGA.

The 1,4-lactones of other 3,6-anhydro acids are available in a similar manner.

LAGA is a rigid diol with a melting point of 117°C (large prisms from ethyl acetate). It has two non-equivalent hydroxyl groups in endo-5 and exo-2 positions.

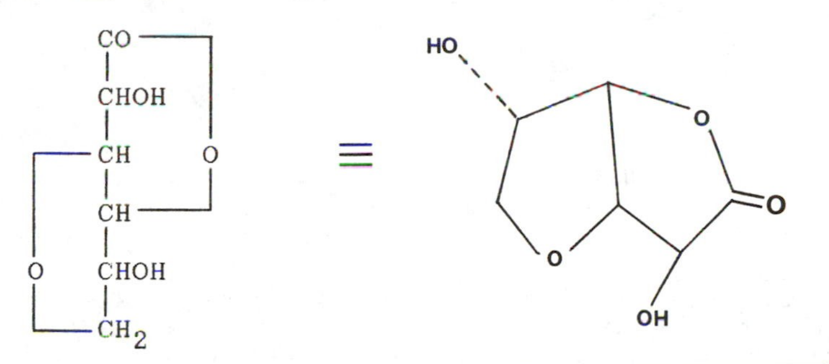

LAGA is a monocarbonyl derivative of isosorbide. In contrast to isosorbide, however it is not hydroscopic and easily opens its lactone ring with the formation of potential cross-linking sites.

A literature search does not indicate any publications concerning the use of LAGA in polymer applications.

We have prepared two linear soluble polyurethanes: P(LAGA-MDI) and P(LAGA-HMDI) by solution polymerization of LAGA with MDI and HMDI respectively in dimethylacetamide as a solvent using dibutyltin dilaurate as a catalyst at 75°C over 24 hours according to the procedure used in the preparation of the isosorbide polyurethanes as described in the experimental section. Both polymers have been isolated in quantitative yield in their "lactone" form by precipitation from chloroform (15).

As obtained, P(LAGA-MDI) and P(LAGA-HMDI) are soluble in polar solvents such as dimethylacetamide, dimethyl sulfoxide, and dimethylformamide, and insoluble in non-polar solvents as chloroform and carbon tetrachloride. These polymers possess high molecular weight in the range of 20-25,000.

The properties of the polyurethanes depend on the nature of the diisocyanate used in their preparation.

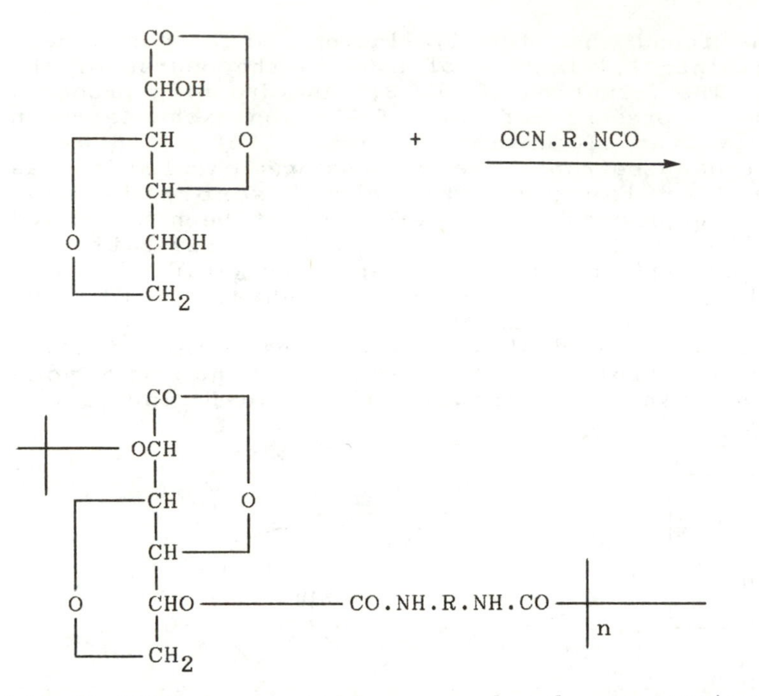

P(LAGA-MDI) is a rigid macromolecule possessing a glass transition temperature above the onset of degradation which is observed at about 240°C. It forms brittle films and brittle compression moldings. In contrast, the aliphatic polyurethane, P(LAGA-HMDI) is a more flexible macromolecule. It is a thermoplastic with a glass transition temperature of 85°C and a softening temperature of 150°C. Its degradation onset is observed around 240°C. This polyurethane forms good films from solution and colourless transparent compression moldings.

These results show that in general, LAGA polyurethanes in their lactone form have practically the same transitions and properties as the corresponding isosorbide polyurethanes.

In contrast to isosorbide polymers precipitation of the initially formed LAGA polymers in water instead of chloroform results in slow opening of the lactone rings with formation of hydroxyl and carboxyl groups. This hydrolysis results in a dramatic change in solubility, TGA, and IR, etc.. Although the rate of the lactone ring opening is very slow at the beginning formation of the carboxyl groups with the advancement of hydrolysis lowers the pH of the aqueous media and accelerates the hydrolysis of the remaining lactone rings. As a result, the solubility of the polymer in water sharply increases with hydrolysis.

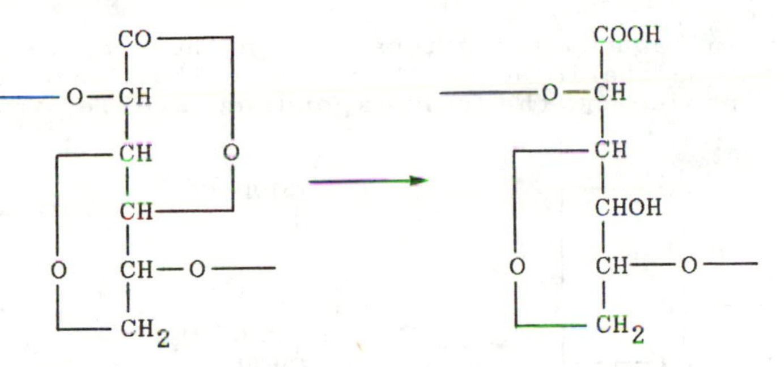

The infrared spectra of the hydrolyzed polymer show a strong broad absorption band in the range of 2500-3500 cm^{-1} which is not observed in the spectra of the initial non-hydrolyzed polymer. This absorption band corresponds to the O-H vibrations of the free hydroxyl and carboxyl groups. Its intensity increases with the degree of hydrolysis. The broad character of this absorption without distinguised separate band indicates that the free hydroxyl and carboxyl groups form intermolecular hydrogen bonding.

The TGA curves of the initial P(LAGA-HMDI) polymer in the lactone form show a weight loss onset at approximately 240°C. In contrast, the hydrolyzed polymer in the so-called acidic form is very hydrophilic, absorbs large amounts of water from the air at room temperature and decreases the temperature at which weight loss occurs. This weight loss which occurs below 180°C, probably corresponds entirely to the release of absorbed water. The glass transition temperature of the hydrolyzed polymers is not observed in DSC due to the intermolecular hydrogen bonding between the hydroxyl and carboxyl groups.

These polymers, which slowly dissolve in water, might be suitable for control release and preparation of biodegradable polymers. LAGA homo- and copolymers open the lactone rings of their LAGA monomeric units with formation of hydroxyl and carboxyl groups and thus, increase the polymer solubility, hydrophilicity and the rate of their biodegradation. In this sense LAGA offers a new approach for polymer controlled biodegradation.

The partially hydrolyzed LAGA polymers have free hydroxyl and carboxyl groups which act as sites for cross-linking. Their cross-linked polymers have been obtained in the presence of additional amount of diisocyanate.

1,4:3,6-DILACTONE OF MANNOSACCHARIC ACID

Another group of diol monomers of a similar type is based on the dilactones of the saccharic acids (DLSA) (XIII) which exist in two forms. Their acidic form (XIV) is formed only in aqueous solutions. Upon the evaporation of water, the saccharic acids (XIV) close their two lactone rings producing the corresponding stable diols (XIII).

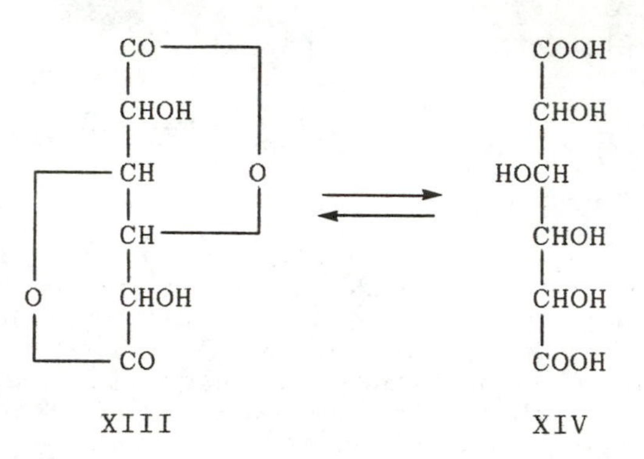

DLSA are rigid diols. They have the same skeleton as isosorbide and LAGA. For instance, 1,4:3,6-dilactone of mannosaccharic acid is a dicarbonyl derivative of isomannide and monocarbonyl derivative of the lactone of the anhydromannonic acid. Therefore, DLSA could find applications as diols and their polymers are expected to have similar properties to that of isosorbide and LAGA polymers.

There are two possible routes for DLSA preparation: 1). They are available by oxidation of the corresponding monosaccharides with nitric acid (16). This reaction generates relatively low yields in the range of 40%.

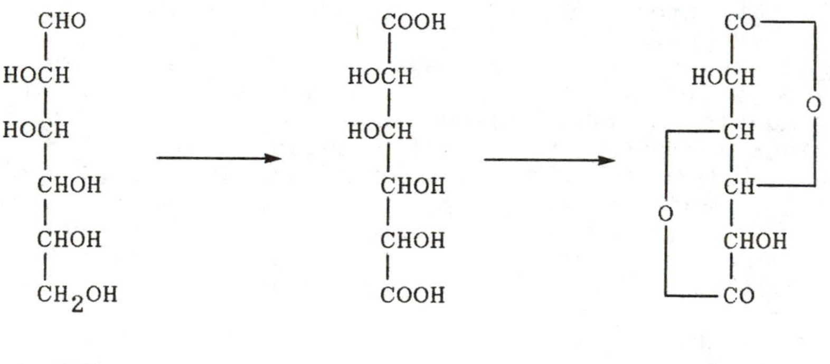

2). An alternative method for their preparation is by oxidation of uronic acids (XV) which are available from alginic acids (from kelp).

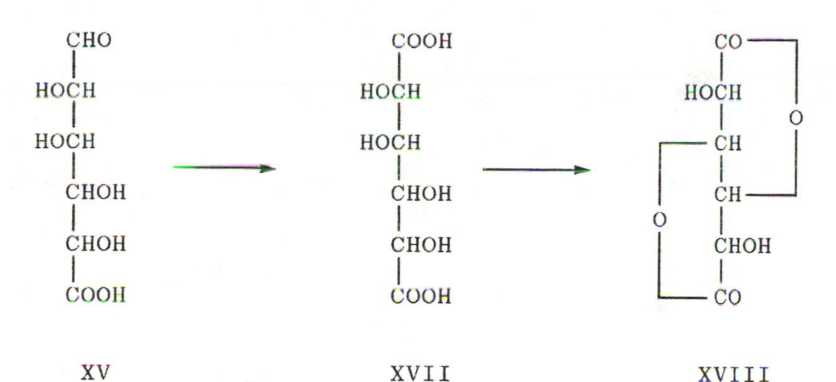

Our experience indicates that DLSA are less available than isosorbide and LAGA. We have prepared manno-saccharic acid (XVII) and its 1,4:3,6-dilactone (XVIII) by oxidation of mannose (XVI) with nitric acid according to (16). Pure dilactone with a melting point of 187°C has been isolated in 26% yield.

Its polyurethanes are under investigation.

EPOXY RESINS: 1,2:5,6-DIANHYDRO-3,4-O-ISOPROPYLIDENE-D-MANNITOL

One would also think that saccharides are more attractive starting raw materials for direct preparation of epoxy resins than hydrocarbons since they already possess oxygen atoms needed for the epoxy rings. Such diepoxy derivatives of hexitols are known compounds, for instance 1,2:5,6-dianhydro-3,4-O-isopropylidene-D-mannitol (XIX). Their preparation, however is difficult and requires 15 to 18 steps which proceed in very low overall yield (17-19).

As a first step in this direction, we have been able to prepare 1,2:5,6-dianhydro-3,4-O-isopropylidene-D-mannitol as a model diepoxy hexitol derivative in four steps with overall 50% yield according to the following scheme:

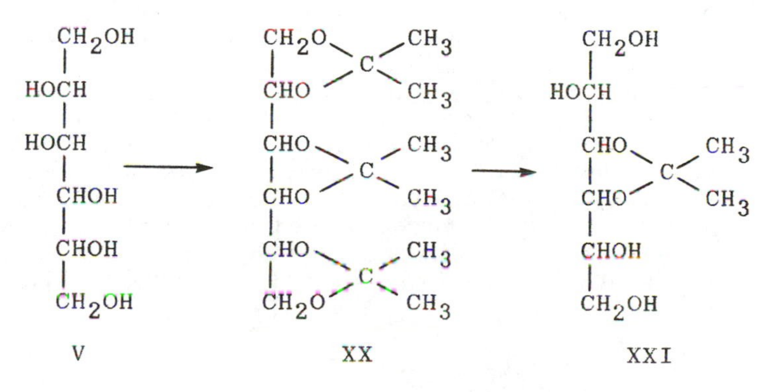

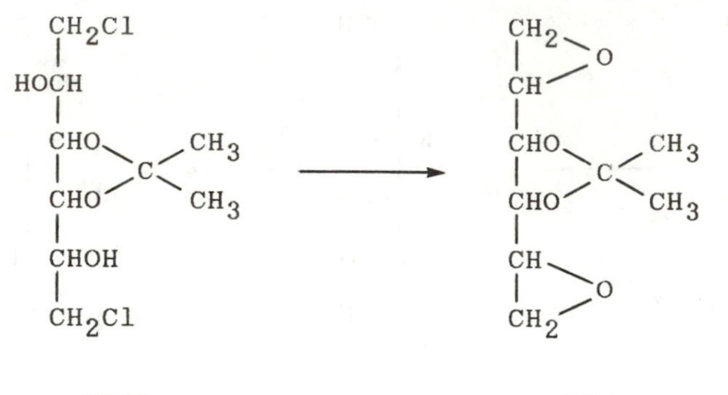

The first step which proceeds in 75% yield to form (XX) involves the blocking of mannitol (V) with acetone at room temperature in the presence of 1% concentrated sulfuric acid as a catalyst (20).

The second step involves unblocking of the primary hydroxyl groups in 70% acetic acid at 40°C for 1.5 hours (21). By this procedure, 3,4-O-isopropylidene-D-mannitol (XXI) has been obtained in 80% yield.

Then, selective halogenation has been carried out at mild conditions under which only the halogenation of the two primary hydroxyl groups occurs. The secondary hydroxyl groups remain intact under these conditions. The reaction proceeds smoothly with carbon tetrachloride and triphenylphosphine in anhydrous pyridine at 5°C for 18 hours (22). 1,6-Dichloro-1,6-dideoxy-3,4-O-isopropylidene-D-mannitol (XXII) has been obtained as described in the experimental part in quantitative yield according to the NMR spectrum of the reaction product (23). Its purification, however, is difficult and requires tedious chromatographic separation on a silica gel column. The purification conditions, however, have not been optimized and this results in a lower yield.

The NMR and IR spectra confirm the structure of 1,6-dichloro-1,6-dideoxy-3,4-O-isopropylidene-D-mannitol (XXII). The resonance signals of the $C\underline{H}_3$ protons of the isopropylidene residues of both compounds XXII and XIX appear at about 8,60 ppm. The signal for the $C\underline{H}OH$ protons of XXII at 5,45 ppm is not observed in the NMR spectra of XIX. Instead, two new multiplets centered at 6,93 and 7,30 ppm for the methine and the methylene protons of the epoxy groups of XIX appear. At the same time, the relative intensity of the $>C\underline{H}-O$ and $-C\underline{H}_2-Cl$ protons of XXII at 6,15 ppm decreases with six protons in the spectrum of XIX. The infrared spectrum of XXII shows an absorption band at 3500 cm^{-1} which corresponds to its hydroxyl groups. This band is not observed in the spectrum of XIX.

Finally, epoxy ring formation has been carried out in methanol at room temperature with sodium methoxide as a catalyst (24). 1,2:5,6-Dianhydro-3,4-O-isopropylidene-D-mannitol (XIX) has been obtained in near quantitative yield.

This diepoxy mannitol derivative cures with amines at room or elevated temperatures. When reacted with Versamid 140 polyamide resin (Henkel Corporation) with an amine value of 370-400, commonly used for curing of commercial epoxy resins, at epoxy/amine molar ratio of 1:1, it produces epoxy resins with good "physico-mechanical" properties and excellent adhesion to glass. The homogeneous mixture is cured as a thin layer between two (glass) microslides. The curing is carried out either at room temperature for 24 hours and then at 150°C for 2 hours, or at room temperature for 7 days. In both cases, transparent cured epoxy resins are obtained. Attempts to separate the two microslides always result in breaking the microslide which indicates strong epoxy resin/glass adhesion. The curing process is nearly complete since extraction with different solvents does not give any extractables.

Other renewable resources: sorbitol (I), cellobiose (XXV), etc. can be used in similar ways as starting materials in the preparation of epoxy resins. We are now working on a direct two-step preparation of diepoxy sorbitol derivatives by direct halogenation of the two primary hydroxyl groups with the formation of 1,6-dichloro-1,6-dideoxy-sorbitol, (XXIII), followed by epoxy ring formation (1,2:5,6-dianhydrosorbitol (XXIV):

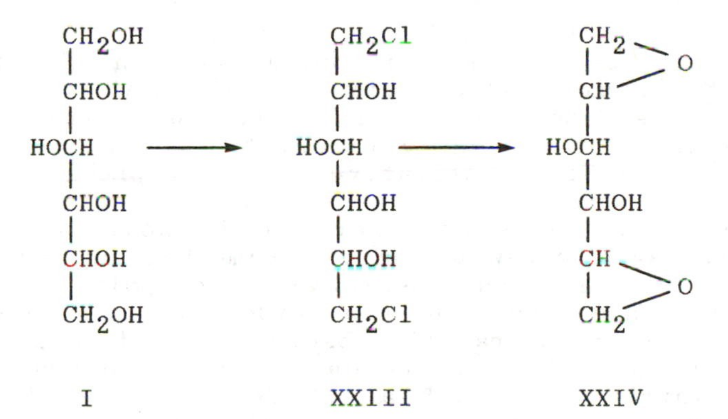

Diepoxy derivative of sorbitol is a known stable compound.

CELLOBIOSE

Cellobiose (XXV), the basic unit of cellulose, is a potentially inexpensive monomer obtained in 90-95% yield by bacterial hydrolysis of cellulose. It is another attractive disaccharide with many potential applications in polymer chemistry.

The two primary hydroxyl groups of cellobiose have higher reactivity than the remaining six secondary hydroxyl groups. Kurita and co-workers have recently reported the preparation of soluble polyurethanes by direct polyaddition of cellobiose to diisocyanates without blocking the excess hydroxyl groups of cellobiose by using the difference in reactivity between its primary and secondary hydroxyl groups (25,26):

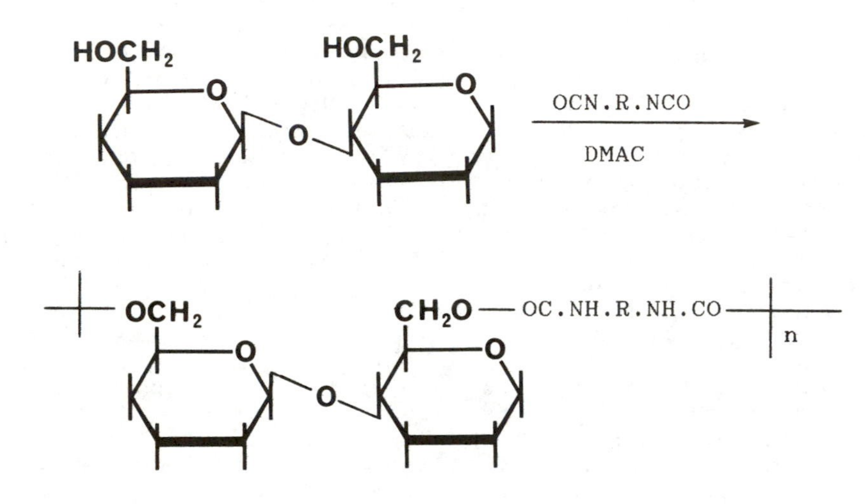

We have carried out the polymerization of cellobiose with MDI in dimethylacetamide at 17°C with stirring for 24 hours as described by Kurita (25,26). The polymerization proceeds without gel formation. The polymer has been isolated in quantitative yield by precipitation in chloroform.

This cellobiose polyurethane is soluble in polar solvents as dimethylacetamide, dimethyl sulfoxide, and dimethylformamide, and insoluble in non-polar solvents as chloroform and carbon tetrachloride. It is a very hydrophilic polymer and rapidly absorbs water from air. TGA analysis shows that it releases absorbed water up to approximately 200°C with a 12% decrease in weight. It has unexpectedly high average molecular weight in the range of 100,000. Polymers prepared from several different polymerization runs give reproducible the same molecular weight. This high molecular weight indicates that some branching probably occurs during the polymerization by the participation of small amount of the secondary hydroxyl group of cellobiose in addition to the main type of polymerization of its two primary hydroxyl groups.

The pendant hydroxyl groups of cellobiose have been confirmed to be useful for crosslinking of its polymers in the presence of additional amount of diisocyanate. Cross-linked insoluble films have been obtained by casting a polymer solution in dimethylacetamide containing 7% of additional amount of MDI.

The polymer limiting oxygen index (LOI) value is 26,5 which corresponds to that of the isosorbide polyurethanes. It is again higher than that of conventional MDI polyurethanes with similar structure based on 1,4-cyclohexanedimethanol (LOI = 20) and indicates lower flammability of cellobiose polymers.

TGA analysis shows that polymer degradation starts at about 235°C which corresponds to the temperature of decomposition of the cellobiose monomer (m.p. 239°C with decom.). Torsion Braid analysis and differential scanning calorimetry measurements show that this polymer is very rigid and does not exhibit any transition in the range of -100 to +250°C, e.g. the polymer decomposition occurs below any transition temperature. This result is expected since both of the monomers, cellobiose and MDI, have rigid molecules and because cellobiose units of the polymer form intermolecular hydrogen bondings. Cellobiose polyurethanes based on aliphatic diisocyanates, e.g. HMDI, are expected to be more flexible.

D-GLUCOSE METHACRYLATE

D-Glucose methacrylate (XXVI) is the last monomer. Its polymers have many potential applications in medicine, secondary oil recovery, etc (27,28).

The industrial production and applications of D-glucose methacrylate are limited by its difficult multi-step preparation with blocking and unblocking procedures of the secondary hydroxyl groups of glucose according to the procedures described in the literature (28).

The primary hydroxyl group of glucose has higher reactivity than the remaining four secondary hydroxyl groups. Our initial results indicate that direct preparation of glucose methacrylate is possible from glucose (II) (dried over P_2O_5) and freshly distilled (in vacuum) methacroyl chloride in 1:1 molar ratio in anhydrous dimethylacetamide and pyridine (in molar ratio as a hydrochloride scavenger) at room temperature under nitrogen with stirring for 20 hours, by a procedure similar to that described for the preparation of cellobiose polyurethanes, without blocking/deblocking procedure of the excess secondary hydroxyl groups, just by using the difference in reactivity between glucose primary and secondary hydroxyl groups.

A spontaneous polymerization of glucose methacrylate, however, is observed during the monomer preparation without gel formation and a very hydrophilic water soluble polymer with high viscosity is obtained by precipitation in acetone. This procedure requires further improvement.

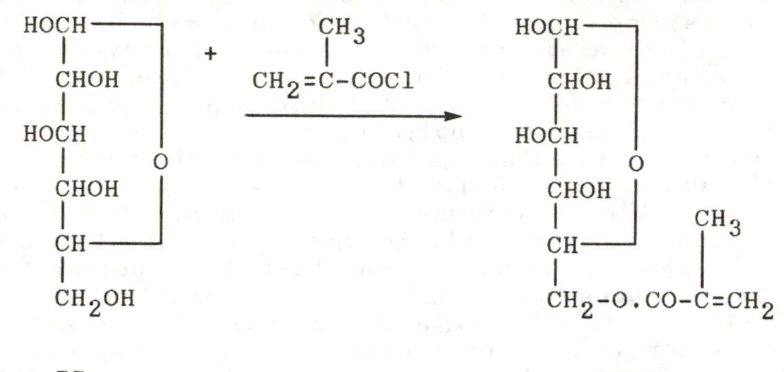

EXPERIMENTAL

All monomers and intermediates have been prepared according to the references given in the text. This experimental part contains only procedures which have not been previously published.

1,4-Lactone of 3,6-Anhydrogluconic Acid (LAGA) (X). 50.0 g of 1,5-lactone of gluconic acid (Sigma), 5.0 g of DOWEX 50WX4 and 500 ml of toluene were loaded into a flask equipped with a reflux condensor and a Dean-Stark adapter for water separation. The mixture was refluxed for 5 hours and cooled. Toluene was separated by decantation and the residue dissolved in 500 ml of methanol. The solution was filtered several times for separation of DOWEX resin and methanol distilled on a Rotavapor. 15.7 g of pure LAGA was isolated by vacuum distillation using a Kugelrohr distillation apparatus (Aldrich) at 160-170° C/0.2 mm Hg (temperature of the heating bath) (37% yield).

1,6-Dichloro-1,6-Dideoxy-3,4-O-Isopropylidene-D-Mannitol (XXII). 100 ml of carbon tetrachloride was gradually added to a solution of 11.1 g (5 mmole) of 3,4-O-isopropylidene-D-mannitol (XXI) and 26.3 g (10 mmole) of triphenylphosphine in 500 ml of anhydrous pyridine at 0°C. After holding the resulting solution at 5°C for 18 hours, methanol was added, and the mixture was evaporated to a crystalline residue which was chromatographically separated on a silica gel column. Elution first with chloroform and then with 6/1 v/v chloroform/acetone gave XXII fraction. An additional recrystallization from toluene/heptane produced 6.50 g of pure XXII (50% of yield) with m.p. 75-76°C (lit. 24: m.p. 76°C).

Poly(Isosorbide-Hexamethylene Diisocyanate). HMDI (Polysciences) was purified prior to use by vacuum distillation through a short Vigreux column (10 cm), b.p. 127°C/10 mm. Isosorbide was prepared according to method (1). It was purified by vacuum distillation using a Kugelrohr Distillation Apparatus (Aldrich) at 115-130°C (temperature of the heating bath) and 0.2 mm vacuum, followed by vacuum drying at 50°C over phosphorus pentoxide. N,N-Dimethylacetamide (DMAc) (Aldrich, 99+%, GOLD label) was stored over molecular sieve 3A and used without further purification. Dibutyltin dilaurate (Polysciences) was used without purification. The polymerization was carried out in a nitrogen atmosphere in 250 ml four-neck flask equipped with a dropping funnel, reflux condenser, mechanical stirrer and nitrogen inlet tube.

16.82 g (0.1 mole) of HMDI was added dropwise at room temperature from the dropping funnel into the rapidly stirred solution of 14.62 g (0.1 mole) of isosorbide in 100 ml of DMAc. After the addition was completed, 0.2 ml of dibutyltin dilaurate as a catalyst was added and the polymerization was carried out at 75°C for 24 hours. A very viscous completely transparent solution was obtained. It was cooled to room temperature, diluted with 500 ml of DMAc and precipitated by dropwise addition into 5 L of methanol (Note 1). The suspension was left overnight, filtered, washed with methanol and dried in vacuum at 25°C. A white powder of P(I-HMDI) was obtained in 100% yield.

All other polyurethanes have been prepared in a similar manner.

Measurements. The melting points of the monomers and the glass transition temperatures and softening points of the polymers have been determined on capillary melting point apparatus, DuPont differential scanning calorimeter (with 10°C/min.), and Dennis thermal bar, respectively. Temperature of decomposition is measured as first onset of weight loss on the thermogravimetric analysis curve in air at 10°C/min.. Limiting Oxygen Index is measured on a home-made instrument. Molecular weight of the polymers is determined by gel permeation chromatography with a DuPont Zorbax Bimodal PSM 1000s column and dimethylsulfoxide as a mobile phase. Infrared (in KBr) and NMR spectra (in deuterochloroform) are taken on Perkin-Elmer infrared and 60 MHz Varian NMR spectrometers, respectively.

CONCLUSION

Mono- and disaccharides appear to be very attractive renewable resources for preparation of broad variety of unique monomers and polymers.

REFERENCES

1. Montgomery, R.; Wiggins, L.F., J. Chem. Soc., 1946, 390.
2. Courtaulds Ltd. Neth. Appl. 6405497, 1964; Chem. Abstr. 1965, 62, 10588.
3. Medem, H.; Schreckenberg, M.; Dhein, R.; Nouvertne, W.; Rudolph, H.; Ger. Offen 3002762, 1981; Chem. Abstr. 1981, 95, 151439n.
4. Medem, H.; Schreckenberg, M.; Dhein, R.; Nouvertne, W.; Rudolph, H.; Ger. Offen. 2938464, 1981; Chem. Abstr. 1981, 95, 44118k.
5. Dirlikov, S.; Schneider, C., U.S. Patent 4443563.
6. Montgomery, R.; Wiggins, L.F., J. Chem. Soc. (London), 1948, 2204.
7. Montgomery, R.; Wiggins, L.F., J. Chem. Soc. (London), 1946, 393.
8. Cope, A.C.; Shen, T.Y., J. Amer. Chem. Soc., 1956, 78, 3177, 5912, and 5916.
9. Chle, H.; Dickhauser, E., Chem. Ber., 1925, 58, 2593.
10. Chle, H.; Von Vargha, L.; Erlbach, H., Chem. Ber., 1928, 61, 1203.
11. Haworth, W.; Owen, L.; Smith, F., J. Chem. Soc. (London), 1941, 88.
12. Fischer, E.; Zach, K., Chem. Ber., 1912, 45, 456.
13. Fischer, E.; Zach, K., Chem. Ber., 1912, 45, 2068.
14. Dirlikov, S.; Schneider, C., U.S. Patent 4581465
15. Dirlikov, S.; Schneider, C., U.S. Patent 4438226
16. Hayworth, W.; Heslop, D.; Salt, E.; Smith, F., J. Chem. Soc. (London), 1944, 217.
17. Jarman, M.; Ross, W., Carbohydr. Res., 1969, 9, 139.
18. Kuszmann, J., Carbohydr. Res., 1979, 71, 123.
19. Institoris, L., et al.: Arzneimittel-Forsch., 1967, 17, 145.
20. Fischer, E., Chem. Ber., 1895, 28, 1167.
21. Wiggins, L.F., J. Chem. Soc. (London), 1946, 13.
22. Anisuzzaman, A.; Whistler, R., Carbohydr. Res., 1978, 61, 511.
23. Dirlikov, S.; Schneider, C., U.S. Patent 4709059.
24. Wiggins, L.F., J. Chem. Soc. (London), 1946, 384.
25. Kurita, K. et. al., Makromol. Chem., 1979, 180, 855.
26. Kurita, K. et. al., Makromol. Chem., 1980, 181, 1861.
27. Klein, J.; Kulicke, W.M., Polymer Pre-prints, ACS symposium, March 1982, 88pp.
28. Colquhoun, J.A.; Dewar, E.T., Process Biochem., 1968, 31.

RECEIVED January 22, 1990

Chapter 17

Polymers and Oligomers Containing Furan Rings

Alessandro Gandini

Ecole Française de Papeterie (INPG), BP 65, 38402 Saint Martin d'Hères, France

Recent work on the synthesis, structure and some properties of macromolecules bearing furan rings is discussed. Two basic sources of monomers are considered, viz. furfural for monomers apt to undergo chain polymerization and hydroxymethylfurfural for monomers suitable for step polymerization. Within the first context, free radical, cationic and anionic systems are reviewed and the peculiarities arising from the presence of furan moieties in the monomer and/or the polymer examined in detail. As for the second context, the polymers considered are polyesters, polyethers, polyamides and polyurethanes. Finally, the chemical modification of all these oligomers, polymers and copolymers is envisaged on the basis of the unique reactivity of the furan heterocycle.

Agricultural and forestry resources, currently called "the biomass", provide a valuable starting point for the chemical exploitation of hemicelluloses, polysaccharides and simple sugars to give certain furanic compounds. Whether in the form of wastes (oat lulls, sugar-cane bagasse, corn cobs,...), by-products of paper mills, eluents from a biomass refinery (steam explosion, organosolv extraction,...) or simply foodstuffs (saccharides, starch,...), one deals with renewable starting materials for the preparation of two basic non-petroleum chemicals: 2-furancarboxyaldehyde (F, commonly known as furfural), arising from the acid-catalysed dehydration of pentoses, and 5-hydroxymethyl-2-furancarboxyaldehyde (HMF), obtained from the corresponding reaction on hexoses and commonly called hydroxymethylfurfural. Whereas the former has been an industrial commodity for decades (the present world production of F is close to 200,000 t per year), the latter is only produced at pilot plant scale, but the situation is likely to evolve in the near future.

0097–6156/90/0433–0195$06.00/0

The chemistry of furan derivatives is well documented: after the classical and monumental book by Dunlop and Peters (1) in the fifties, several monographs have brought the subject up to date (2-5). The specific topic of furanic polymers has received much less attention, except for the resinification of 2-furanmethanol (furfuryl alcohol) which is at present the only major industrial process involving the polymerization of a furan monomer (6). The author has been actively engaged for the past twenty years in research dealing with the synthesis and characterization of macromolecules bearing furan rings and has reviewed the field on two occasions (7,8). The present paper discusses recent advances mostly in the realm of well-defined systems capable of providing reliable information on mechanisms, polymer structures, etc.

The possibility of making monomers from F and HMF and of studying their polymerization and copolymerization behaviour, as well as the properties of the ensuing materials, is an attractive proposition considering (i) the ubiquitous and non-depletive character of the sources of F and HMF and (ii) the unique and useful chemical properties of the furan heterocycle with a view to possible structural modifications of the polymers.

Some of the transformations of F into furanic monomers are shown in Scheme 1. In recent years renewed interest in these syntheses has produced much progress in terms of yields, simplicity and economy. Thus, for example, very convenient routes leading to 2-vinylfuran (9) and 2-furyl oxirane (10) have been reported. The structures of the monomers in Scheme 1 simulate those of aliphatic and aromatic counterparts which are the basis of most polyaddition reactions and which are prepared in typical petrochemical operations. The only (but major) difference stems from the presence of the furan ring in each of the structures.

The use of HMF or the corresponding dialdehyde as precursors obviously applies to the synthesis of monomers for polycondensation reactions as shown by the examples given in Scheme 2. These difunctional structures again mimic the corresponding well-known aliphatic and aromatic counterparts used in the preparation of polyesters, polyamides, polyurethanes , etc.

The purpose of this paper is to highlight the achievements and drawbacks related to the use of furanic monomers in recent investigations, including work in progress.

CHAIN POLYMERIZATIONS

Homo- and copolymerizations involving the monomers depicted in Scheme 1 are chain reactions which can be initiated, at least potentially, by typical free-radical, cationic or anionic promoters. The object of the studies reported below is to establish first of all which monomers adapt best to each type of initiation, then what peculiarities (if any) are caused by the presence of the furan ring, compared to the known behaviour of the corresponding aliphatic and/or aromatic homologues and finally to establish the structure-properties relationships of the materials obtained.

Radical Polymerization

The reactions of free radicals with furan and its derivatives can give both addition and substitution products depending on the specific system (11-13). With 2-substituted furans, the attack takes place predominantly at C5 and leads, by additon, to the corresponding furyl radicals which must be viewed as relatively stabilized interemediates because of the dienic-aromatic character of the furan heterocycle. These premises are essential to the understanding of the varied responses of furan monomers to free-radical activation.

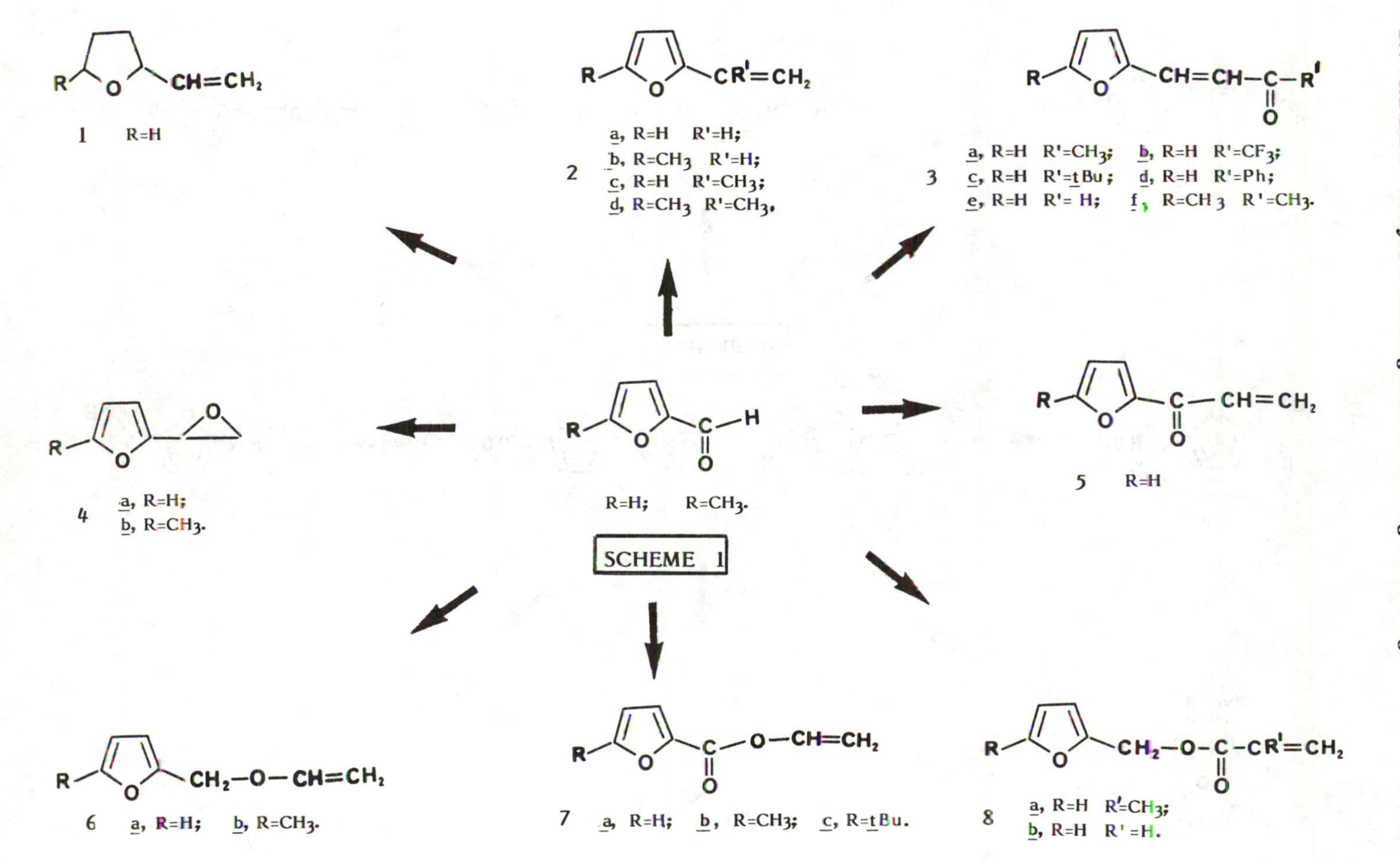
1 R=H
2 a, R=H R'=H; b, R=CH3 R'=H; c, R=H R'=CH3; d, R=CH3 R'=CH3.
3 a, R=H R'=CH3; b, R=H R'=CF3; c, R=H R'=tBu; d, R=H R'=Ph; e, R=H R'=H; f, R=CH3 R'=CH3.
4 a, R=H; b, R=CH3.
R=H; R=CH3.
SCHEME 1
5 R=H
6 a, R=H; b, R=CH3.
7 a, R=H; b, R=CH3; c, R=tBu.
8 a, R=H R'=CH3; b, R=H R'=H.

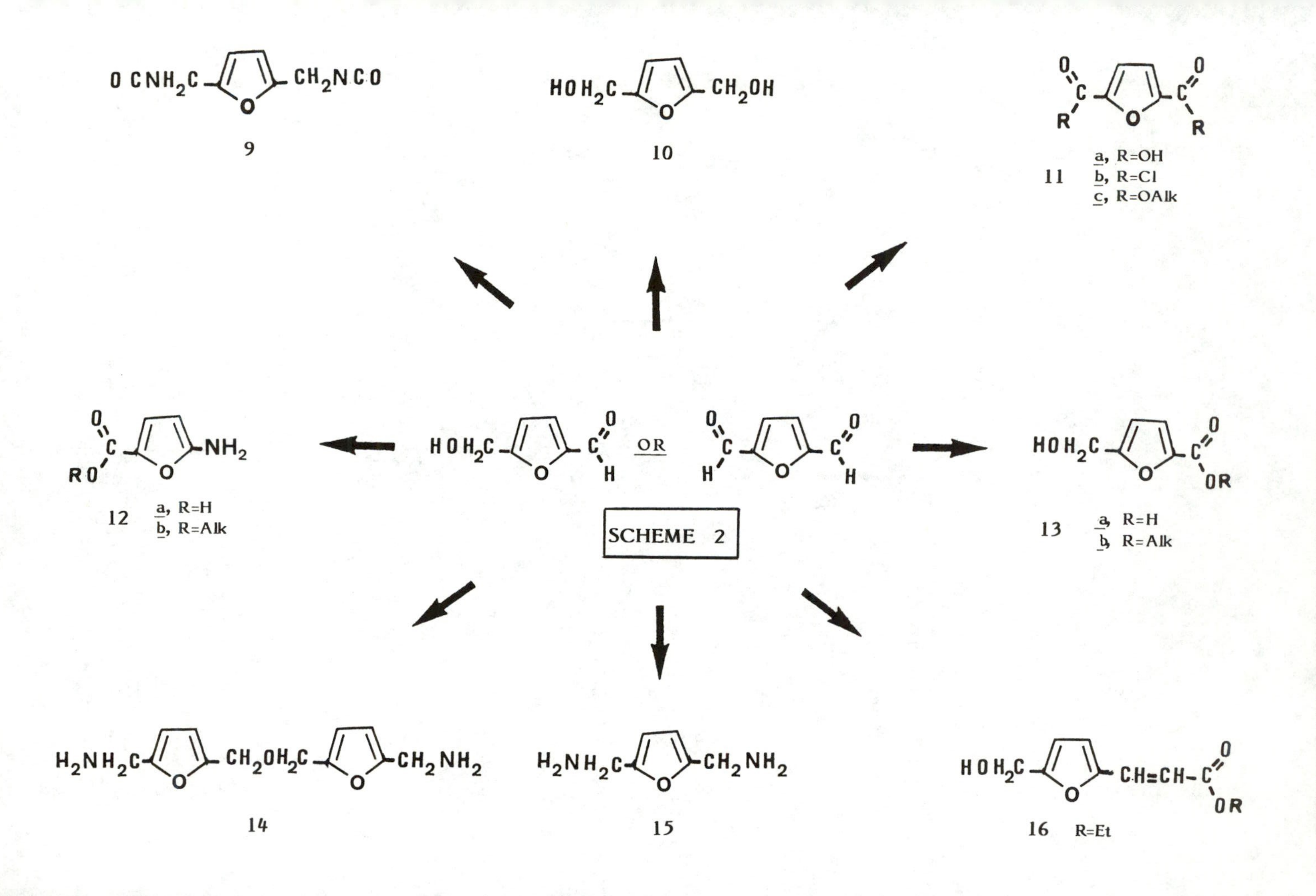
9
10
11 a, R=OH
b, R=Cl
c, R=OAlk
OR
12 a, R=H
b, R=Alk
SCHEME 2
13 a, R=H
b, R=Alk
14
15
16 R=Et

All monomers shown in Scheme 1 have been tested in this context. In principle, primary and/or macromolecular radicals can react with all of them according to two alternative pathways: (i) addition to the external double bond (or oxirane ring), i.e. the classical initiation or propagation reactions, or (ii) addition to the C5 position of the furan ring to give the furyl radical. The relative importance of these alternative events will depend on the relative degree of stabilization of the corresponding radicals formed. Thus, with monomers giving a "normal" radical (addition to the polymerizable function) which is more stabilized than the furyl radical arising from addition at C5, polymerization should proceed smoothly, whereas if the inverse is true, the furanic monomer should encounter serious difficulties in polymerizing and in giving regular polymer structures.

This is precisely what is observed (14). Indeed, with three typical monomers from Scheme 1 different responses are obtained as a function of the structure of the substituent at C2:

- 2-vinyl furoate **7a** does *not* polymerize under normal conditions of free-radical initiation because the primary radicals add predominantly onto the C5 position of the ring rather than onto the vinyl group. This occurs because the furoate-type radical is much less stabilized than the furyl radical. When the amount of initiator is increased substantially, oligomers are formed in low yields and their structure bears dihydrofuran rings, indicating that the very occasional correct initiation leads rapidly to the "wrong" propagation, i.e. the oligoradicals attack the C5 position of a monomer molecule eventually giving dihydrofuranic structures (11). The use of 2-vinyl furoates substituted at C5, e.g. with methyl or t-butyl groups as in monomers **7b** and **7c**, reduces the probability of addition of radicals onto the ring : the yields and DPs of the polymers are increased, but polymerizations are still very sluggish with respect to a "correct" behaviour (11,15).
- 2-furfuryl methacrylate **8b** polymerizes readily and without complications from additions onto the C5 position of the furan ring. In fact, it is well known that the free radicals arising from the methacrylic moiety are stabilized entities, and in the present system more stabilized than the alternative furyl radical structure.
- Finally, 2-vinyl furan **2a** displays an intermediate behaviour in that it polymerizes slowly (because "normal" radicals formed from addition to the vinyl group are relatively stabilized), but gives modest DPs and limiting yields due to the fact that the furan rings pendant to the polymer chains act as radical traps which retard the polymerization and inhibit it above a certain concentration (equivalent to a given polymer yield).

The same general features are encountered in copolymerization experiments (14). 2-vinyl furoates inhibit or retard the polymerization of vinyl acetate, vinyl chloride, acrylonitrile and styrene, but do not interfere with the polymerization of methyl methacrylate although they are not incorporated in the polymer to any appreciable extent. 2-vinylfuran copolymerizes with styrene, acrylonitrile, vinilydene chloride and butadiene but with a marked tendency to homopolymerization and again these reactions do not reach high yields. 2-furfuryul methacrylate copolymerizes with styrene and methyl methacrylate giving complete conversions and high DPs: the reactivity ratios are close to unity with both systems, as expected from the relative stabilization of the corresponding "normal" macroradicals.

In conclusion, the response of a given radical polymerization or copolymerization involving a furanic monomer or comonomer depends largely on the relative degree of stabilization of the two free radicals which that monomer can generate, viz. the "normal" vinylic-type radical and the furyl radical from the C5 addition. If the former is strongly favored, the monomer will homopolymerize readily and copolymerize following conventional criteria. If the latter predominates, the furanic monomer will not homopolymerize and will moreover inhibit or retard the polymerization of comonomers giving poorly stabilized macroradicals.

A consequence of the above observations is that highly conjugated furan compounds should display a strong inhibiting role (radical traps) in free-radical reactions. This has indeed been verified even for structures as simple as:

O CH=CH–C–R
‖
O

17

R=H,OH,CH_3,OAlk

Furanic black resins arising from furfural, furfuryl alcohol and furfurylidene acetone **3a** are characterized by highly conjugated structures (6-8): their inhibiting power as radical scavengers was tested and turned out to be extremely high even in heterogeneous conditions, e.g. with the resin suspended in a monomer solution.

Cationic Polymerization

Among the monomers given in Scheme 1, several possess sufficient nucleophilicity to undergo cationic polymerization (7,8), namely 2-vinyltetrahydrofuran **1**, the alkenylfurans **2**, the furfurylidene ketones **3**, 2-furyl oxiranes **4** and the furfuryl vinyl ethers **6**. Moreover, furfural and its 5-methyl derivative can act as comonomers in certain cationic copolymerizations.

The ineluctable side reaction encountered with monomers *unsubstituted* at C5 is a typical feature of 2-alkyl furans, i.e. electrophilic substitution at C5. The structural consequence of these events in cationic polymerization is branching and in some instances even crosslinking.

Thus, 2-furfuryl vinyl ether **6a** is extremely sensitive to cationic activation (16) because of its very pronounced nucleophilic character, but the polymerization is accompanied by some gel formation due to abundant alkylation of the furan rings pendant to the macromolecules. This structural anomaly is not encountered with the 5-methylated monomer **6b** (16) precisely because electrophilic substitutions take place predominantly at C5 and are therefore impossible with this monomer. A similar difference of phenomenology was observed with the 2-furyl oxiranes **4a** and **4b** (17).

The 2-alkenylfurans **2** have been studied in considerable depth both mechanistically and kinetically (18). With 2-vinylfuran, the alkylation reactions occurred as expected at C5 and produced branched structures as **18** below, but another important side reaction was detected leading to polyunsaturated sequences in some polymer chains. In fact, a chain carrier can abstract a hydride ion from a dead polymer molecule bearing a terminal double bond. The mobile hydrogen atoms are those on the —CH(Fu)— groups next to the unsaturation, i.e. on the tertiary carbon atoms of the penultimate monomer unit. The resulting allylic carbenium ion can lose a proton and give a doubly unsaturated (conjugated) polymer chain. The repetition of the hydride-ion-abstraction/proton-expulsion cycle on the same chains brings about a third conjugated terminal unsaturation as shown in structure **19** below, and eventually polyunsaturated sequences which are responsible for the dark colours of the materials.

The cationic polymerization of 2-vinylfuran with strong acids can be allowed to proceed to black crosslinked resins which display a remarkable proton affinity when swelled in organic solvents (19). Their very high Lewis basicity can be exploited to scavenge Brönsted acids: the insoluble resin is easily removed by filtration at the end of the operation and readily regenerated by neutralization with a strong Brönsted base.

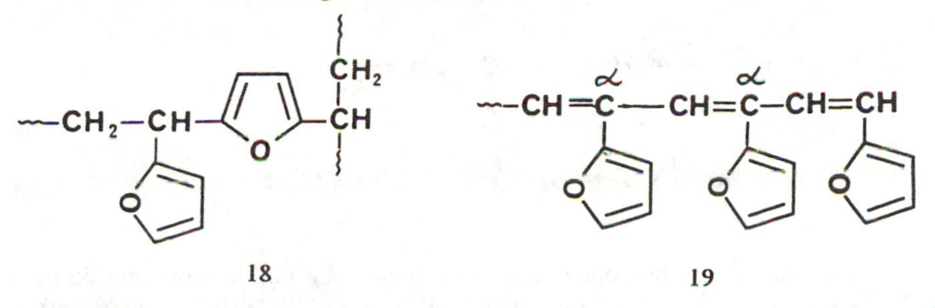

18 19

Monomer **2b** is "protected" against electrophilic substitution at C5 and therefore does not give any branching; it is however susceptible to form polyunsaturated sequences because it bears a mobile hydrogen atom at Cα. Monomer **2c** can give branched structures (free C5 position), but its polymer is refractory to hydride-ion abstraction bacause of the methyl group on the carbon atom bearing the furan ring (Cα). The cationic polymerization of monomer **2d** proceeds undisturbed by side reactions and yields linear chains with a regular structure.

The propensity of the C5 site towards electrophilic substitution has been exploited to prepare functionalized oligomers by cationic polymerization. Thus monomers like isobutene, styrene, the vinyl ethers, etc. polymerize in the presence of simple furan derivatives such as 2-methyl furan to give essentially short chains (DP between 2 and 100 depending on the specific experimental conditions) with a terminal furan ring as a result of predominant transfer onto the C5 position of the added furan compound (20).

Other interesting systems responding to electrophilic activation are the copolymerizations of furancarbonyl compounds such as furfural and 5-methyfurfural (which cannot homopolymerize because of thermodynamic restrictions, i.e. too low ceiling temperature) with alkenes like α-methylstyrene, indene and vinyl ethers (16,21). By choosing the appropriate experimental conditions, alternating structures can be obtained, i.e. polyethers resulting from the opening of the C=O and C=C bonds from each comonomer.

The cationic polymerization of 2-vinyltetrahydrofuran **1** is marred by side reactions due to the allylic structure of the substituent and the participation of the ring (22). The oligomers obtained have therefore a complex structure.

Anionic Polymerization

Monomers **1** (22), **2** (23) and **6** do not polymerize with anionic initiators or give low yields of ill-defined oligomers. Vinyl 2-furyl ketone **5** (24) is sensitive to anionic activation, but the products have low molecular weights. The most interesting results have been obtained with **3** and **4**.

Whereas the cationic polymerization of furfurylidene acetone **3a** engenders crosslinked structures (25), the use of anionic initiators results in linear structures (26). However, the propagation is preceded by an isomerization of the active species which eliminates the steric hindrance to propagation arising from the 1,2-disubstitution in the monomer structure. A proton shift from the 4- to the 2-position places the negative charge at the extremity of the monomer unit and the incoming monomer can add onto this anion without major restrictions. The polymer structure thus obtained is :

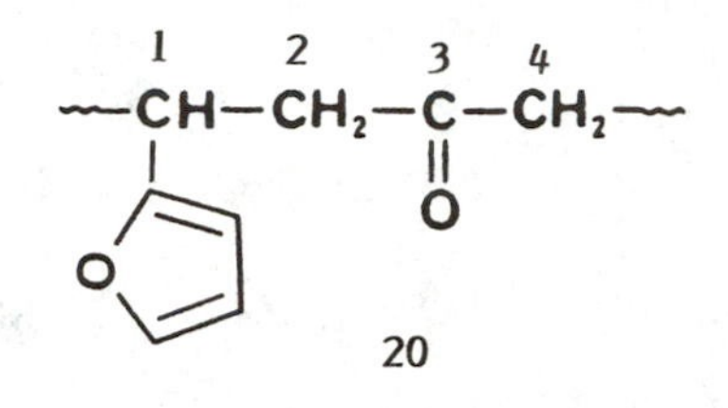

20

The key role of the hydrogen atoms of the methyl group in monomer **3a** in allowing the occurrence of the reorganization of the chain carrier is proved by the fact that similar structures bearing CF_3, Ph or tBu instead of CH_3, i.e. monomers **3b**, **3c** and **3d**, do not polymerize anionically because they do not have a mobile hydrogen atom at the 4-position to insure the 4-to-2 proton shift. On the other hand, the non-intervention of the C5 position in this isomerization polymerization is confirmed by the fact that monomer **3f** behaves exactly like **3a**.

The anionic polymerization of 2-furyl oxirane **4a** (17) proceeds smoothly and can be made to yield oligomers bearing a hydroxyl group at each end:

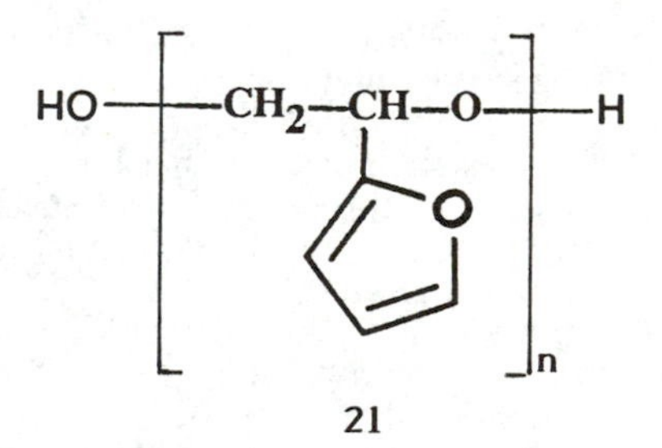

21

Apart from conventional initiators such a tBuOK, MeONa or BuLi, this monomer can be activated by much milder bases such as amines, alcohols and even water (27). Conversely, the carbazyl anion or the naphthalene radical anion are inactive. The initiation mechanism by water is novel and does not operate with other oxiranes like styrene oxide. The first step is a slow hydrolysis of the three-member ring to give the corresponding diol. Thereafter the diol adds onto the oxirane ring of the monomer to provide the propagation step. Thus, the active species are hydroxyl functions. Initiation by alcohols such as MeOH or tBuOH proceeds likewise by the addition of the OH group onto the oxirane ring to form the furanic alcohol which insures propagation; this mechanism is also new in that it does not require catalysis by a strong base. These peculiarities obviously arise from the presence of the furan ring in the monomer structure **4a**; they are presently being investigated further.

STEP POLYMERIZATIONS

Apart from the well-known mechanistic differences between chain and step polymerizations, another important aspect is relevant here, namely the fact that polyaddition (chain) reactions of monomers like those given in Scheme 1 result in macromolecular structures bearing *pendant* furan rings, whereas polycondensation (step) reactions with monomers like those depicted in Scheme 2 provide macromolecules with furan rings *within* the main chain. Naturally, the structure-properties relationships will be sensitive to this major topological difference in the positioning of the heterocycle. In other words, one would expect the polymers formed by polycondensation to have stiffer chains and therefore higher glass transition temperatures and other improved physical properties.

This section deals with the preparation of such structures, including all-furanic compositions and mixed ones calling upon furanic and aromatic or aliphatic monomers.

Polyesters

Early studies on furanic polyesters are few and far between. A systematic investigation of these polymers was carried out just over a decade ago (28) and showed that only those structures arising from furan-2,5-dicarboxylic compounds **11** and aliphatic diols or diphenols were interesting in terms of materials. Indeed, all-furanic polyesters or polyesters derived from bis-2,5-hydroxymethyfuran **10** prepared in those studies seemed much more fragile with respect to heat or weathering and had low DPs. In part at least this problem must stem from the rather severe conditions of synthesis which probably induced the parallel resinification of the furanic diol and the onset of some polymer degradation.

More recently two different approaches have given successful results in the preparation of similar structures. One of these was based on the use of a single monomer, viz. 5-hydroxymethyl-2-ethyl furanacrylate **16**. This compound is readily obtained from HMF (29) and polymerizes in bulk or in solution by transesterification under mild catalytic conditions and at temperatures below 100°C (30). The polymers crystallize during the synthesis and precipitate out of the reaction medium. They possess the regular structure **22** and melt at about 180°C:

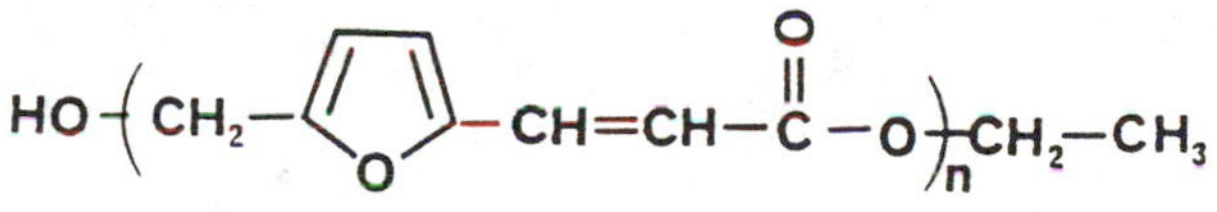

$$HO\left(CH_2-\text{C}_4\text{H}_2\text{O}-CH{=}CH-\overset{O}{\overset{\|}{C}}-O\right)_n CH_2-CH_3$$

22

Photoirradiation of this polyester in the near UV induces its crosslinking through interchain cycloaddition of the acrylic unsaturations.

The other approach called upon phase-transfer catalysis (31). The potassium salt of 2,5-furandicarboxylic acid **11a** was treated with primary aliphatic dihalides in a typical solid/liquid system in the presence of a crown ether. The resulting furanic-aliphatic polyesters had molecular weights of several thousands and a regular structure as shown by NMR spectroscopy and the fact that they crystallized quite readily.

Finally, a novel investigation on furanic polyesters is being carried out with the aim of arriving at polymers with regular structure by solution reactions involving the diacid dichloride **11b** or similar compounds and various diols and diphenols (32). Thus, for example, with 1,8-octandiol a medium-DP polyester was obtained with a melting point of about 190°C.

Polyamides

The most extensive research on furanic polyamides is recent (33) and deals essentially with furanic-aromatic structures, although an important effort was also devoted to all-furanic compositions. The reaction of the diacid **11a** with various aromatic diamines leads to high-molecular weight polymers with good thermal stability and crystallinity. Structure **23**, obtained with p-phenylenediamine, exhibited features resembling closely those of polyaramides:

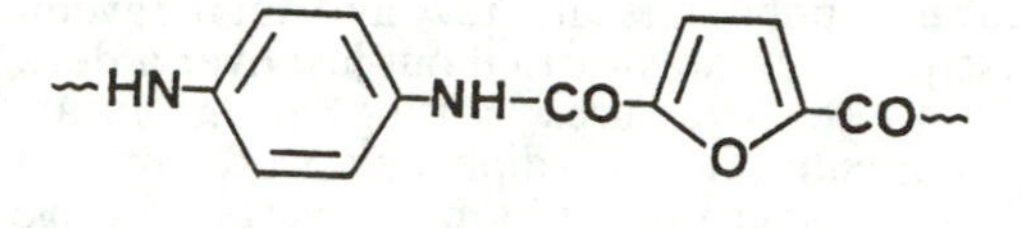

This polymer suffered thermal degradation only above 400°C, before melting. For molecular weights higher than about 20,000, it showed a lyotropic liquid-crystal behaviour in N-methypyrrolidone containing LiCl. This is the first reported example of a furanic polymer with mesogenic properties.

Two polyamides bearing only furanic moieties were prepared, as shown by the following structures:

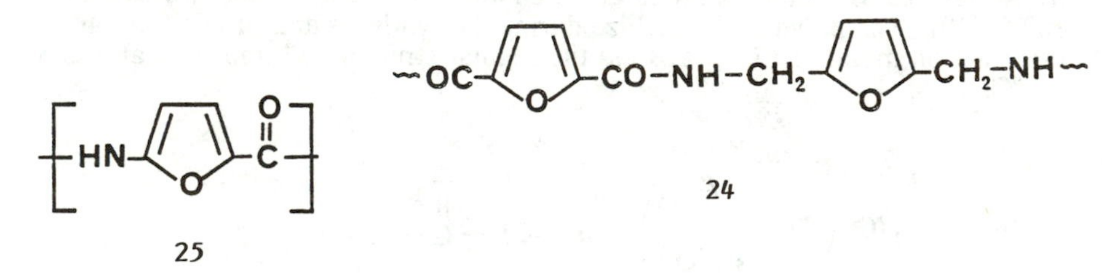

Polymer **24** was synthesized from the acid chloride **11b** and the diamine **15**. It had lower DPs than the furanic-aromatic counterparts and was less resistant to thermal degradation. This must stem from the relative instability of the diamine and the lability of the –Fu–CH_2–NH– group. Polymer **25**, obtained from the selfcondensation of aminoester **12b** seems more promising, but more work is needed to improve its preparation and assess its properties.

Polyethers

2,5-bis(hydroxymethy)lfuran **10** can be made to react with primary dihalides under typical liquid/liquid phase transfer conditions involving an aqueous

NaOH solution of the diol and an organic solution of the dihalide in the presence of a phase-transfer catalyst (31). The best results were obtained with dichloro-p-xylene as comonomer and gave polymers with the furanic-aromatic structure **26**:

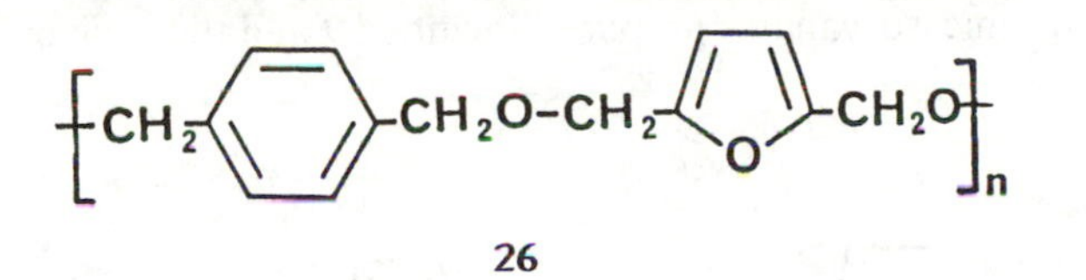

26

This is the first published example of a polyether of regular structure bearing furan rings in the chain backbone. Molecular weights varied between 2,000 and 4,000 and it was possible to obtain narrow DP distributions and $-CH_2Cl$ end groups to avoid ageing of the polymers due to $-FuCH_2OH$ terminal moieties.

Polycarbonates

Using a synthetic approach similar to that described above for the polyethers and based on liquid/liquid phase-transfer catalysis, the first furanic polycarbonate shown below was prepared (31):

$$\left[\overset{O}{\overset{\|}{C}}-O-CH_2-CH_2-O-CH_2-CH_2-O-\overset{O}{\overset{\|}{C}}-O-CH_2-\text{(furan)}-CH_2O \right]_n$$

27

Its NMR and IR spectra confirmed the structure proposed, its Tg was about 10°C and its decomposition threshold around 150°C.

Polyurethanes

Surprisingly little information is available on simple furanic urethanes and only one investigation on thermoplastic polyurethanes containing the heterocycle (34). The latter made use of difurylic diisocyanates with aliphatic diols to arrive at structures like:

$$\sim\overset{O}{\overset{\|}{C}}-O-(CH_2)_n-O-\overset{O}{\overset{\|}{C}}-NH-CH_2-\text{(furan)}-CH_2-\text{(furan)}-CH_2-NH\sim$$

28

These materials were characterized in terms of thermal and mechanical properties, but not by a full structural analysis or DP and DP distribution. Also, no clear-cut evidence was given of the long-term stability of the urethane linkages near a furan ring.

In order to gain a better insight into these problems, a fundamental study of simple mono- and di-urethanes of the furan series was undertaken. This included the determination of their structure, properties and stability (35) and the mechanism and kinetics of their formation. The combinations investigated were furan alcohols and diols with aliphatic, aromatic and furanic isocyanates and the latter mono- and bis- derivatives with aliphatic and arylalkyl alcohols and diols. The furanic isocyanates prepared included **9** and those given below:

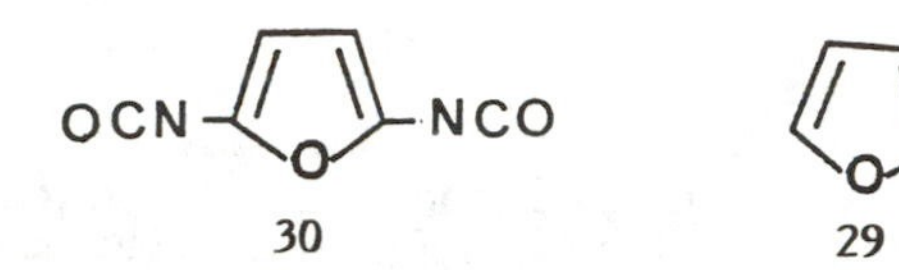

30 29

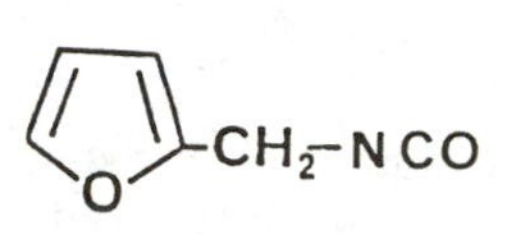

31

but also difuranic diisocyanates like those reported previously (34). Compound **31** was not listed in the Chemical Abstracts before this investigation and the diisocyanate **9** had only been mentioned in a patent.

It was found that whereas 2-furyl isocyanate **29** and its difunctional 2,5-homologue **30** are unstable and resinify on standing at room temperature even if kept under nitrogen, the corresponding urethanes formed with aliphatic and furanic alcohols are stable. In more general terms, both the kinetics and the mechanism of the formation of furanic urethanes and diurethanes could be rationalized on the basis of established criteria combined with the specific effects of the furan moiety (35). All the products were fully characterized and proved to have the expected structure. No side reactions involving the heterocycle were detected.

These results stimulated the continuation of the research towards the synthesis of polyurethanes. Various furanic diols and polyols were selected for this purpose and made to react with aliphatic and aromatic diisocyanates. Diol **10** gave regular linear polymers which crystallized readily (36). Crosslinked structures were obtained with furanic tetrafunctional polyols derived from saccharose (37). Polymers were also prepared using the macrodiols **21** (38). Some of these materials were tested as barriers for the confinement of radioactive wastes arising from the dismantling of nuclear power stations, i.e. building and structural materials possessing low radioactivity (38).

CHEMICAL MODIFICATION OF POLYMERS

Two distinct working hypotheses have been considered. The first involves the use of furan compounds as additives in a given polymerization system and the second deals with a specific reaction on the polymer after its synthesis. Both, however, make use of particular chemical specificities of the furan heterocycle.

An example of the first type of study is the cationic polymerization of alkenes and heterocyclic monomers in the presence of 2-alkylfurans. As discussed above, electrophilic substitution at C5 is quite facile with these compounds and one can therefore prepare monofunctional oligomers bearing a furanic end-group. By a judicious choice of experimental conditions this transfer reaction will predominate over all other chain-breaking events and virtually all the chains will have the same terminal structure, i.e. a 5-oligomer-2-alkylfuran. Structure **32** illustrates this principle with isobutyl vinyl ether oligomers capped by 2-methylfuran:

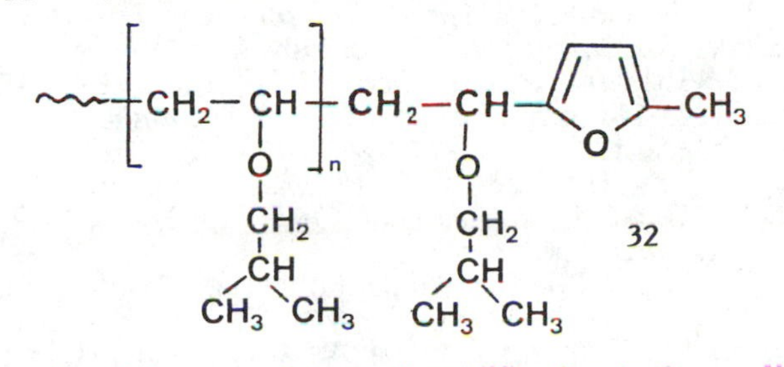

An example of the second type of modification is the application of the Diels-Alder cycloaddition reaction to polymers and copolymers containing pendant or backbone furan moieties. The use of bis-dienophiles such as propiolic acid and its esters or bis-maleimides provides a means of crosslinking based on multiple bridging by the double interchain cycloadditions. The thermal reversibility of these reactions allows the return to the original linear structure (thermoplastic material) by simply heating the gel.

AKNOWLEDGMENTS

Special thanks are due to Q.O. Chemicals for a generous gift of diol **10** and to Drs. Borredon, Delmas, Gaset and Rigal for providing precious samples of several monomers among those cited in this work.

LITERATURE CITED

1. Dunlop, A.P.; Peters, F.N. *The Furans*; ACS Monograph Series 119; Reinhold Publishing Corporation: New York, 1953.
2. Bosshard, P.; Eugster, C.H. *Adv. Heterocyc. Chem*. 1966, **7**, 377.
3. McKillip, W.J.; Sherman, E. *Furan Derivatives*, in Kirk-Otmer Encyclopedia of Chemical Technology; Grayson, M., Ed.; John Wiley and Sons, Inc.: New York, 1981; Vol. 11, p 499.
4. Dean, F.M. *Adv. Heterocyc. Chem*. 1982, **30**, 167 and 1983, **31**, 237.
5. Bird, C.W.; Cheeseman, G.H.W., Eds.; *Comprehensive Heterocyclic Chemistry*; Pergamon Press: Oxford, 1984; Vol. 4, Chapters 1, 2,3,10,11,12.
6. McKillip, W.J. in *Adhesives From Renewable Resources*; Hemingway, R.W.; Conner, A.H.; Branham, S.J., Eds.; ACS Symposium Series 385; American Chemical Society: Washington, DC, 1989; p 408.

7. Gandini, A. *Adv. Polym. Sci.* 1977, **25**, 47.
8. Gandini, A. in *Encyclopedia of Polymer Science and Engineering*; John Wiley and Sons: New York, 1986; Vol. 7, p.454.
9. Arekion, J.; Delmas, M.; Gaset, A. *Biomass* 1983, **3**, 59.
10. Borredon, M.E.; Delmas, M.; Gaset, A. *Tetrahedron* 1987, **43**, 3945 and refs therein.
11. Gandini, A.; Rieumont, J. *Tetrahedron Lett.* **1976**, 2101.
12. Lunazzi, L.; Placucci, G.; Grossi, L. *J. Chem. Soc., Perkin II*, **1982**, 875.
13. Burkey, T.J. *J. Org. Chem.* 1983, **48**, 3704.
14. Gandini, A.; Rieumont, J *Prepr. IUPAC Int. Symp. Free Radical Polymerization*, S.ta Margherita Ligure, Italy, May 1977, p. 93.
15. Rieumont, J.; Espinosa, R. *Rev. Cienc. Quim. (Cuba)* 1983, **14**, 293.
16. Brunache, M.H. *Doctoral Thesis*, National Polytechnic Institute, Grenoble, France, 1984.
17. Salon, M.C.; Amri, H.; Gandini, A *Polymer Comm.*, in press.
18. Martinez, R.; Gandini, A. *Makromol. Chem.* 1982, **183**, 2399; 1983, **184**, 189; and *Acta Polym.* in press.
19. Salon, M.C. *Doctoral Thesis*, National Polytechnic Institute, Grenoble, France, 1985.
20. Razzouk, H.; Bouridah, K.; Gandini, A.; Cheradame, H. in *Cationic Polymerization and Related Processes* Goethals, E.J., Ed.; Academic Press: New York, 1984, p.355.
21. Gandini, A.; Rieumont, J. *Brit. Polym. J.* **1977**, 28.
22. Salon, M.C.; Gandini, A.; Gey, C. *Makromol. Chem., Rapid Comm.* 1988, **9**, 539.
23. Gandini, A.; Hernandez, C. *Polym. Bull.* 1978, **1**, 221.
24. Labidi, A.; Salon, M.C.; Gandini, A.; Cheradame, H. *Polym. Bull.* 1985, **14**, 271.
25. Rodriguez, V.J.; Gandini, A. *Rev. CENIC, Cienc. Fis. (Cuba)* 1974, **5**, 29 and 1975, **6**, 155.
26. Salon, M.C.; Gandini, A.; Cheradame, H. *Polym. Bull.* 1984, **12**, 441.
27. Amri, H; Gandini, A. *Unpublished results*, 1989.
28. Moore, J.A.; Kelly, J.E. *Macromolecules* 1978, **11**, 568; *J. Polym. Sci., Polym. Chem. Ed.* 1978, **16**, 2407 and 1984, **22**, 863; *Polymer* 1979, **20**, 627.
29. Mouloungui, Z.; Delmas, M.; Gaset, A. *Synth. Comm.* 1984, **14**, 701.
30. Roudet, J.; Gandini, A. *Submitted to Polymer.*
31. Boileau, S.; Dorigo, R.; Majdoub, M.; Mechin, F.; Gandini, A. *Submitted to European Polym. J.*
32. Maccio, F.; Valenti, B.; Costa, G.; Gandini, A. *Unpublished results*, 1988.
33. Mitiakoudis, A.; Cheradame, H; Gandini, A. *Polym. Comm.* 1985, **26**, 246. Mitiakoudis, A. *Doctoral Thesis*, National Polytechnic Institute, Grenoble, France, 1986.
34. Cawse, J.L.; Stanford, J.L.; Still, R.H. *Makromol. Chem.* 1984, **185**, 697 and 708.
35. Quillerou, J.; Belgacem, M.; Gandini, A.; Rivero, J.; Roux, G. *Polym. Bull.* 1989, **21**, 555; *European Polym. J.* 1989, **25**, 1125.
36. Belgacem, M.; Quillerou, J.; Gandini, A., *Unpublished results*, 1989.
37. Roux, G.; Rivero, J.; Gandini, A. *French Patent*
38. Signoret, C., *Doctoral Thesis*, National Polytechnic Institute, Grenoble, France, 1989.

RECEIVED January 18, 1990

AGRICULTURAL POLYMER UTILIZATION
Alternate Crop Strategies

Chapter 18

Microbial Fructan

Production and Characterization

Y. W. Han[1] and M. A. Clarke[2]

[1]Southern Regional Research Center, U.S. Department of Agriculture, P.O. Box 19687, New Orleans, LA 70179
[2]Sugar Processing Research, Inc., 1100 Robert E. Lee Boulevard, New Orleans, LA 70179

As part of an ongoing study to develop new products from the agricultural resources provided by sugar-producing crops, a search was initiated for microorganisms to produce polymeric compounds for industrial use.

A fructan-producing bacterium was isolated from soils and characterized for polysaccharide synthesis. The composition and properties of the polysaccharide produced were studied. The organism, identified as a strain of Bacillus polymyxa, produced a large quantity of polysaccharide when grown on sucrose. The polysaccharide consisted entirely of fructose: methylation analysis showed that the primary fructose linkages were β(2→6) fructofuranosyl linkages. Carbon 13 nmr showed the product to be a levan type fructan.

As part of an ongoing study to develop new products from the agricultural resources provided by sugar-producing crops, a search was initiated for microorganisms to produce polymeric compounds for industrial use.

Polysaccharides were the first group of polymers considered. Dextrans, polymers of glucose synthesized from sucrose, are important industrial polysaccharides (1).

Fructans are natural polymers of fructose. Depending on the linkage types, fructans are classified into two groups: the levans, with mostly β-(2→6) linkages and the inulins with β-(2→1) linkages. Many fructans of both types have branched chains. Levans and inulins of low molecular weight are abundantly found in plants, while high molecular weight fructans are produced by many microorganisms (2-4).

0097–6156/90/0433–0210$06.00/0

A variety of microorganisms produce extracellular polysaccharides in the form of capsules attached to the cell wall, or as slime secreted into the growth medium. These materials are used in the organism's defense mechanism, or as a food reservoir. Some bacteria produce fructan, among which *Bacillus* spp. predominate. Oral bacteria such as *Rothis dentocariosa*, *Streptococcus salivarius* and *Odontomyces viscosus* accumulate fructan in human dental plaque (5-7). Several species of yeast and fungi are also known to produce levan (8,9). Most research on the biosynthesis of fructan has been conducted using *Bacillus subtilis*, *Aerobacter levanicum*, and *Streptococcus salivarius* (10-21).

Microbial fructans or levans, like dextran, were first found in sugar factories (16,2). These polysaccharides caused difficulties in the beet sugar manufacturing process by increasing the viscosity of the processing liquor. Since their discovery (22), fructans have received little attention and have never been exploited for industrial applications. Recently the sugar industry has faced intense competition from high fructose corn syrup, which is used as a low cost alternative sweetener. In search of new products from sucrose, the possibility of producing microbial fructan from sucrose for industrial applications has been investigated. In this paper, we report the isolation of a levan-producing bacterium and characteristics of the fructan it produces.

Isolation of a Fructan-Producing Bacterium

Figure 1 shows the isolation scheme for a levan producing bacterium. About 1 g of rotting sugarcane stalks and the adhering soil particles were added to 100 ml of basal medium and incubated at 30°C with constant shaking. The isolation medium consisted of sucrose 150g; peptone, 2g; yeast extract, 2g; K_2HPO_4, 2g; $(NH_4)_2SO_4$, 0.3g; in a liter of water. The growth culture was then transferred to fresh media every 7-10 days. After several successive transfers, the culture was plated on agar media and the bacterial colonies with gummy appearance were selected. These are organisms for which sucrose is the sole carbon source, and which thrive in high osmotic pressure. The organisms that produce polysaccharide (alcohol precipitate) having negative rotation of polarized light was tentatively selected as levan producers. The levan was finally confirmed by ^{13}C nmr, infrared, and methylation analyses. A detailed isolation procedure was reported elsewhere (23) and the organisms has been registered at USDA, Northern Regional Research Center, Peoria, Illinois, and identified as NRRL B-18475.

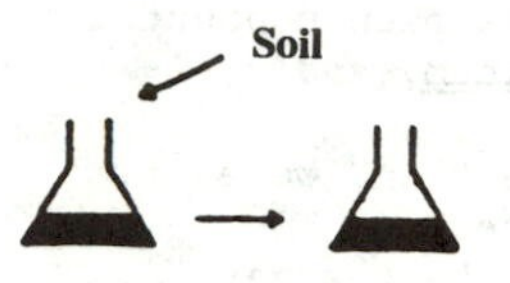

Figure 1. Isolation of a fructan-producing bacterium.

Production of Fructan

The B. polymyxa (NRRL B-18475) produces a large quantity of fructan when grown on 4%-16% sucrose solution. The production scheme is shown in Fig. 1. The organism converted the fructose moiety of sucrose to fructan; of the remaining glucose, most were used as the carbon source for microbial growth and a small amount accumulated in the growth medium. Some acids were produced as evidenced by decrease in pH in the growth medium.

The composition of the products was monitored by HPLC (Sugar Analyzer, Waters Associates; HPX-87C column, BioRad Corp. with deionized water, 40 ppm in calcium acetate as mobile phase). During fermentation, the sucrose levels dropped and fructan started to appear in 2 days; thereafter, sucrose level gradually decreased as fructan increased. Glucose was the major by-product. A small amount of fructose and other unidentified fermentation products smaller in molecular weight were also observed. The pH of the growth medium was controlled, and fell from 7.0 to 4.7 due to acid production. In reports of other fructan production, maintaining pH above 5.5 was important because the optimum pH for fructansucrase is between 5.5-7.0 and fructan may be hydrolyzed at a lower pH (2). Optimum temperature for growth and fructan production was around 30°C.

Aeration has been shown to be important in the biosynthesis of fructansucrase (24). Polysaccharide production was especially pronounced when the culture was gently shaken during cultivation, but vigorous agitation and aeration inhibited fructan production (23). A small amount of microbial polysaccharide (detected by alcohol precipitation) was also produced when the organism was grown on lactose, maltose, and raffinose, but no polysaccharide was produced on glucose or fructose. The organism produced polysaccharide from sugarcane juice, but the yield was much less than that from sucrose. High sucrose concentration has been reported to lower the average molecular weight of the fructan synthesized (25).

Fructan was harvested by precipitation from the culture broth by addition of ethanol or isopropanol. Acetone and methanol can also be used. The yield and consistency of the product varied depending on the amount of alcohol added. The fructan started to precipitate at the medium/alcohol v/v ratio of 1:1.2, and the yield peaked at about 1:1.5. Further increase in the ratio hardened the fructan and made the product less fluid. Slightly less isopropanol was needed than ethanol to precipitate levan (fructan). Although most of the bacterial cells, unfermented sugars, and other solubles remained in the aqueous alcohol phase, pre-removal of microbial cells by centrifuging was needed to obtain a pure form of fructan. The product was further purified by repeated precipitation and dissolution in water, followed by dialysis or ultrafiltration. The final product was an off-white, flaky or powdery material that could be freeze- or vacuum-dried, or alternately dried by trituration and pulverization in a high-speed blender with absolute alcohol. In a typical fermentation, B. polymyxa produced about 3.6g of levan (fructan) in 100 ml of 15% sucrose in 10 days (about 46% yield on available fructose, where 7.89 g fructose are available from 15 g sucrose).

Composition and Properties of the Fructan

GLC analysis of the oxime derivatives of TFA-hydrolyzed fructan shows over 93% fructose with a small amount of glucose and traces of degradation products. GLC analysis of oxime derivatives (26), after hydrolysis of the fructan by trifluoracetic acid, was performed on a Hewlett Packard chromatograph model 5880, with a fused silica capillary column. The initial molecule in fructan chain formation is sucrose, therefore terminal glucose groups will be present in fructan. Some free glucose may have been adsorbed on the crude sample analyzed, because essentially no glucose was observed on methylation analysis. At the molecular weight observed (see below), there is a very small percentage of glucose units as terminal groups.

X-ray crystal structure analysis showed no crystallinity. Fructan is amorphous. X-ray analysis was performed on a General Electric X-ray diffraction refractometer.

A 5% aqueous solution of crude fructan, after dialysis through a membrane with 12,000 daltons cut-off, gave a single, sharp clean peak just below 2 x 10^6 daltons on Sephacryl S-500. The compound is stable in aqueous solution at pH 4.5 for up to 36 hours, when monitored by HPLC analysis. Fructan is readily hydrolyzed by 0.5% oxalic acid (19). It is not decomposed by amylase enzymes.

The fructan has an optical rotation $[\alpha]^{25}$ - 47.2. It is non-hygroscopic, unusual in view of its high solubility. Lyophilized sheets of fructan have been maintained under condition of 25°-30°C and 70%-85% relative humidity for up to 6 months. The solubility of fructan is very high: up to 30% in cold water, with no apparent viscosity increase. It is extremely soluble in hot water. This high solubility is characteristic of β-(2→6) linked fructans.

Structure. The ^{13}C nmr spectra, shown in Figure 2, indicates that essentially all fructose molecules in the polymers are in the same conformation. In Table I, nmr peaks from fructan are compared to peaks from known inulin (β-(1→2) linked) and bacterial levan (β-(2→6) linked). Data clearly show the fructan to be of the β-(2→6) type (27). (See Table II.)

Nmr ^{13}C spectroscopy was performed at 100 MHz with a JEOL GX-400 instrument, at 70°C, with internal standard 1,4-dioxane (δ67.40).

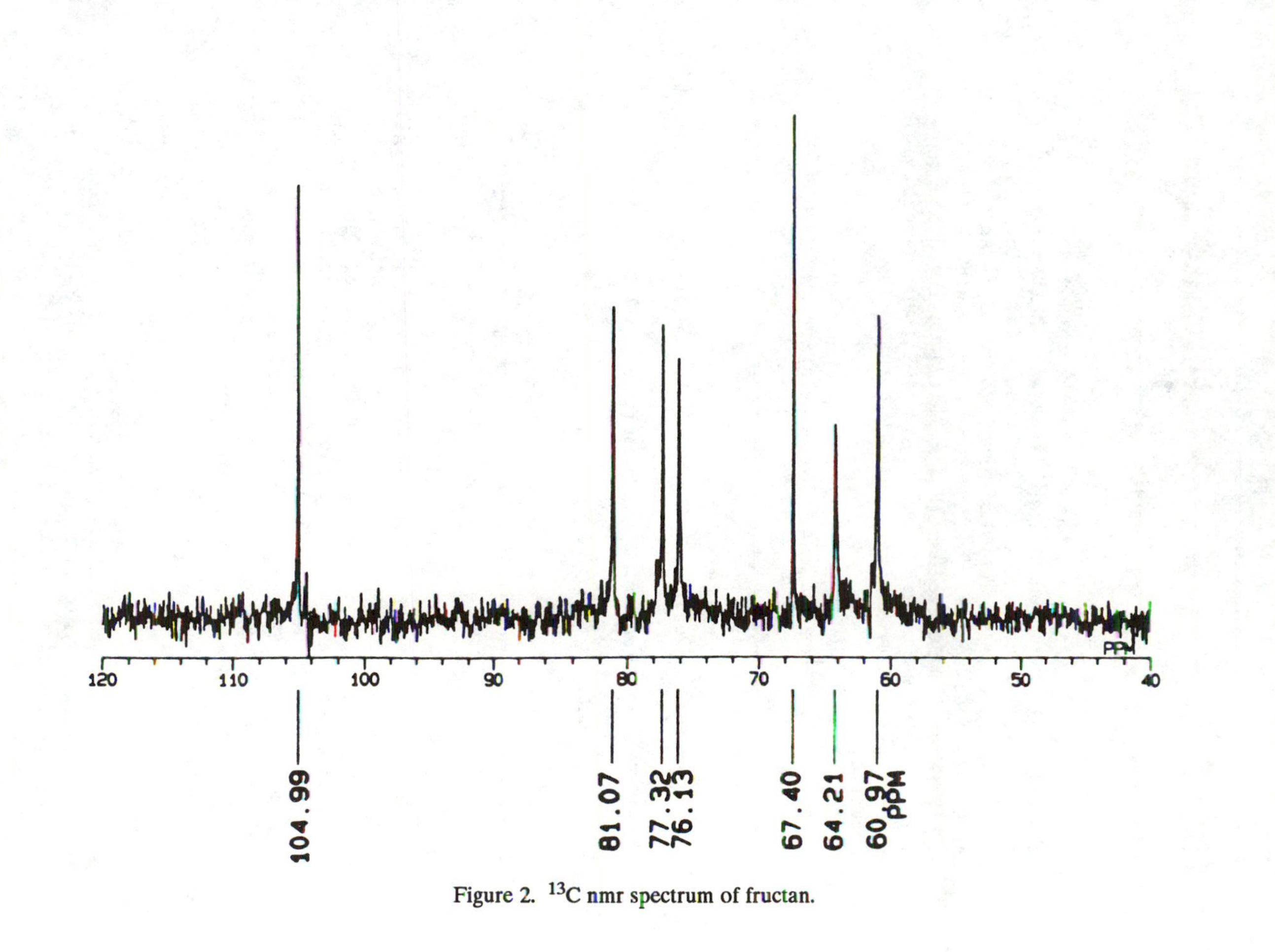

Figure 2. ^{13}C nmr spectrum of fructan.

Table I. Assignment of nmr peaks

	1	2	3	4	5	6
Inulin (1-2)	62.2	104.5	78.5	76.6	87.4	63.4
Levan (2-6)	61.4	105.1	77.5	76.6	81.3	64.6
Fructan	61.4	105.0	77.8	76.4	81.1	64.2

Methylation analysis was run by the method of Hakomori (28), followed by hydrolysis with trifluoracetic acid, sodium borohydride reduction, and acetylation. GLC was performed on a Hewlett-Packard 5970, used as an inlet for a mass spectrometer. Molecular weight was determined on a Sephacryl S-500 column (2.6 x 70 cm), using deionized water as solvent, upward flow, 2.75 ml/min, and detection by refractive index monitor, Model R-401 (Waters Associates).

Table II. Linkages indicated by methylation analysis

β-(2→6) linked fructose	71%
Branch points (at 1, 2, 6)	12%
Terminal groups (1 or 2 position)	13%
Undissolved material	4%

Branch points are indicated by the presence of 3,4 dimethyl substituted fructose, and the degree of branching of 12% is supported by the observation of 13% terminal groups, indicated by tetramethylated fructose residues, substituted at the 1- or 2-positions. The branches are formed by β-(1→2) linkages with side-chains of β-(2→6) linked residues. The degree of branching in fructans has been shown to range from 5-20% (28). The free hexose probably results from material that was not dissolved during methylation. (See Figure 3.)

Summary

Fructans (levans) are natural polymers of fructose, found in many plants and microbial products. Like dextrans, they can be formed as an undesirable microbial byproduct in the processing of sugar juice and have deleterious effects on processing. On the other hand, fructans, which

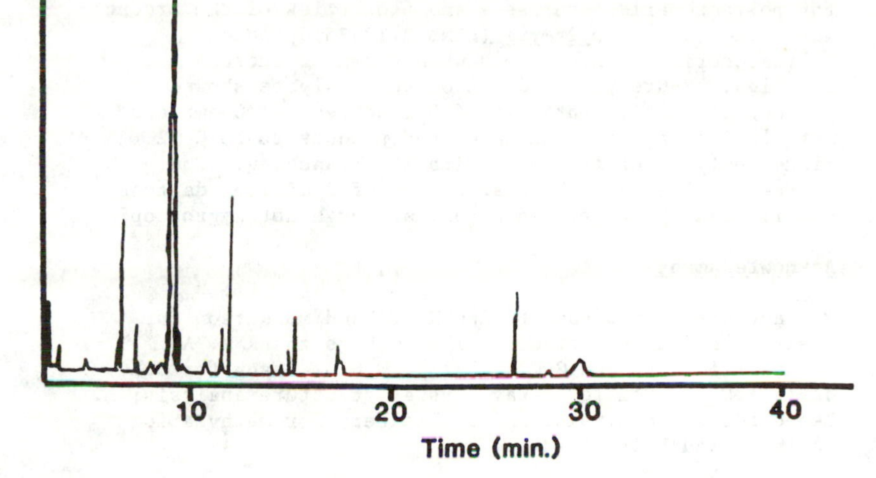

Figure 3a. GLC of alditol acetate derivatives of the hydrolyzed methylated fructan.

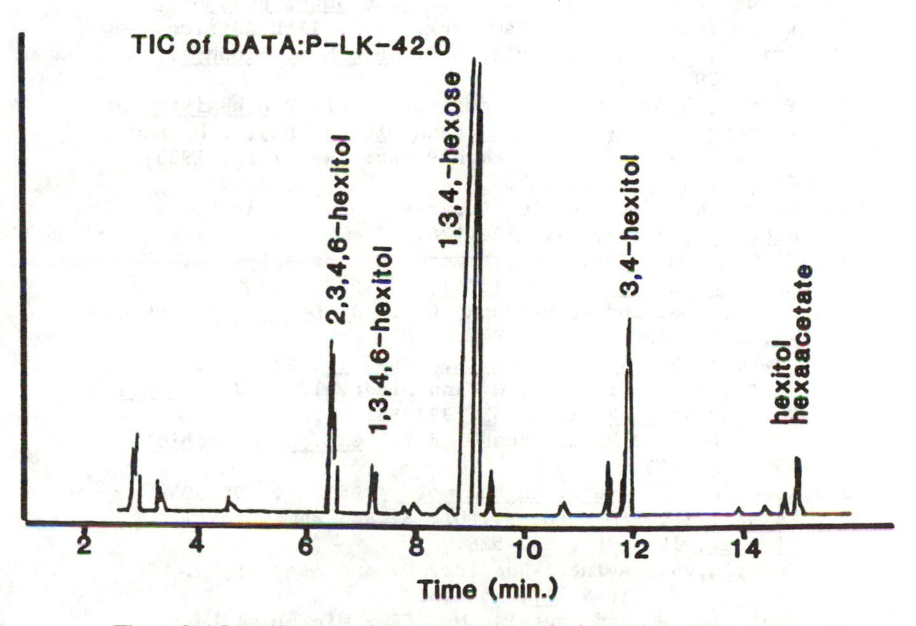

Figure 3b. Identification of alditol acetate peaks (expanded section of 3a).

can be produced only from sucrose, have potential industrial applications as thickeners and encapsulating agents and could provide products of added value from sugar crops. In this study, a fructan-producing bacterium was isolated from soil and its characteristics for polysaccharide synthesis and properties of the product were studied. B. polymyxa (NRRL B-18475) produced polysaccharide in high yield when grown on sucrose solution. Hydrolysis and subsequent analyses showed the product to consist entirely of D-fructose. ^{13}C-nmr and methylation analyses indicated the product to be β-(2→6) linked polymer of fructose, with 12% branching. The polysaccharide has molecular weight of 2 million daltons and is readily soluble in water, although not hygroscopic.

Acknowledgment

The authors acknowledge L. Ban-Koffi and M. Watson for their technical assistance. The authors thank M. A. Godshall for GLC and GPC analysis; W.S.C. Tsang for HPLC data; A.D. French for x-ray crystal structure analysis; L. Kenne for nmr analysis, and B. Lindberg for methylation analysis and helpful advice.

Literature Cited

1. Meade, G.P. and Chen, J.P.C. Cane Sugar Handbook, Wiley Interscience: New York, 1985, 11th Edition.
2. Avigad, G., and Feingold, D.S. Biochem. Biophys. 1965, 70, 178.
3. Pontis, H.G., and Del Campillo, E. In Biochemistry of storage carbohydrates in green plants; Dey, P.M. and Dixon, R.A., Eds.; Academic Press: New York, 1985; Chapter 5, pp. 205-227.
4. Vandamme, E.J. and D.G. Derycke. Adv. in Appl. Microbiol. 1983, 29, 139-176.
5. Higuchi, M., Iwami, Y., Yamada, T. and Araya, S. Arch. Oral Biol. 1970, 15(6), 565-567.
6. Manly, R.S. and Richardson, D.J. J. Dent. Res., 1968, 47, 1080-1086.
7. Newbrun, E. J. Dent. Child, 1969, 14, 239-248.
8. Fuchs, A., DeBruijn, J.M. and Niedeveld, C.J. Antonie Van Leeuwenhoek, 1985, 51, 333-351.
9. Loewenburg, J.R. and Reese, E.T. Can. J. Microbiol. 1957, 3, 643.
10. Dedonder, R. Meth. in Enzymol. 1966, 8, 500-505.
11. Tanaka, T., Yamamoto, S., Oi, S. and Yamamoto, T. J. Biochem, 1981, 90, 521-526.
12. Hestrin, S., Avineri-Shapiro, D. and Aschner, M. Biochem. J., 1943, 37, 450-456.
13. Mantsala, P. and Puntala, M. FEMS Microbial Lett., 1982, 13, 395-399.

14. Perlot, P. and Monson, P. Enz. Eng. 7th Int'l. Conf. Annal. 1984, New York Academic Science, 434, 468-471.
15. Yamamoto, S., Iizuka, M., Tanaka, T. and Yamamoto, T. Agric. Biol. Chem. 1985, 49, 343-350.
16. Fuchs, A. Doctoral thesis, Rijksuniversiteitte Leiden, Waltman Ed., Delft, 1959.
17. Robeiro, J.C.C. Guimarues, Borges, W.V., Silv, A.C., D.O. and Crug, C.D. Rev. Microbiol. 1988, 19(2), 196-201.
18. Lyness, E.W. and Doelle, H.W. Biotechnol. Lett. 1983, 5(5), 345-350.
19. Evans, T.H. and Hibbert, H. Adv. Carbohydr. Chem. 1946, 2, 253-277.
20. Feingold, D.S. and Gehatia, M. J. Polymer Sci. 1957, 23, 783-790.
21. Takeshita, M. J. Bacteriol. 1973, 116, 503-506.
22. Lippman, E.O. Chem. Ber., 1881, 14, 1509.
23. Han, Y.W. J. Indus. Microbiol. 1989, In press.
24. Tkachenco, A.A. and Loitsyankaya, M.S. Appl. Biochem. Microbiol. 1979, 14(4), 502-505.
25. Dedonder, R. and Peaud-Lenoel, C. Bull Soc. Chim. Biol. 1957, 39, 483.
26. Schaffler, K.J. and Morel du Boil, P.G. J. Chromatog. 1981, 207, 221-229.
27. Barrow, K.D., Collins, J.G., Rogers, P.L. and Smith, G.M.C. Eur. Jour. Biochem. 1984, 145, 173-179.
28. Lindberg, B., Loungren, J. and Thompson, J.L. Acta Chem. Scand. 1973, 27, 1819-1821.

RECEIVED December 29, 1989

Chapter 19

Coatings Based on Brassylic Acid (An Erucic Acid Derivative)

David E. Chubin, James P. Kaczmarski, Zeying Ma, Daozhang Wang, and Frank N. Jones

Polymers and Coatings Department, North Dakota State University, Fargo, ND 58105

Erucic acid, $HOOC(CH_2)_{11}CH{=}CH(CH_2)_7CH_3$, can be economically obtained from rapeseed and *crambe abyssinica* oils and is potentially a major source of industrial materials. It can be ozonized to brassylic acid, $HOOC(CH_2)_{11}COOH$, which is known to impart flexibility and moisture resistance to nylons. Here preliminary results of a study of brassylic acid as a monomer for polyester resin/melamine resin coatings are described. It is demonstrated that brassylic acid imparts good flexibility to such coatings. It is also shown that brassylic acid is polymorphic.

The United States Department of Agriculture has initiated an "Alternative Crops Program" to promote cultivation of new types of crops grown primarily to yield industrial materials. Several crops are being studied, each study involving the many steps required to get from farm to factory. A component of the program is focusing on the high-erucic vegetable oils, primarily rapeseed and crambe oils. The authors are involved in a subcomponent of the high-erucic oils project -- a study of potential uses of brassylic acid, a derivative of erucic acid, as a monomer for industrial coatings and elastomers. Here we offer a preliminary report of our results with coatings. First we will summarize background on high-erucic oils in order to put the subject in perspective.

BACKGROUND INFORMATION

The 22-carbon, monounsaturated fatty acid erucic acid, $HOOC(CH_2)_{11}CH{=}CH(CH_2)_7CH_3$, is obtained from high-erucic rapeseed oil, a commodity that is now produced mainly in eastern Europe. High-erucic rapeseed is raised on a modest scale in U.S., and it could be more widely grown. Its oil contains about 45 weight per-cent of erucic acid.

0097–6156/90/0433–0220$06.00/0

A potentially richer source of erucic acid is the oil of crambe abyssinica, a crop which is being cultivated in the U.S. on an experimental basis. Crambe oil contains 55 to 61 per-cent erucic acid, and there is hope that new strains can be developed that will be contain a higher proportion. Crambe grows and matures rapidly under relatively dry conditions. Test plots of current strains of crambe in varied U.S. locations (several thousand acres in all) gave highly encouraging results. For example, ten-acre test plots of crambe grown in North Dakota in 1988 yielded 1000 to 2000 lbs of seed per acre. As crambe seed is 30% to 40% oil, it is possible to project yields of oil comparable to yields produced by present day temperate-climate oilseed crops. Because crambe matures quickly, various double-cropping schemes can be envisaged. While many agricultural, regulatory, marketing and processing details will need to be worked out before crambe can be cultivated on a large scale, it seems very probable that large-scale cultivation of crambe in relatively arid parts of the U.S. is feasible.[1]

The question, then, is whether substantial markets for high-erucic oils can be developed. The answer depends on the industrial demand for erucic acid and its derivatives. Potential end uses can be separated into two categories: uses for derivatives with the C-22 chain intact and uses for derivatives made by breaking the C-22 chain. Today the only major end uses are in the first category. By far the largest is for erucamide, a lubricant additive (a "slip agent") for polyolefins. The U.S. market is about 9×10^6 kg/year. Other potential uses for erucic acid and its derivatives with the C-22 chain intact include lubricants, flotation aids, surfactants, plasticizers, waxes and dielectric fluids.[1] However, other potential markets for C-22 erucic acid derivatives have been slow to develop despite the fact that rapeseed oil has been available for a long time at prices below $1.00/kg.

Beside being a source of C-22 chemicals, erucic acid is a natural source of C-13 and C-9 chemicals made by cleavage at the site of the double bond. Ozonolysis of erucic acid followed by oxidative cleavage of the ozonide to form brassylic acid (BA) and pelargonic acid (Chart I) has been demonstrated on the pilot plant scale.[2,3] Yield of 99% pure brassylic acid was 72 to 82% of theoretical.

Chart I

$$HOOC(CH_2)_{11}CH{=}CH(CH_2)_7CH_3 \xrightarrow[\text{2. [O], } H_2O]{\text{1. } O_3} HOOC(CH_2)_{11}COOH + HOOC(CH_2)_7CH_3$$

Brassylic acid (BA) (1,11-undecanedicarboxylic acid) — Pelargonic acid (nonanoic acid)

It seems probable, but not certain, that the reaction in Chart I could be a basis for economical large scale production. The fact that an analogous ozonolysis process is now used on a large scale to convert oleic acid to azeleic [$HOOC(CH_2)_7COOH$] and pelargonic acids [4] strongly indicates that commercial production is feasible. Azeleic acid is sold at about $3.30/kg and is widely used as a monomer for adhesives, coatings, plasticizers, elastomers and specialty plastics,[4] and pelargonic acid has well established markets as a component of plasticizers and synthetic lubricants. In 1977 Carlson et al. projected cost of producing brassylic acid at about $1.50/kg,[2] but more recent projections have been higher. In undertaking this study we assumed that brassylic acid could, if produced on large scale, be marketed at a price no higher than dodecanedioic acid, [$HOOC(CH_2)_{10}COOH$], now marketed at about $4.30/kg, and possibly at a price competitive with azeleic acid. Thus performance advantages of BA over azeleic will probably need to be identified to motivate substitution.

Possible alternatives to the ozonolysis process for production of brassylic acid include oxidation with nitric acid or with potassium permanganate; these routes are feasible,[5] but yields were inferior to ozonolysis.

Because only one substantial market for C-22 derivatives of erucic acid has appeared, it is likely that the economic viability of large-scale high erucic oil production very probably depends on the cost of converting it to C-13 and C-9 derivatives and on development of large-scale applications for them. As the market for pelargonic acid is already well established, brassylic acid (BA) and its derivatives are key. At present the only commerical use for BA or its derivatives is small: The 17-member ring cyclic ester, ethylene brassylate, is a commercial fragrance. However, potential large-scale uses are mainly for acyclic diesters -- fluids, oligomers and polymers.

Several polymer-related uses of brassylic acid (BA) have been investigated. For example, a BA/1,3-butanediol/lauric acid oligomer is an effective plasticizer for polyvinyl chloride,[6] BA/ethylene glycol and BA/propylene glycol polymers function as polyester based polyurethane elastomers,[7] and BA has been patented as a cross-linker for glycidyl methacrylate copolymer powder coatings.[8] However, the most detailed studies have involved polyamides; selected data from these studies are summarized in Table I.

Table I: SELECTED PHYSICAL PROPERTIES OF BA-CONTAINING NYLONS[10]

PROPERTY	N-13	N-13/13	N-11	N-6/10
Tensile strength, psi	5550	5700	6310	7680
Elong. at break, %	130	130	110	160
Impact rest., ft-lb/in	2.1	2.6	2.6	1.7
Hardness, Shore D	74	72	74	81
Heat defl. T, 264 psi, °C	51	53	54	57
Water absorbt., % @ 23°C	1.04	0.75	1.61	3.1
Moisture regain, % @ 50 % RH, 23 °C	0.36	0.29	1.05	1.75

As indicated in Table I, most properties of polyamide derivatives of BA, nylons 13, and 13/13, are predictable from properties of commercial engineering plastics such as nylon-11 and nylon-6/10 -- the BA based nylons are have lower moduli and most physical properties are unexceptional.[9,10] However, the BA based nylons have one exceptional property -- their very low capacity to absorb moisture. This property suggests that these materials may be less affected by water plasticization than other nylons, and it has attracted interest in developing BA-based nylons commercially. Development has been impeded by the fact that BA is not produced on a sufficient scale to make it cost-competitive, and apparently the attractive markets are not large enough to justify investment in development of BA processes, creating a "chicken-or-egg" problem.

Our program is concerned with finding additional uses for BA to help overcome the "chicken or egg" problem that has impeded commercial development. We undertook to investigate BA-based polyester resins for coatings, an area that seems not to have been investigated previously. Of course if we could show that BA polyesters can be designed that will perform as well as polyesters made from competing materials, such as dodecanedioic acid [$HOOC(CH_2)_{10}COOH$], sebacic [$HOOC(CH_2)_8COOH$], and azaleic acid, [$HOOC(CH_2)_7COOH$] market development would depend on producing BA at competetive cost.

Further, if BA polyesters could be shown to have performance advantages BA could command a premium price. There are at least two places to look for premium performance in coatings -- one, suggested by the performance of BA nylons, is the area of low water plasticization and perhaps resistance to hydrolytic degradation. Moisture resistance is highly desirable for coatings as it is related to weatherability, corrosion resistance and retention of properties under humid conditions. The latter problem can be severe -- moduli of polyester resin/polyurethane resin films is reduced by more than 50% by water plasticization.[11] A second potential area of premium performance is in high solids coatings. It is theoretically possible that the flexible C-13 chain may impart very low viscosity to polyesters, reducing the amount of volatile solvent required to reach application viscosity and helping coatings producers meet the new, very stringent U.S. environmental regulations.

Thus our first objective was to learn how BA polyesters could be designed to make resins which will perform as well as commercial polyesters and the second was to find ways in which they might be superior. Here we present initial results of our project, showing how BA polyesters can be designed for good performance in industrial coatings. We also demonstrated, apparently for the first time, that brassylic acid is polymorphic, undergoing phase transitions at about 112 ^{o}C and 114 ^{o}C.

EXPERIMENTAL DETAILS

Materials. The following materials were used as received: adipic acid 99% (AA, Aldrich), azeleic acid 90% (AZA, Emerox 1144, Huls), sebacic acid 99% (SA, Aldrich), dodecanedioic acid 99% (DDA, DuPont), isophthalic acid (IPA, 99%, Aldrich), 2,2-dimethyl-1,3-propanediol 97% (NPG, Aldrich). 4-methyl-2-pentanone (MIBK, 98%) and other solvents (reagent grades) were purchased from chemical supply houses. K-Flex 188 polyester polyol (King). Brassylic acid (BA) assayed at about 95% purity by gas chromatography of methyl esters was supplied by Dr. Kenneth Carlson, U.S. Department of Agriculture, Peoria IL. 1,4-Cyclohexane-dimethanol (CHDM, Eastman), was a 99%, commmercial grade; it is a 70/30 mixture of cis and trans isomers which tend to separate during storage; the material was homogenized by melting and stirring before use. Melamine resins were the hexa(kis)methoxymethyl-melamine (HMMM) type, contributed by Monsanto as Resimenes 746 & 747. Catalysts were "Fascat 4100" a soluble tin catalyst supplied by M & T Chemical said to be 95% $H_9C_4Sn(O)OH$ and p-toluenesulfonic acid monohydrate (p-TSA, Lancaster).

Brassylic acid for polymerization was recrystallized twice from 50/50 v/v ethanol/water; this material had two melting points in the range 111-113.5 °C by DSC at a heating rate of 1 °C/min. Brassylic acid for melting point studies (see below) was recrystallized twice from toluene (91% overall recovery) or twice from from ethyl acetate (84% overall recovery).

Synthesis of Polyester Resins Described in Table II. The monomers and 3 wt % of xylene were placed in a 1-L breakaway reaction flask equipped with a thermometer, a Liebig condensor, a mechanical stirrer, an N_2 line and a heating mantle. The mixture was slowly heated under N_2 to 210 °C, with stirring being started as soon as possible. Temperature was kept at 210 °C for 7.5 to 15 hr as water, probably contaminated with NPG, was collected. The volume of water/NPG collected ranged from 50 to 200% of the theoretical amount of water. When the acid number approached the target of 10 mg of KOH/g of resin the reaction mass was cooled to 80 °C and was diluted to 70% solids with xylene. Final acid numbers ranged from 6 to 11 mg of KOH/g of resin.

Synthesis of CHDM/BA and CHDM/DDA Polyester Resins Described in Tables III and IV. CHDM, brassylic acid (or dodecandioic acid) and "Fascat 4100" in a mol ratio of 2/1/0.005 and 5 wt % of xylene were placed in a three-necked, 250-mL round-bottomed flask fitted with N_2 line, a thermometer, an Ace temperature controller, a heating mantle, a mechanical stirrer, a steam-heated Allihn partial condenser with a xylene-filled Barrett moisture test receiver below the Allihn condensor and a total condensor above it. The mixture was heated to about 140 °C without stirring to melt the reactants and stirring was started. Heating was continued to 205 °C. A slow N_2 sweep was maintained throughout the reaction. About 95 to 100% of the theoretical water was collected in the Barrett receiver. Upon cooling the reaction mixtures partially solidified to white waxy solids which liquified when heated to about 40 °C. Acid numbers were < 1 mg of KOH/g of resin.

Preparation of unpigmented coatings. Coatings of the resins described in Table II were prepared by dissolving polyester resin and HMMM resin (Resimene 746) in a 3/1 weight ratio in xylene, adding 0.3

wt % p-TSA (resin solids basis) and casting films on 24 Ga. Bonderite 1000 (Parker Chemical) phosphate treated, cold rolled steel panels using a #36 wire-wound drawdown bar. These coatings were baked at 133 ^{o}C for 30 min. Film thickness was about 25 um. Coatings described in Tables III (CHDM/BA) and IV (CHDM/DDA) were prepared similarly except that catalyst level was 0.5 % p-TSA, the solvent was MIBK, the HMMM resin was Resimene 747, the polyester resin/melamine was varied from 68/32 to 53/47 as indicated and the coatings were baked for 15 min at 175 ^{o}C.

Coatings from the resins in Tables III and IV: The resin (5.0 g, warmed with steam to just liquify it) and a solution of 0.042 g of p-toluenesulfonic acid in 1.0 g of 2-heptanone were mixed until homogeneous, and then HMMM (3.33 g) was added and mixed. In other experiments the resin/HMMM ratio was varied. The mixtures were cast onto 3"x9"x24 ga "Bonderite 1000" steel panels. A #36 wire drawdown bar was used, leading to dry film thicknesses of 18 to 24 µm. The coated panels were baked in a forced air oven at 133 ^{o}C for 15 min or at 150 ^{o}C for 30 min.

Melting points were recorded under N_2 using a DuPont Differential Scanning Calorimeter Model 910 using a heating rate of 1 ^{o}C/min. over a 90°-120 ^{o}C range. Samples of brassylic acid recrystallized twice from toluene were subjected to different thermal histories in the following manner. Samples were placed in test tubes under a steady N_2 flow and melted in an oil bath. One sample was cooled to room temperature at a rate of 2-3 ^{o} C/min. while the other sample was rapidly cooled to room temperature by placing in an ice water bath. DSC scans of the quenched and annealed samples were recorded.

Coating Test Procedures: Pencil hardness was measured according to ASTM D-3363. Solvent resistance was measured by double rubbing with acetone-saturated disposable laboratory wipers ("Kimwipes"). The criterion used for acceptablility was imperviousness to the solvent and absence of film surface marring. Conical mandrel bend testing was performed according to ASTM D-522. Knoop hardness was determined according to ASTM D-1474 using a Tukon hardness tester, Model MO, (Wilson Inst.). Crosshatch adhesion was measured according to ASTM D-3359. Dry film thickness was measured using a General Electric Type B film thickness gauge. Impact resistance was measured according to ASTM D-2794-84 using a Paul N. Gardner Co. Model 172 impact tester. Hydroxyl numbers were determined by the method of Fritz and Schenk.[12]

RESULTS AND DISCUSSION

Initially we synthesized model polyester resins by substituting C-9, C-12 and C-13 diacids in a formula developed for the C-6 diacid by Belote and Blount.[13] This type of resin is made by esterifying a mixture of three mols of a diol (neopentyl glycol, NPG) with two mols of a mixture of diacids, in this case an equimolar mixture of isophthalic acid (IPA) and the aliphatic acid being studied. The products are oligomeric diols; such oligomers are generally called "polyester resins" in coatings terminology although they are oligomers with $\overline{M}_n$'s ranging from 500 to a few thousand. Coatings were made by cross-linking these polyesters at 133 °C with a hexa(kis)methoxymethyl melamine (HMMM) type resin, a widely used co-reactant for polyols in industrial enamels. The chemistry of cross-linking is complex, as discussed in recent reviews.[14,15] The principle cross-linking reaction is an acid catalyzed ether interchange which liberates volatile methanol. All the coatings made in this way had excellent adhesion and flexibility, but they varied widely in hardness. The results are summarized in Table II.

Table II. SUBSTITUTION OF LONGER CHAIN DIACIDS FOR ADIPIC ACID IN MODEL POLYESTER COATINGS

Polyester Resin Composition	$\overline{M}_n$	OH#[a]	Hardness[b]
NPG/IPA/adipic acid (3/1/1)	850	120	B
NPG/IPA/azelaic acid (3/1/1)	520	207	2B
NPG/IPA/dodecanedioic acid (3/1/1)	910	113	B
NPG/IPA/brassylic acid (3/1/1)	500	215	5B

All coatings had excellent adhesion and reverse impact resistance.
[a]OH# in mg KOH/g of resin. [b]Pencil hardness (ASTM D-3363).

While variations in $\overline{M}_n$ made it impossible to draw precise conclusions from study of these resins, it was evident that coatings made from the BA polyester were far too soft to be satisfactory. Accordingly, work with this series of materials was discontinued, and we began seeking polyols which, when co-oligomerized with BA, would give stiffer structures and harder coatings. Stiffer polyester resins were prepared by oligomerizing BA with cyclohexanedimethanol (CHDM), a commercially available diol, to yield oligomers of the type:

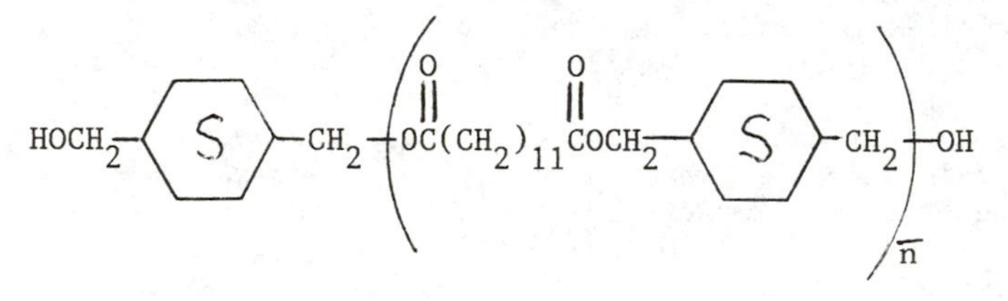

$\overline{n}$ = about 1

Analogous oligomers were made from dodecanedioic acid (DDA, n=10) and for brassylic acid (BA, n=11) for comparison studies. Resins made from the oligomeric diols derived from DDA and BA were made into similar coatings in which the diol/HMMM ratio was varied from 68/32 to 53/47 with p-TSA catalyst. The coated panels were baked at 133 °C for 15 min and, in the case of the n = 11 materials at 150 °C for 30 min. Results are provided in Tables III and IV.

Table III: PROPERTIES OF COATINGS BASED ON BRASSYLIC ACID AT VARIOUS HMMM LEVELS, BAKED AT 133 °C AND 150 °C

A: Baked at 133 °C for 15 min:

Diol/HMMM, w/w	68/32	63/37	58/42	53/47
Tukon hardness (KHN)	6	8	9	9
Pencil hardness	H	H	H	H-2H
Fwd. impact (in-lb)	100	130	140	140
Rev. impact (in-lb)	60	48	80	60
Acetone rub	>100	>100	>100	>100
X-Hatch adhesion	76%	86%	82%	86%
1/8" Mandrel bend	Pass	Pass	Pass	Pass
Dry film thinkness (μm)	19	18	17	19

B: Baked at 150 °C for 30 min:

Tukon Hardness (KHN)	6	7	9	9
Pencil Hardness	H	F-H	H	2H
Fwd. impact (in-lb)	100	100	100	120
Rev. impact (in-lb)	20	20	32	20
Acetone Rub	>100	>100	>100	>100
X-Hatch Adhesion	100%	100%	100%	100%
1/8" Mandrel bend	Pass	Pass	Fail	Pass
Dry film thickness	28	28	32	23

Hardness in the H to 2H range is adequate for many industrial coatings; thus the data in Table III show that it is possible to make coatings with adequate hardness from the BA/CHDM polyester. Further, these coatings have fair impact resistance and flexibility. Both the 133 and the 150 °C bakes are probably sufficient to drive the co-dondensation reaction of polyol with melamine resin close to completion, and some degree of self-condensation of the excess melamine resin can be expected, especially at 150 °C.[15] Self-condensation often causes hardening and embrittlement, and modest property changes in this direction are indeed observed for the coatings formulated at high melamine resin levels.

Table IV: PROPERTIES OF COATINGS BASED ON DODECANEDIOIC ACID AT VARIOUS HMMM LEVELS

A: Baked at 133 °C for 15 min:

Diol/HMMM, w/w	68/32	63/37	58/42	53/47
Tukon hardness (KHN)	6	6	5	9
Pencil hardness	H	H	H	H-2H
Fwd. impact (in-lb)	120	120	140	140
Rev. impact (in-lb)	48	60	120	40
Acetone rub	>100	>100	>100	>100
X-Hatch adhesion	91%	91%	98%	100%
1/8" Mandrel bend	Pass	Pass	Pass	Fail
Dry film thickness (μm)	23	24	18	18

Comparison of the data in Tables III-A and IV-A suggests that the BA/CHDM enamels are slightly harder and less elastic than the DDA/CHDM enamels. However, considerable more data would be needed to confirm this observation, which is mildly surprising in view of the slightly longer chain length of BA. The tests used here are imprecise, and they can be affected by variables such as location in the baking oven, film thickness and variations in test panels.

In the course of this work we noted ambiguous information in the literature about the melting point of brassylic acid;[16] it is variously reported to be 111 to 114 ^{o}C. Differential scanning calorimetry of 95% brassylic acid showed two endothermic peaks, one at about 110 and one at 112 ^{o}C, the second about twice as high as the first. In comparison, DDA showed a single peak at 130.2 ^{o}C. Recrystallization of BA raised the temperature of these twin peaks but did not substantially change their relative sizes or spacing (Table V). The heat history of the samples sometimes has modest effects, shifting the peaks by as much as 0.5 ^{o}C.

Table V: MELTING POINTS, ^{o}C, OF BRASSYLIC ACID MEASURED DIFFERENTIAL SCANNING CALORIMETRY

Brassylic acid sample	1st peak	2nd peak
As received (ca 95 % pure)	110.90	112.80
Recryst. 1x, toluene	111.45	113.15
Recryst. 2x, toluene	111.58	114.10
Recryst. 2x, toluene, annealed	112.19	114.18
Recryst. 2x, toluene, quenched	112.20	113.86
Recryst. 1x, ethyl acetate	111.89	113.42
Recryst. 2x, ethyl acetate	112.29	114.02

It was concluded that brassylic acid is inherently polymorphic, with melting points at about 112 and 114 ^{o}C. While several other diacids in the series $HOOC(CH_2)_nCOOH$ have been reported to exist in more than one crystal form,[17] polymorphism of brassylic acid has no to our knowledge been reported. A recent study demonstrated that erucic acid itself is polymorphic.[18]

CONCLUSIONS

Our results indicate that brassylic acid has potential utility as a monomer for polyester resins designed for industrial coatings. While substitution of brassylic acid for adipic or azeleic acids in conventional polyester coating resin formulations can be expected to yield coatings that are too soft to be servicable, combining brassylic acid with a relatively rigid diol yields "polyesters" that afford coatings with a reasonable balance of hardness and flexibility for commerical use. Systematic comparisons of resins made from the series of diacids, $HOOC(CH_2)_nCOOH$, ($n = 4$ to 11) will be reported subsequently. More extensive testing of the properties, especially those related to moisture absorbtion, of optimized BA and DDA based resins are planned.

ACKNOWLEDGMENTS

Financial support by the U.S. Department of Agriculture Alternative Crops Program is gratefully acknowledged. Very helpful discussions with Dr. K. D. Carlson of the U.S.D.A. Regional Laboratory at Peoria. IL are appreciated; Dr. Carlson also helped by providing the brassylic acid used in this work.

REFERENCES

[1] Various Authors. Presented at the Marketing Workshop--High Erucic Acid Development Effort, Lincoln, NE, January 26, 1989.

[2] Carlson, K. D.; Sohns, V. E.; Perkins, R. B., Jr.; Huffman, E. L. Ind. Eng. Chem., Prod. Res. Dev. 1977, 16(1), 95-101.

[3] Jaskierski, J.; Beldowicz, W.; Szczepanska, H.; Szelejewski, W. Zesz. Probl. Postepow Nauk. Roln. 1981, 211, 159-165.

[4] Quantum Chemicals Technical Bulletin 165A, Cincinnati, OH 1988.

[5] Gidanian, K.; Howard, G. J. J. Macromol. Sci., Chem. 1976, A10(7), 1391-1398.

[6] Gidanian, K.; Howard, G. J. J. Macromol. Sci., Chem. 1976, A10(7), 1399-1414.

[7] Gidanian, K.; Howard, G. J. J. Macromol. Sci., Chem. 1976, A10(7), 1415-1424.

[8] Labana, S.S.; Theodore, A.N. US Patent 4,359,554 (1982).

[9] Nieschlag, H. J.; Rothfus, J. A.; Sohns, V. E.; Perkins, R. B., Jr. Ind. Eng. Chem., Prod. Res. Dev. 1977, 16(1), 101-107.

[10] Perkins, R. B., Jr.; Roden, J. J.; Tanquary, A. C.; Wolff, I. A. Mod. Plast. 1969, 46(5), 136-142.

[11] Wedgewood, A. R.; Seferis, J. C.; Beck, T. R. J. Appl. Poly. Sci. 1985, 30(1), 111-133.

[12] Fritz, J.S.; Schenk, G.H., Anal. Chem., 1959, 31, 1808-1812.

[13] Belote, S. N.; Blount, W. W. J. Coat. Technol. 1981, 53(681), 33-37.

[14] Bauer, D. R. Prog. Org. Coat. 1986, 14, 193-218.

[15] Nakamichi, T. Prog. Org. Coat. 1986, 14, 23-43.

[16] "Dictionary of Organic Compounds", 5th ed.; Chapmann and Hall: New York, 1982; Vol. 5, pp. 5463-5464.

[17] Morgan, P., in "Encyclopedia of Chemical Technology," 3rd Ed., Vol. 7, pp 614-628.

[18] Suzuki, M.; Sato, K.; Yoshimoto, N.; Tanaka, S.; Kobayashi, M. J. Am. Oil Chem. Soc. 1988, 65(12), 1942-1947.

RECEIVED January 18, 1990

Chapter 20

Synthesis and Characterization of Chlorinated Rubber from Low-Molecular-Weight Guayule Rubber

Shelby F. Thames and Kareem Kaleem[1]

Polymer Science Department, University of Southern Mississippi, Hattiesburg, MS 39406–0076

The chlorination of low molecular weight natural rubber from Guayule (Parthenium Argentatum Grey) has been accomplished. The structure of the chlorinated product is consistent with that of chlorinated Hevea rubber. The use of Azo-bis-isobutyronitrile was as a catalyst resulted in increased chlorine content with a concomitant reduction in molecular weight, thereby allowing the preparation of lower viscosity grades of chlorinated rubber.

Recently, considerable attention has been given the development of a domestic source for natural rubber (NR).(1) Among the rubber bearing plants, Guayule (Parthenium Argentatum Grey) is known to provide good quality and high molecular weight NR, poly(cis-isoprene). The physical and mechanical properties of guayule NR are similar to that of Hevea (Malaysian rubber).(1,2) In contrast to Hevea, guayule rubber (GR) is isolated from the guayule shrub by a selective solvent extraction process and in addition to high molecular weight NR (Mn$\sim 10^6$), other isolated by-products include low molecular weight rubber, (Mn$\sim$75,000), organic soluble resins, a water soluble resin fraction and bagasse. The value and quantity of the high molecular weight rubber alone is insufficient to offset the cost of the planting, cultivating, harvesting and the extraction process; thus coproduct commercialization is required if the production of GR is to be a viable commercial domestic industry. Hence, it is necessary to explore the opportunities offered by guayule coproducts in an effort to improve the economic outlook of GR production. Thus, investigators have evaluated the organic soluble resin fraction for a variety of applications

[1]Current address: 7831 Village Drive, Apt. D, Cincinnati, OH 45242

0097–6156/90/0433–0230$06.00/0

including anti-termite activity (3) and adhesion modifiers for epoxy coatings.(4) Similarly, low molecular weight GR is of interest in that many of its properties are identical to that of high molecular weight rubber.(5) However, since mechanical properties are largely dependent on molecular weight, low molecular weight GR cannot be used for tire production and other conventional rubber products. In fact, there are reports that NR having molecular weight of 10^5 gm/mole are of no commercial value.(2) We have found, however, that low molecular weight GR rubber is an attractive feedstock for the synthesis of coatings grade chlorinated rubbers. For instance, the process of chlorination of NR typically involves two steps: mastication of high-molecular weight rubber to a reduced molecular weight and then chlorination. Since viscous solutions of NR pose manufacturing difficulties (such as gel formation and heat build up) the mastication process is necessary to reduce solution viscosity to a manageable level. The low molecular weight of naturally occurring GR is therefore an advantage as mastication is not required prior to chlorination.This communication describes, for the first time, the synthesis and characterization of coatings grade chlorinated rubber from low molecular weight GR.

Experimental

Materials and Methods.

Guayule resin was supplied by Firestone Tire and Rubber Company, Akron, Ohio. Low molecular weight GR was isolated by treating the resin with 90% ethanolic solution in the following manner. Typically, 900 gm of 90% ethanol was added to 100 gm of guayule resin in a 1500 ml beaker. The low molecular weight rubber separates as a solid and the dark green supernatant solution was decanted from the low molecular weight rubber. The raw rubber was purified by dissolving in carbon tetrachloride (5% solution) followed by precipitation with the addition of 90% ethanolic solution. The purification was monitored by ^1H NMR and ^{13}C NMR.

High purity, research grade chlorine gas purchased from Matheson Gas Products was used, as was research grade carbon tetrachloride, toluene, and ethanol.

Chlorination of Guayule Rubber.

A dilute solution of guayule rubber (5%) in CCl_4 was added to a three neck flask fitted with a water condenser, gas dispersion tube and an adaptor. The gas dispersion tube was connected via Teflon tubes to a chlorine cylinder through two gas traps. The exit port of the condenser was connected via Teflon tubing to ice-cooled traps containing 2 N sodium hydroxide solution. The reaction flask was immersed in an oil bath for temperature control. Nitrogen was purged through the system for 10 minutes to insure complete removal of oxygen. A blanket of nitrogen was maintained over the reaction throughout the chlorination process. The solution was allowed to reflux at

79^{0} C with constant stirring via a magnetic stirrer. A slight excess of chlorine (10% mole excess) gas was bubbled through the solution with the liberated hydrogen chloride being trapped in the sodium hydroxide solution. The chlorinated rubber product was isolated as a white precipitate with addition of 90% ethanol.

FTIR spectra were recorded on a Mattson, Polaris spectrometer. Films were cast by the evaporation of a toluene solution of chlorinated rubber. The films were dried in a vacuum oven to insure removal of all solvent.

^{1}H NMR and ^{13}C NMR were obtained from a 300 MHz Brucker fourier transform spectrometer. The solutions (20% w/w) were prepared by dissolving the chlorinated rubber (CR) in $CDCl_3$ and C_6D_6 for the ^{13}C NMR spectra analysis with tetramethylsilane as an internal standard.

Gel permeation chromatograms were generated from a Waters Associates, Inc. GPC equipped with a refractive index detector. The following operating conditions were employed: mobile phase, THF; flow rate; 1 ml/min., columns 10^{6}, 10^{4}, 500, 100 A^{0}. Sample concentrations were prepared at 0.2% (w/w); a 100 microliter aliquot was used for molecular weight analysis. Standard polystyrene samples (Polymer Laboratories, Inc.) were used to create a calibration curve.

Thermal analyses were performed on a Dupont Model 9900 thermal analyzer under nitrogen atmosphere. A heating rate of $10^{0}C$/min. was employed for glass transition temperature (T_g) determinations.

Duplicated elemental analysis of CR samples were carried out by the MHW Laboratory, Phoenix, Az.

Results and Discussion.

The synthesis of Chlorinated Rubber (CR) from low molecular weight Guayule Rubber (GR) is reported. Other investigators (5) have performed similar studies on the formation of chlorinated rubber from Hevea rubber. They found the empirical formula of the chlorinated product to be $C_5H_8Cl_{3.5}$, indicating that chlorination involves more than one isoprene unit. Their products were soluble in organic solvents thereby lending additional support to the thesis that cyclization rather than crosslinking is the predominant reaction.(6) The chlorine content was found to be 65% and is consistent with a combination of substitution and addition followed by cyclization.(5) The concept of cyclized units along the polymer backbone have also been supported by FTIR and ^{13}C NMR analysis.(7)

The physical and mechanical properties of CR are often determined by its chlorine content and molecular weight. For instance, lower molecular weight CR is used for printing inks while higher molecular weight CR's are required for coatings applications. In our work we found that the chlorination of low molecular weight GR principally yielded coating grade CR.

Since GR is contaminated with wax and other hydrocarbons it must be purified before chlorination. In our case purification was monitored by proton NMR spectroscopy (Figure 1). The peak assignments representing satisfactory purification are as follows:

1.67 ppm (Cis double bond methyl protons)
2.00 ppm (methylene protons)
5.12 ppm (vinyl proton)

The small peak at 1.25 ppm is representative of an impurity(ies) that was not removed during the extraction process. Further purification from an ethanolic solution reduces the impurity(ies) to an insignificant level. These peak positions are in excellent agreement with literature values for natural rubber (Hevea).

The results of various experimental chlorinating conditions are summarized in Table I.

TABLE I. Experimental Conditions Used In The Synthesis Of Chlorinated Rubber

Batch No.	Amount of Guayule Rubber Used, g	Amount of Chlorinated Rubber Isolated, g	Cl%
1[a]	3.200	6.00	59.81
2[b]	5.00	11.80	60.00
3[c]	5.00	13.14	63.67

[a] Solvent; CCl_4, 200 ml
[b] Solvent; CCl_4, 100 ml
[c] Solvent; CCl_4, 100 ml and 37.5 mg of AIBN

The radical initiator, Azo-bis-isobutyronitrile (AIBN) was used in catalytic amounts to determine its efficacy, if any, in affecting changes in the chlorine content of the products. Recent studies of rubber chlorination have reported that AIBN can be used to increase the rate and amount of chlorination. In our case, use of purified and partially purified GR provided essentially identical levels of chlorination at 60%. The lower yield obtained (6.00 gm) in the first instance is attributed to the presence of impurities in the starting materials. However, purified GR (Figure 2), provided for a significant increase in product yield. The use of AIBN increased the chlorine content to a value approaching the theoretical value for fully chlorinated rubber (64.7%). While AIBN was instrumental in increasing the degree of chlorination its mechanistic role is not fully understood.

The FTIR spectra of Figure 3 comparing a commercial grade rubber (Alloprene CR-20 from ICI) with guayule CR shows the two materials to be essentially identical. Absorption bands characteristic of CR appear near 780 cm^{-1} and 736 cm^{-1} and represent the secondary C-Cl and the CH_2 rocking

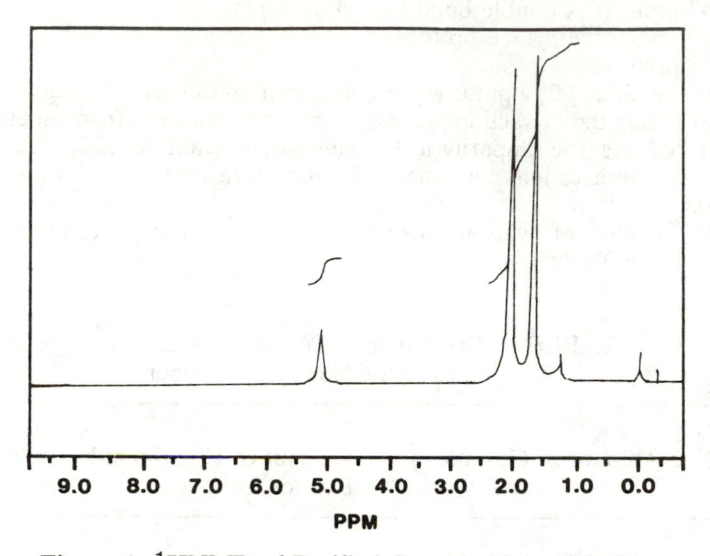

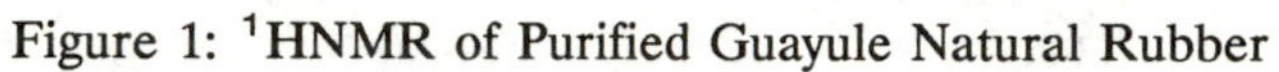

Figure 1: ^{1}HNMR of Purified Guayule Natural Rubber

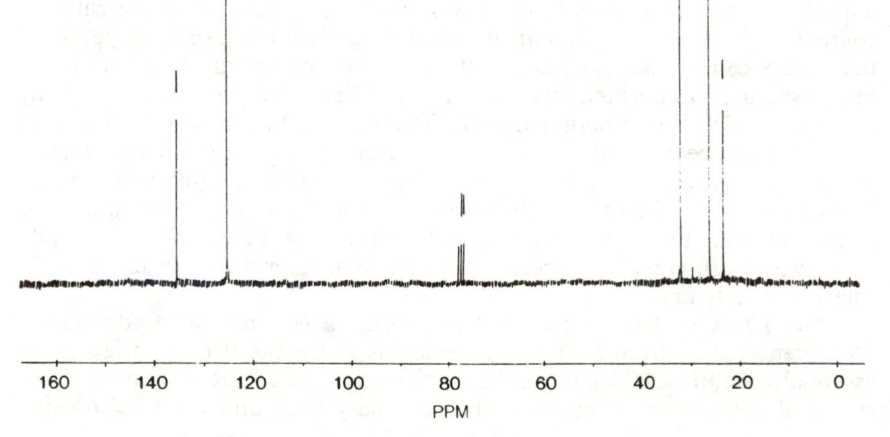

Figure 2: ^{13}C NMR Spectrum of Guayule Rubber in $CDCL_3$

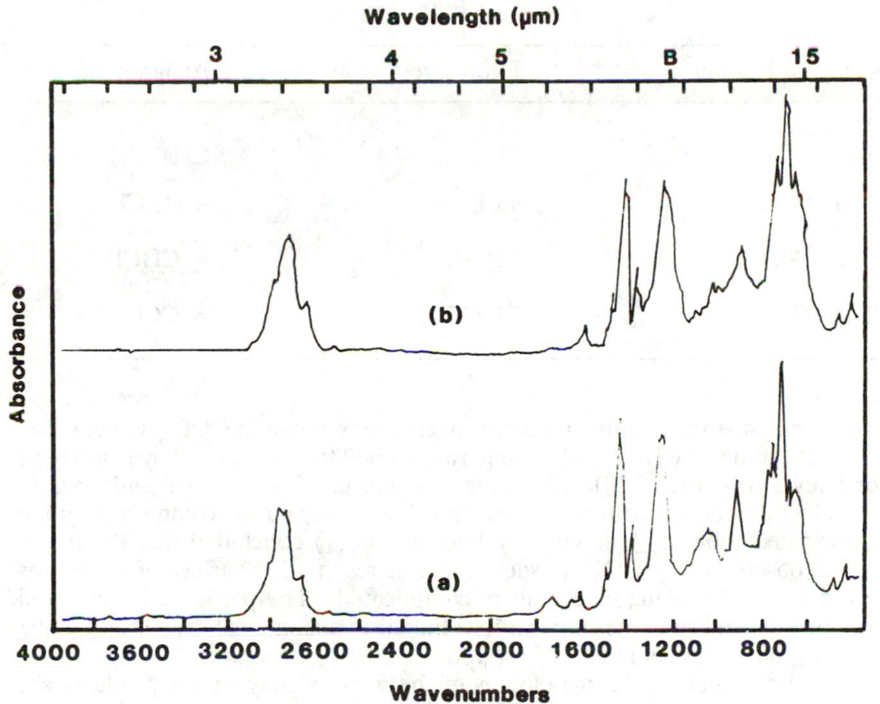

Figure 3: FTIR Spectrum of Chlorinated Rubber (a) Guayule CR (b) Commercial CR

frequency, (7) respectively, while the absorbance bands at 2939 cm^{-1}, 1440 cm^{-1} and 1260 cm^{-1} are due to the C-H stretching and bending absorptions, respectively. The weak absorbance near 1630 cm^{-1} indicates residual unsaturation.

The major ^{13}C NMR chemical shifts of guayule CR are shown in Table II; the spectra are displayed in Figure 4.

TABLE II. Characteristics of ^{13}C NMR Spectra of Guayule CR And Commercial Grade CR

Chemical Shifts (PPM)		
Guayule Rubber	Literature	Assignment
21.5, 28.7, 34.7, 37.6	21.5, 28, 34.7, 37.4, 21.5, 38, 37.4	CH_2, CH_3
45.4, 48.01	45.4, 48	$-CH_2Cl$
62.25, 64.37	62 - 64	= CHCl
75.1, 77.18	74 - 77	= CCl

To identify the unsaturated carbons, spectra were taken in $CDCl_3$ solvent, and peak assignments were based on spectra obtained for chlorinated hydrocarbons of known structure.(8-10) The values for guayule CR chemical shifts are in excellent agreement with the literature values reported for commercial grade chlorinated rubber.(7) Makani and coworkers (7) concluded that the broad peaks observed in ^{13}C NMR spectra are a result of a variety of structures present in CR making its chemistry complicated. The peaks at 74 ppm and 77 ppm are due to quaternary carbons linked to a single chlorine atom. The CHCl group appears at 63 and 64 ppm.

Gel permeation chromatograms of the various guayule CR products are shown in Figures 5 and 6, with a chromatogram of commercial grade Alloprene CR-20 (CR-20) chlorinated rubber being included for comparative purposes. The guayule CR and commercial grade CR-20 (10) are of the same molecular weight range, and molecular weight distributions (Figure 5) with the exception that the commercial CR-20 exhibited a low molecular weight shoulder. Since CR-20 is used primarily in the coatings industry and especially in traffic paints and marine coating these data suggest a similar use for guayule CR. Guayule CR seems particularly suited for this purpose as it forms continuous, transparent films from toluene solutions (20% solution). In contrast, guayule CR, obtained in the presence of AIBN, was found to have a lower molecular weight than that prepared without AIBN. Its molecular weight corresponds to that of commercial grade Alloprene CR-5 (10) (CR-5)

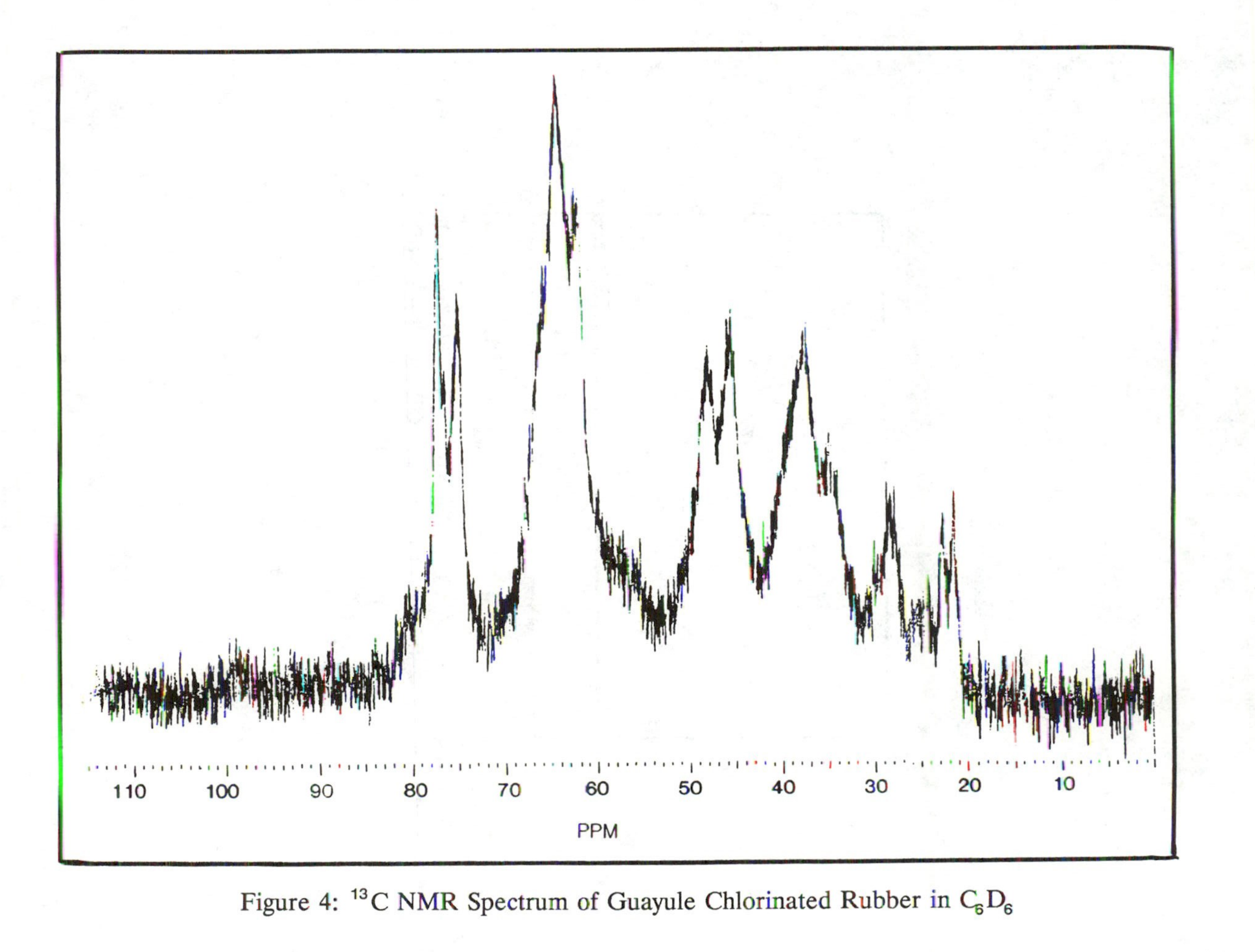

Figure 4: ^{13}C NMR Spectrum of Guayule Chlorinated Rubber in C_6D_6

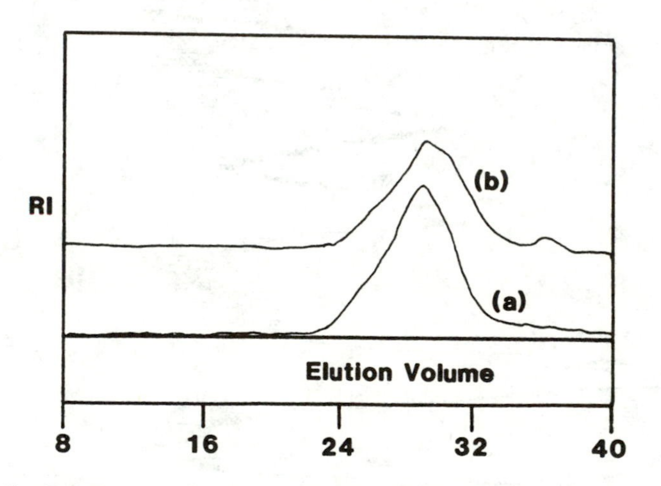

Figure 5: Gel Permeation Chromatograms of Chlorinated Rubber (a) Guayule CR, Cl ~ 60% (b) Commercial CR-20, Cl ~ 64-65%

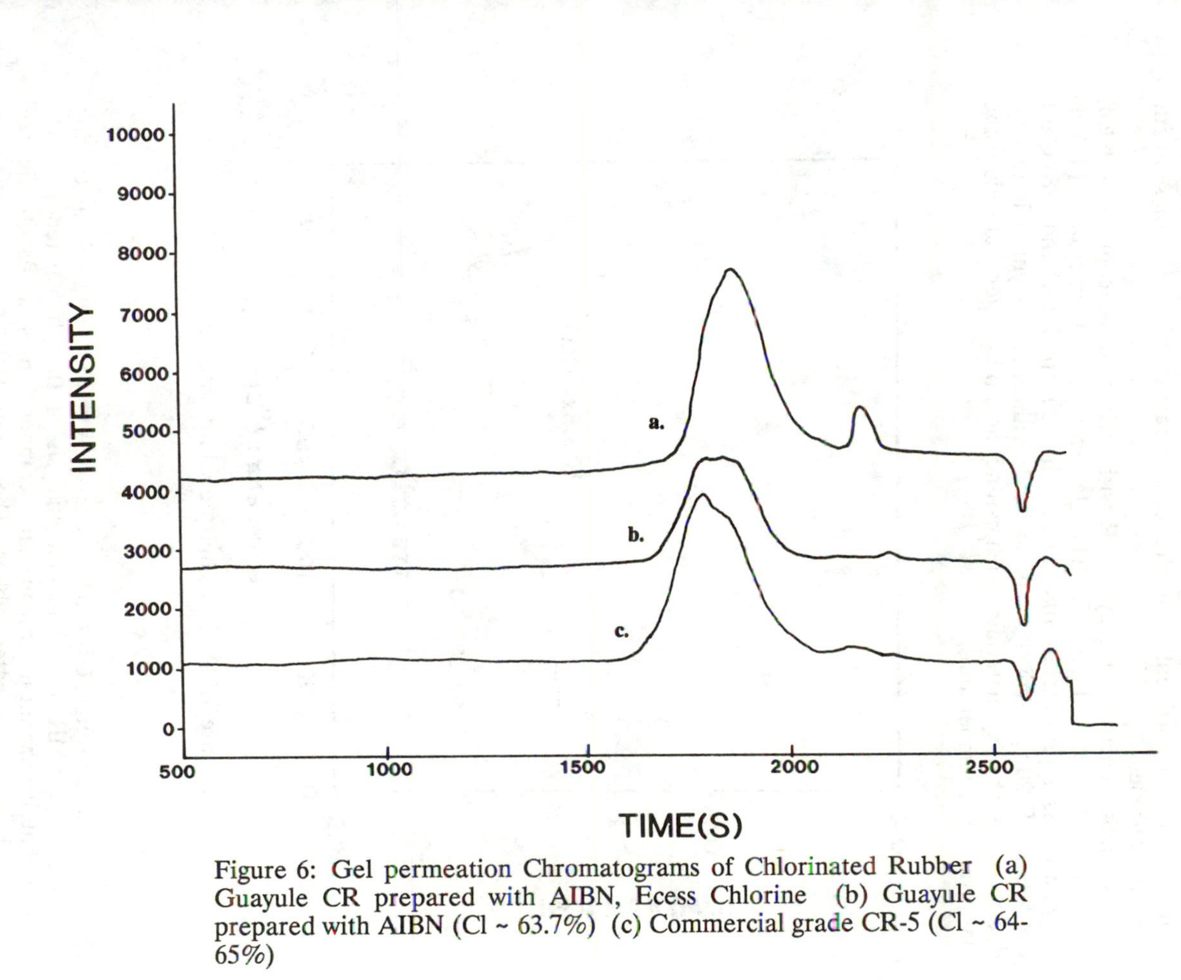

Figure 6: Gel permeation Chromatograms of Chlorinated Rubber (a) Guayule CR prepared with AIBN, Ecess Chlorine (b) Guayule CR prepared with AIBN (Cl ~ 63.7%) (c) Commercial grade CR-5 (Cl ~ 64-65%)

(Figure 6), which is used primarily in the printing ink industry. It is clear therefore that the use of AIBN results in a slight increase in the chlorine content yet with a concomitant and significant reduction in molecular weight. Films prepared from CR-5 and AIBN derived guayule CR were found to be brittle and difficult to remove from the steel panels. These results indicate that guayule low molecular weight NR can be successfully chlorinated in a fashion similar if not identical to that of Hevea natural rubber or synthetic poly *cis*-isoprene.

The DSC spectrum of guayule CR with 60% chlorine content (Table I, batch 1), is shown in Figure 7. The glass transition temperature (T_g) of guayule CR is approximately 108°C, while the T_g's are 126°C and 128° C for CR-5 and CR-20 respectively. The lower T_g values for guayule CR may be due to traces of waxy materials which can act as a plasticizer and reduce the glass transition temperature.

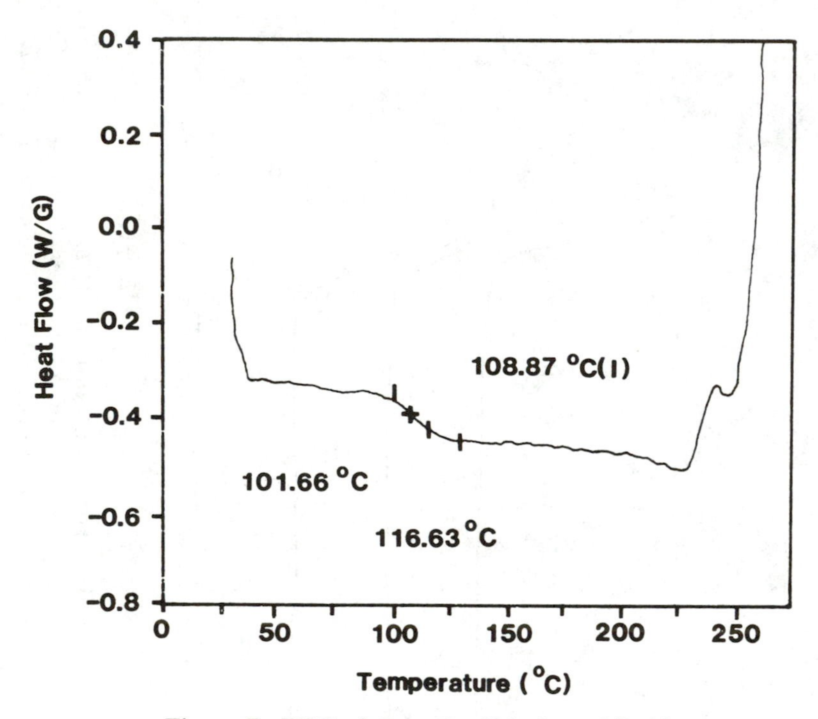

Figure 7: DSC of Guayule Chlorinated Rubber

Conclusions

The low molecular weight GR can be successfully chlorinated to obtain coating grade CR. ^{13}C NMR and elemental analysis of the chlorinated products confirm that its chemical structure and composition is similar to that of commercial grade chlorinated rubber. The use of AIBN during chlorination significantly reduces the molecular weight and in turn allows the preparation of lower viscosity grade chlorinated rubbers.

Acknowledgments

Financial support from the United States Department of Agriculture Grant No. 89-COOP-1-4218 is gratefully acknowledged. We are particularly thankful to Mr. George Donovan and Dr. s Richard Wheaton and Daniel Kugler for their encouragement and support.

Literature Cited

1. Hammond, B. L.; Polhamus, L. G. Research on Guayule (Parthenium Argentatum Grey): 1942-1959, U.S.D.A, Agr. Res. Ser. Tech. Bull. No. 1327, Washington, D. C., 1965, 143.
2. Backhaus, R. A.; Nakayama, F. S. Rubber Chemistry and Technology, 1986, 61, 78-85.
3. Bultman, J. D.; Gilbertson, R. L.; Amburgey, T. L.; Bailey, C. A. "Guayule resin - a new wood preservative." In Annual Meeting of the Wood Preservers' Association, Minneapolis, MN, (1988).
4. Thames, S. F.; Kaleem, K. Communication to "Biomass," 1989.
5. Bloomfield, G. F. J. Chem. Soc., 1943, 289.
6. Rubber Chemistry, J. A. Brydson, Ed.: Applied Science publisher Ltd, London, (1978), pp 178-179.
7. Makani, S.; Brigodiot, M.; Marechal, E.; Dawans, F.; Durand, J. P., Journal of Applied Polymer Science, 1984, 29, 4081-4089.
8. Torosyan, K. A., Voskanyan, E. S., Mkryan, G. M., and Karapetyan, N. G. arm. Khim. Zh., 1973, 26, 413. (C.A. 79, 93181d); 871 (1973) (C.A. 80, 97016x); Hlevca, B., Gheorghe, G. Rom. Pat. 67,006 (C.A. 95, 170767j (1981).
9. Velichko, F. K.; Chukovskaya, E. G.; Dostovalova, V. Il; Kuzmina, N. A.; Freidlina, R. Kh. Org. Mag. Res., 1975, 7.
10. Commercial Grade CR-20, CR-5 were obtained from Polyvinyl Chemical Co. (ICI Resins US), Wilmington, MA.

RECEIVED January 18, 1990

Chapter 21

Chemical Modification of Lignocellulosic Fibers To Produce High-Performance Composites

Roger M. Rowell

Forest Products Laboratory, Forest Service, U.S. Department of Agriculture, Madison, WI 53705–2398

The performance properties of composites made from wood and other lignocellulosic materials can be greatly improved by changing the basic chemistry of the cell wall polymers. This paper reviews published research on reducing dimensional instability and susceptibility to degradation by biological organisms, heat, and ultraviolet radiation to produce high-performance lignocellulosic composites based on acetylation of the furnish before product formation.

Wood and other lignocellulosic materials are three-dimensional, polymeric composites made up primarily of cellulose, hemicelluloses, and lignin. These polymers make up the cell wall and are responsible for most of the physical and chemical properties of these materials. Wood and other lignocellulosic materials have been used as engineering materials because they are economical, renewable, and strong and have low processing-energy requirements. They have, however, several undesirable properties, such as dimensional instability due to moisture sorption with varying moisture contents, biodegradability, flammability, and degradability by ultraviolet light, acids, and bases. These properties are all the result of chemical reactions involving degradative environmental agents. Because these types of degradation are chemical in nature, it should be possible to eliminate them or decrease their rate by modifying the basic chemistry of the lignocellulosic cell wall polymers.

Most research on chemical modification of lignocellulosic materials has focused on improving either the dimensional stability or the biological resistance of wood. This paper reviews the research on these properties for wood and other lignocellulosic composites and describes opportunities to improve fire retardancy and resistance to ultraviolet degradation.

Reaction Chemistry

Reactive organic chemicals can be bonded to cell wall hydroxyl groups on cellulose, hemicelluloses, and lignin. Much of our research has involved simple epoxides (1-3) and isocyanates (4), but most of our recent effort has focused on acetylation. Acetylation studies have been done using fiberboards (5,6), hardboards (7-11), particleboards (12-20), and flakeboards (21-23), using vapor phase acetylation (8,24-26), liquid phase acetylation (1,27), or reaction with ketene (28).

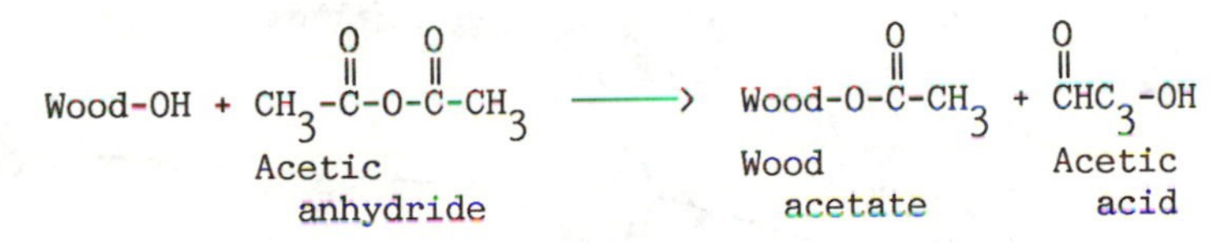

If this acetylation system does not include a strong catalyst or cosolvent, only the easily accessible hydroxyl groups will be acetylated. We developed an acetylation system that uses no strong catalyst or cosolvent and probably acetylates only easily accessible hydroxyl groups (27).

Several lignocellulosic fibers were acetylated using this procedure; reaction times from 15 min to 4 h were used on Southern Pine, aspen, bamboo (29), bagasse (30), jute (31), pennywort, and water hyacinth (32). All the lignocellulosic materials used were easily acetylated. Acetyl content resulting from acetylation plotted as a function of time shows all data points fitting a common curve (Fig. 1). A maximum weight percent gain (WPG) of about 20 was reached in a 2-h reaction time, and an additional 2 h increased the weight gain only by 2 to 3 percent. Without a strong catalyst, acetylation using acetic anhydride alone levels off at approximately 20 WPG for the softwoods, hardwoods, grasses, and water plants.

Moisture Sorption

Sorption of moisture is due mainly to hydrogen bonding of water molecules to the hydroxyl groups in the cell wall polymers. By replacing some of the hydroxyl groups on the cell wall polymers with acetyl groups, the hydroscopicity of the lignocellulosic material is reduced.

Table I shows the equilibrium moisture content (EMC) of several lignocellulosic materials at 65 percent relative humidity (RH). Reduction in EMC at 65 percent RH of acetylated fiber referenced to unacetylated fiber plotted as a function of the bonded acetyl content is a straight line (Fig. 2). Although the points shown in Figure 2 come from many different lignocellulosic materials, they

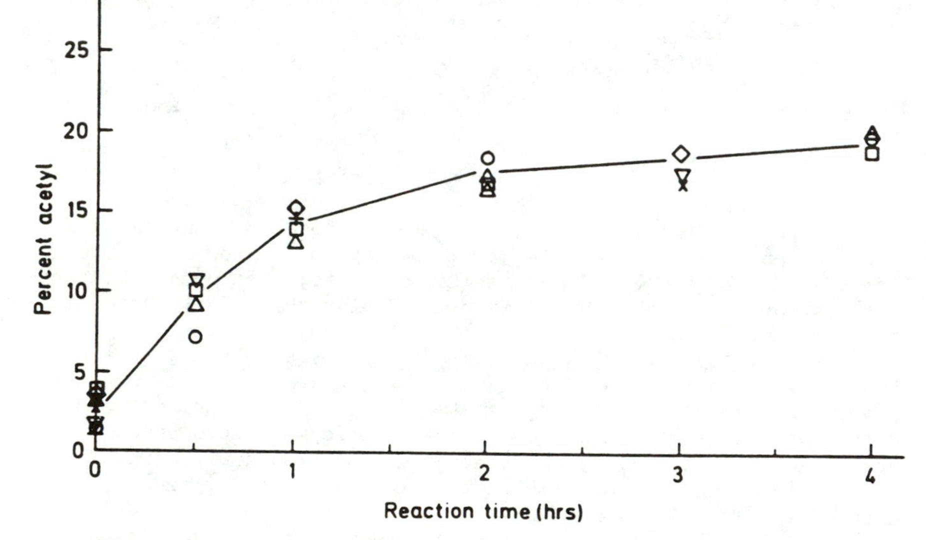

Figure 1--Rate of acetylation of various lignocellulosic materials. ○, Southern Pine; □, aspen; △, bamboo; ◇, bagasse; X, jute; +, pennywort; ▽, water hyacinth.

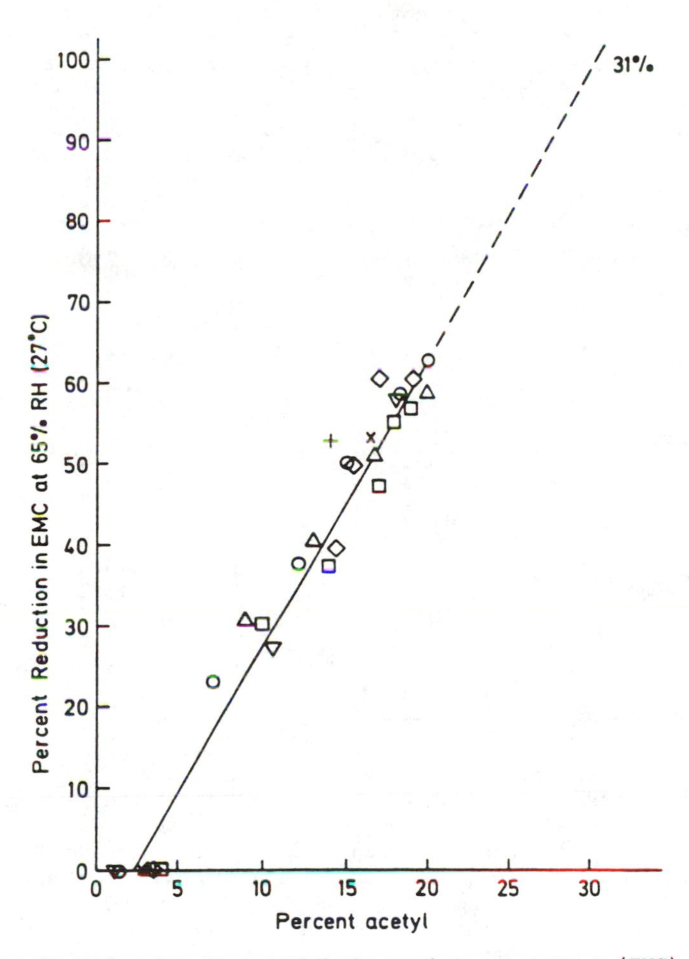

Figure 2--Reduction in equilibrium moisture content (EMC) as a function of bonded acetyl content for various acetylated lignocellulosic materials. ○, Southern Pine; □, aspen; △, bamboo; ◇, bagasse; X, jute; +, pennywort; ▽, water hyacinth.

Table I. Equilibrium Moisture Content (EMC) of Various Acetylated Lignocellulosic Materials (65 Percent RH, 27°C)

Material	Reaction weight gain (percent)	Acetyl content (percent)	EMC (percent)
Southern Pine	0	1.4	12.0
	6.0	7.0	9.2
	14.8	15.1	6.0
	21.1	20.1	4.3
Aspen	0	3.9	11.1
	7.3	10.1	7.8
	14.2	16.9	5.9
	17.9	19.1	4.8
Bamboo	0	3.2	8.9
	10.8	13.1	5.3
	14.1	16.6	4.4
	17.0	20.2	3.7
Bagasse	0	3.4	8.8
	9.4	14.4	5.3
	12.2	15.3	4.4
	17.6	19.0	3.4
Jute	0	3.0	9.9
	15.6	16.5	4.8
Pennywort	0	1.3	18.3
	10.1	14.0	8.6
Water hyacinth	0	1.2	1.7
	8.3	10.8	1.2
	18.6	17.8	0.7

all fit a common line. A maximum reduction in EMC is achieved at about 20 percent bonded acetyl. Extrapolation of the plot to 100 percent reduction in EMC would occur at about 30 percent bonded acetyl. Because the acetate group is larger than the water molecule, not all hygroscopic hydrogen-bonding sites are covered.

The fact that EMC reduction as a function of acetyl content is the same for many different lignocellulosic materials indicates that reducing moisture sorption and, therefore, achieving cell wall stability are controlled by a common factor. The lignin, hemicellulose, and cellulose contents of all the materials plotted in Figure 2 are different (Table II). Earlier results showed that the bonded acetate was mainly in the lignin and hemicelluloses (33) and that isolated wood cellulose does not react with uncatalyzed acetic anhydride (34).

Acetylation may be controlling the moisture sensitivity due to the lignin and hemicellulose polymers in the cell wall but not reducing the sorption of moisture in the cellulose polymer because

Table II. Chemical Composition of Some Lignocellulosic Materials

Material	Composition (percent)										
			Sugars								
	Lignin[a]	Cellulose	Glucose	Xylose	Galactose	Arabinose	Mannose	Uronic acids	Extractives[b]	Ash	Acetyl
Southern Pine	26.6	45	49.0	5.4	2.4	--	19.2	1.6	6.9	0.3	1.4
Aspen	18.5	49	53.3	18.5	1.0	--	1.4	2.5	4.3	0.4	4.1
Bamboo	24.2	42	52.0	21.7	--	0.8	--	0.8	11.0	0.4	3.2
Bagasse	21.1	45	47.4	27.6	--	1.7	--	1.0	8.4	1.4	3.4
Jute	13.7	58	63.8	13.1	1.2	--	0.6	3.2	3.7	1.0	3.0
Pennywort	10.3	49	39.0	3.5	2.8	0.8	2.9	9.8	38.3	11.2	1.3
Water hyacinth	8.5	58	37.2	8.7	5.0	11.4	1.4	7.1	20.5	7.9	1.2

[a] Klason.

[b] 6 h of reflux, benzene ethanol.

1. these materials vary widely in their lignin, hemicellulose, and cellulose content,
2. acetate is found mainly in the lignin and hemicellulose polymer, and
3. isolated cellulose does not acetylate by the procedure used.

Dimensional Stability

Dimensional instability, especially in the thickness direction, is a greater problem in lignocellulosic composites than in solid wood because composites undergo not only normal swelling (reversible swelling) but also swelling caused by the release of residual compressive stresses imparted to the board during the composite pressing process (irreversible swelling). Water sorption causes both reversible and irreversible swelling, with some of the reversible shrinkage occurring when the board dries. Dimensional instability of lignocellulosic composites has been the major reason for their restricted use.

We are in the process of producing fiberboards from various types of acetylated lignocellulosic fibers. Most of our research has been on pine or aspen particleboards or flakeboards, so the data presented here on dimensional stability and biological resistance come mainly from these types of boards.

The rate of swelling in liquid water of an aspen flakeboard made from acetylated flakes and phenolic resin (27) is shown in Figure 3. During the first 60 min, control boards swelled 55 percent in thickness, while the board made from flakes acetylated to 17.9 WPG swelled less than 2 percent. During 5 days of water soaking, the control boards swelled more than 66 percent, while the 17.9-WPG board swelled about 6 percent.

Control boards made from bamboo particles using a phenolic adhesive swelled about 10 percent after 1 h, 15 percent after 6 h, and 20 percent after 5 days. Particleboards made from acetylated bamboo particles swelled about 2 percent after 1 h and only 3 percent after 5 days (35).

Thickness changes in a six-cycle water-soaking/ovendrying test for an acetylated aspen flakeboard are shown in Figure 4 (27). Control boards swelled more than 70 percent in thickness during the six cycles, compared with less than 15 percent for a board made from acetylated flakes. Acetylation greatly reduced both irreversible and reversible swelling.

In a similar five-cycle water-soaking/ovendrying test on bamboo particleboards, control boards swelled more than 30 percent, while boards made from acetylated particles swelled about 10 percent.

Figure 5 (27) shows changes in thickness of aspen flakeboards made from control and acetylated flakes using a phenolic adhesive at different relative humidities. After four cycles of 30 to 90 percent RH, control boards swelled 30 percent in thickness, while acetylated boards at 17.9 WPG swelled about 5 percent.

The results of both liquid water and water vapor tests show that acetylation of lignocellulosic materials greatly improves dimensional stability of composites made from these materials.

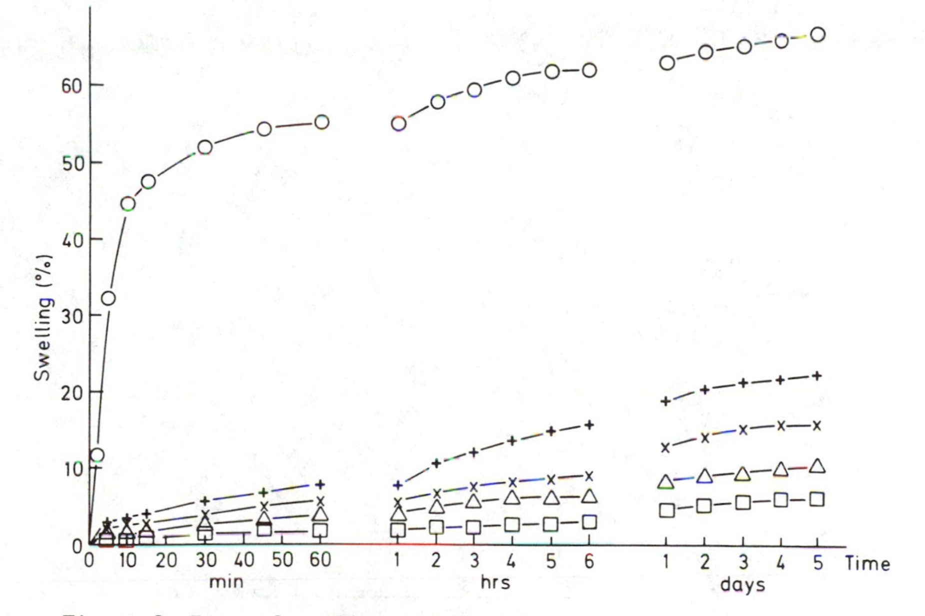

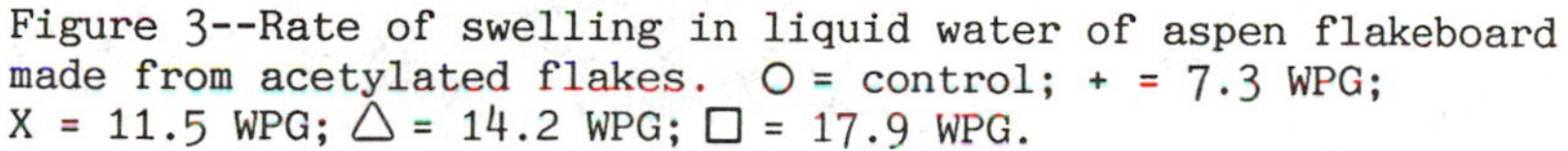
Figure 3--Rate of swelling in liquid water of aspen flakeboard made from acetylated flakes. ○ = control; + = 7.3 WPG; X = 11.5 WPG; △ = 14.2 WPG; □ = 17.9 WPG.

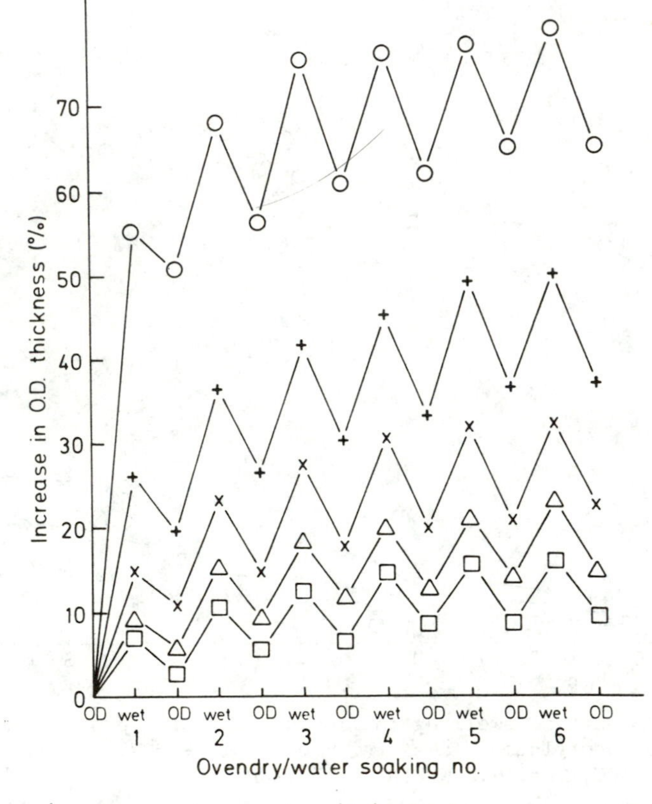

Figure 4--Changes in ovendry (OD) thickness in repeated water-soaking test of aspen flakeboard made from acetylated flakes. ○ = control; + = 7.3 WPG; X = 11.5 WPG; △ = 14.2 WPG; □ = 17.9 WPG.

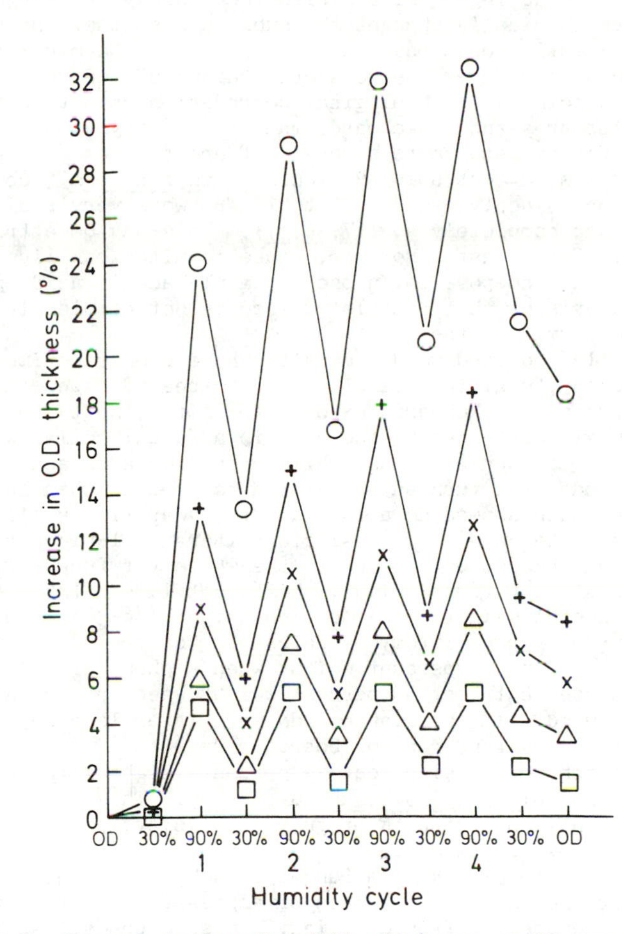

Figure 5--Changes in ovendry (OD) thickness at 30 percent and 90 percent relative humidity of aspen flakeboard made from acetylated flakes (27°C). ○ = control; + = 7.3 WPG; X = 11.5 WPG; △ = 14.2 WPG; □ = 17.9 WPG.

Biological Resistance

Chemical modification of wood composite furnish for biological resistance is based on the theory that the potentially degrading enzymes must directly contact the substrate and that the substrate must have a specific chemical configuration and molecular conformation. Reacting chemicals with the hydroxyl groups on cell wall polymers chemically changes the substrate so that the highly selective enzymatic reactions cannot take place. Chemical modification also reduces the moisture content of the cell wall polymers to a point where biological degradation cannot take place.

Particleboards and flakeboards made from acetylated flakes have been tested for resistance to several different types of organisms. In a 4-week termite test using *Reticulitermes flavipes* (subterranean termites), boards acetylated at 16 to 17 WPG were very resistant to attack, but not completely so (17,36,37). This may be attributed to the severity of the test. However, since termites can live on acetic acid and decompose cellulose to mainly acetic acid, perhaps it is not surprising that acetylated wood is not completely resistant to termite attack.

Chemically modified wood composites have been experimentally exposed to decay fungi in several ways. Untreated aspen and pine particleboards and flakeboards exposed to white-, soft-, and brown-rot fungi and tunneling bacteria in a fungal cellar were destroyed in less than 6 months, while particleboards and flakeboards made from furnish acetylated to greater than 16 percent acetyl weight gain showed no attack after 1 year (Table III) (36-38). In a standard 12-week single-culture soil-block test, control aspen flakeboards exposed to the white-rot fungus *Trametes versicolor* lost 34 percent weight, while acetylated flakeboards (17 percent acetyl weight gain) lost no weight (36-38). In exposure to the brown-rot fungus *Tyromyces palustris*, control aspen flakeboards lost only 2 percent weight when a phenol-formaldehyde adhesive was used but lost 30 percent weight when an isocyanate adhesive was used. If the flakeboards were water leached before the soil-block test was run, control boards made with phenol-formaldehyde adhesive lost 44 percent weight with the brown-rot fungus *Gloeophyllum trabeum* (34-36). This shows that the adhesive can influence results of fungal toxicity, especially in a small, closed test container.

Weight loss resulting from fungal attack is the method most used to determine the effectiveness of a preservative treatment to protect wood composites from decaying. In some cases, especially for brown-rot fungal attack, strength loss may be a more important measure of attack since large strength losses are known to occur in solid wood at very low wood weight loss (39). A dynamic bending-creep test has been developed to determine strength losses when wood composites are exposed to a brown- or white-rot fungus (40).

Using this bending-creep test on aspen flakeboards, control boards made with phenol-formaldehyde adhesive failed in an average of 71 days with *T. palustris* and 212 days with *T. versicolor* (41). At failure, weight losses averaged 7.8 percent for *T. palustris* and 31.6 percent for *T. versicolor*. Isocyanate-bonded control

Table III. Fungal Cellar Tests on Aspen Flakeboards Made from Control and Acetylated Flakes[a,b]

Acetyl WPG[c]	Rating at intervals[d]					
	2 mo	3 mo	4 mo	5 mo	6 mo	12 mo
0	S/2	S/3	S/3	S/3	S/4	--
7.3	S/0	S/1	S/1	S/2	S/3	S/4
11.5	0	0	S/0	S/1	S/2	S/3
13.6	0	0	0	0	S/0	S/1
16.3	0	0	0	0	0	0
17.9	0	0	0	0	0	0

[a]Nonsterile soil containing brown-, white-, and soft-rot fungi and tunneling bacteria.

[b]Flakeboards bonded with 5 percent phenol-formaldehyde adhesive.

[c]Weight percent gain.

[d]Rating system: 0, no attack; 1, slight attack; 2, moderate attack; 3, heavy attack; 4, destroyed; S, swollen.

flakeboards failed in an average of 20 days with T. palustris and 118 days with T. versicolor, with average weight loss at failure of 5.5 percent and 34.4 percent, respectively (41). Very little or no weight loss occurred with either fungi in flakeboards made using phenol-formaldehyde or isocyanate adhesive with acetylated flakes. None of these specimens failed during the test period.

Deflection-time curves for flakeboards are shown in Figure 6. The curves show an initial increase of deflection for both control and acetylated flakeboards, then a stable zone, and finally, for control boards, a steep slope to failure. The study showed that less than 5 mm of deflection was caused by creep due to moisture alone (41).

Mycelium fully covered the surfaces of isocyanate-bonded control flakeboards within 1 week, but mycelial development was significantly slower in phenol-formaldehyde-bonded control flakeboards. Both isocyanate- and phenol-formaldehyde-bonded acetylated flakeboards showed surface mycelium colonization during the test time, but the fungus did not attack the acetylated flakes, so little strength was lost.

In similar bending-creep tests, both control and acetylated pine particleboards made using melamine-urea-formaldehyde adhesive failed because T. palustris attacked the adhesive in the glueline (42). Mycelium invaded the inner part of all boards, colonizing in both wood and glueline in control boards but only in the glueline in acetylated boards.

After a 16-week exposure to T. palustris, the internal bond strength of control aspen flakeboards made with phenol-formaldehyde

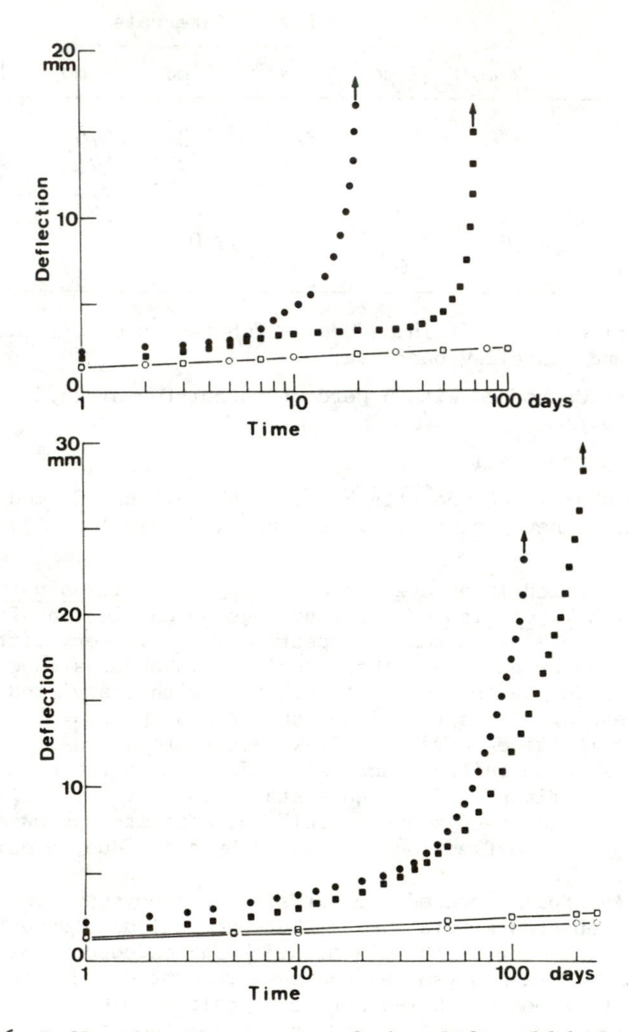

Figure 6--Deflection-time curves of phenol-formaldehyde- (PF-) and isocyanate- (IS-) bonded flakeboards in bending-creep tests under progressive fungal attack by T. palustris (upper) and T. versicolor (lower). ■ = PF control; □ = PF acetylated; ● = IS control; ○ = IS acetylated.

adhesive was reduced more than 90 percent and that of flakeboards made with isocyanate adhesive was reduced 85 percent (43). After 6 months of exposure in moist unsterile soil, the same control flakeboards made with phenol-formaldehyde adhesive lost 65 percent of their internal bond strength and those made with isocyanate adhesive lost 64 percent internal bond strength. Failure was due mainly to great strength reductions in the wood caused by fungal attack. Acetylated aspen flakeboards lost much less internal bond strength during the 16-week exposure to *T. palustris* or 6-month soil burial. The isocyanate adhesive was somewhat more resistant to fungal attack than the phenol-formaldehyde adhesive. In the case of acetylated composites, loss in internal bond strength was mainly due to fungal attack in the adhesive and moisture, which caused a small amount of swelling in the boards.

Acetylated pine flakeboards have also been shown to be resistant to attack in a marine environment (44). Control flakeboards were destroyed in 6 months to 1 year, mainly because of attack by *Limnoria tripunctata*, while acetylated boards showed no attack after 2 years.

All laboratory tests for biological resistance conducted to this point show that acetylation is an effective means of reducing or eliminating attack by soft-, white-, and brown-rot fungi, tunneling bacteria, and subterranean termites. Tests are presently underway on several lignocellulosic composites in outdoor environments.

Other Properties

Other properties of lignocellulosic composites can be improved by changing the basic chemistry of the furnish (45). Acetylation has been shown to improve ultraviolet resistance of flakeboards. Table IV shows that acetylation greatly reduced weight loss and errosion rate due to loss of surface fiber from aspen flakeboards. The rate of errosion for boards made from acetylated flakes is half that of control boards. In outdoor tests, flakeboards made from acetylated pine flakes were still light yellow in color when control boards had turned dark orange to light gray.

Table IV. Weight Loss and Rate of Erosion for Acetylated Aspen Flakeboard After Accelerated Weathering

Sample	Weight loss (percent) after weathering period				Erosion rate at 2,400 h (μm/h)
	140 h	500 h	800 h	1,700 h	
Control	0.9	2.0	2.8	3.7	0.12
Acetylated	0.0	0.5	1.2	2.4	0.06

Acetylation does not change the fire properties of lignocellulosic materials. In thermogravametric analysis, acetylated and control pine sawdust pyrolyzed at the same temperature and rate (46). The heat of combustion and rate of oxygen consumption were also the same for control and acetylated specimens, showing that the acetyl group added to the cell wall has approximately the same carbon, hydrogen, and oxygen content as the cell wall polymers. Reactive fire retardants can be bonded to the cell wall hydroxyl groups. The effect would be an improvement in dimensional stability, biological resistance, and fire retardancy.

Conclusions

Dimensional instability and susceptibility to degradation by biological organisms, heat, and ultraviolet radiation can be greatly reduced by modification of lignocellulosic cell wall polymers. These modifications result in a furnish that can be converted into composites of any desired shape, density, and size and provide an opportunity for a manufacturer to distinguish a product line based on quality, uniformity, and performance. Of the various reaction systems studied to date, a simple noncatalyzed acetylation process appears to be closest to implementation on a commercial basis.

Composites made from lignocellulosic materials have been restricted from many markets because of their moisture sorption, dimensional instability, and to a lesser extent, biological degradation. These negative properties can be overcome, allowing flakes, particles, and fiber from wood and agricultural residues to find markets related to high-performance composites.

For some applications, a combination of materials may be required to achieve a composite with the desired properties and performance. Property-improved lignocellulosic fibers can be combined with materials such as metal, glass, plastic, natural polymers, and synthetic fiber to yield a new generation of composite materials. New composites will be developed that utilize the unique properties obtainable by combining many different materials. This trend will increase significantly in the future.

Literature Cited

1. Rowell, R. M.; Tillman, A-M.; Zhengtian, L. Wood Sci. Technol. 1986, 20, 83-95.
2. Rowell, R. M.; Ellis, W. D. Reaction of Epoxides with Wood. Res. Pap. FPL 451. U.S. Department of Agriculture, Forest Serv., Forest Prod. Lab., 1984. 41 p.
3. Tillman, A-M. J. Wood Chem. Technol. (In press).
4. Rowell, R. M.; Ellis, W. D. Am. Chem. Soc. Symp. Ser. 1981, 172, 263-284.
5. Bekere, M.; Shvalbe, K.; Ozolinya, I. Latvijas Lauksaimniecibas Akademija. Raksti 1978, 163, 31-35.
6. Bristow, J. A.; Back, E. L. Svensk Papperstid. 1969, 72, 367-374.
7. Sudo, K. Mokuzai Gakkaishi 1979, 25, 203-208.
8. Klinga, L. O.; Tarkow, H. Tappi 1966, 49, 23-27.

9. Zhang, G.; Yin, S.; Wang, W.; Xu, R.; Chen, C. J. Nanjing Technol. College Forest Products 1981, 3, 70-75.
10. Ozolinya, I. O.; Shvalbe, K. P.; Bekere, M. R.; Shnyutsinsh, F. A.; Mitsane, L. V.; Karlsone, I. M. USSR Patent 449 822, 1974.
11. Shvalbe, K. P.; Ozolinya, I. O.; Bekere, M. R.; Mitsane, L. V.; Karlsone, I. M.; Dudin'sh, M. M. USSR Patent 478 743, 1974.
12. Rowell, R. M. In Wood Modification; Lawniczak, M., Ed.; Polish Acad. Sci.: Posnan, Poland, 1985; 358-365.
13. Tillman, A.-M.; Simonson, R.; Rowell, R. M. In Wood Modification; Lawniczak, M., Ed.; Polish Acad. Sci.: Posnan, Poland, 1985; 436-446.
14. Nishimoto, K.; Imamura, Y. Mokuzai Kogyo 1985, 40, 414-418.
15. Yoshida, Y.; Kawai, S.; Imamura, Y.; Nishimoto, K.; Satou, T.; Nakaji, M. Mokuzai Gakkaishi 1986, 32, 965-971.
16. Rowell, R. M.; Simonson, R.; Tillman, A.-M. Paperi ja Puu 1986, 68, 740-744.
17. Imamura, Y.; Nishimoto, K.; Yoshida, Y.; Kawai, S.; Sato, T.; Nakaji, M. Wood Res. 1986, 73, 35-43.
18. Rowell, R. M.; Imamura, Y.; Kawai, S.; Norimoto, M. Wood Fiber Sci. 1989, 21, 67-79.
19. Kiguchi, M.; Suzuki, M. Mokuzai Gakkaishi 1985, 31, 200-208.
20. Imamura, Y.; Nishimoto, K. Mokuzai Gakkaishi 1987, 33, 25-30.
21. Youngquist, J. A.; Krzysik, A.; Rowell, R. M. Wood Fiber Sci. 1986, 18, 90-98.
22. Youngquist, J. A.; Rowell, R. M.; Krzysik, A. Holz als Roh- und Werkstoff 1986, 44, 453-457.
23. Rowell, R. M.; Plackett, D. New Zealand J. Forest Sci. (In press.)
24. Arora, M.; Rajawat, J. S.; Gupta, R. C. Holzforschung and Holzverwertung 1981, 33, 8-10.
25. Rowell, R. M.; Tillman, A.-M.; Simonson, R. J. Wood Chem. Technol. 1986, 6, 293-309.
26. Rowell, R. M.; Simonson, R.; Tillman, A.-M. Nordic Pulp and Paper Res. J. 1986, 2, 11-17.
27. Rowell, R. M.; Tillman, A.-M.; Simonson, R. J. Wood Chem. Technol. 1986, 6, 427-448.
28. Rowell, R. M.; Wang, R.H.S.; Hyatt, J. A. J. Wood Chem. Technol. 1986, 6, 449-471.
29. Rowell, R. M.; Norimoto, M. J. Jap. Wood Res. Soc. 1987, 33, 907-910.
30. Rowell, R. M.; Keany, F. Wood and Fiber Sci. (In press.)
31. Rowell, R. M., Simonson, R.; Tillman, A. M. European Patent Application 85850268.5, 1985.
32. Rowell, R. M.; Rowell, J. S. In Cellulose and Wood; C. Schurch, Ed.; Wiley, New York, 1989; 343-356.
33. Rowell, R. M. Wood Sci. 1982, 15, 172-182.
34. Rowell, R. M. Unpublished data.
35. Rowell, R. M.; Norimoto, M. Mokuzai Gakkaishi 1988, 34, 627-629.
36. Rowell, R. M.; Esenther, G. R.; Nicholas, D. D.; Nilsson, T. J. Wood Chem. Technol. 1987, 7, 427-440.

37. Rowell, R. M.; Esenther, G. R.; Youngquist, J. A.; Nicholas, D. D.; Nilsson, T.; Imamura, Y.; Kerner-Gang, W.; Trong, L.; Deon, G.; Proc. Symp. Protection of Wood-Based Composite Products, 1988, 238-266.
38. Nilsson, T.; Rowell, R. M.; Simonson, R.; Tillman, A.-M. Holzforschung 1988, 42, 123-126.
39. Cowling, E. B. Comparative Biochemistry of the Decay of Sweetgum Sapwood by White-Rot and Brown-Rot Fungus, Technol. Bull. 1258, U.S. Department of Agriculture, Forest Serv., 1961, p 50.
40. Imamura, Y.; Nishimoto, K. J. Soc. Materials Sci. 1985, 34, 985-989.
41. Rowell, R. M.; Youngquist, J. A.; Imamura, Y. Wood Fiber Sci. 1988, 20, 266-271.
42. Imamura, Y.; Rowell, R. M.; Simonson, R.; Tillman, A.-M. Paperi ja Puu 1988, 9, 816-829,
43. Imamura, Y.; Nishimoto, K.; Rowell, R. M. Mokuzai Gakkaishi 1987, 33, 986-991.
44. Johnson, B. R.; Rowell, R. M. Mater. und Org. 1988, 23, 147-156.
45. Rowell, R. M. In Chemistry of Solid Wood; Rowell, R. M., Ed.; Advances in Chemistry Series 207; American Chemical Society: Washington, DC, 1984; Chapter 4, pp 175-210.
46. Rowell, R. H.; Susott, R. A.; De Groot, W. G; Shafizadeh, F. Wood and Fiber Sci. 1984, 16, 214-223.

RECEIVED January 22, 1990

Chapter 22

Cellulose and Cellulose Derivatives as Liquid Crystals

Richard D. Gilbert

Fiber and Polymer Science Program, North Carolina State University, Raleigh, NC 27695–8301

The study of mesophases of cellulose and cellulose derivatives is an active field which has expanded rapidly since the initial observation of liquid crystals of hydroxypropyl cellulose in 1976. There are two areas that warrant further investigation: recent observations regarding the influence of solvent and/or substituents on the cholesteric helicoidal twist await a theoretical explanation; there is a lack of careful studies to permit a theoretical treatment of the behavior of ordered cellulose phases. To date, no applications have been developed where the unusual properties of cellulose derivatives are utilized.

The first observation of a liquid crystal solution of a cellulose derivative was by Werbowyj and Gray in 1976 (1). They reported that 20-50% water solutions of hydroxypropyl cellulose (molecular weight, 100,000; four hydroxypropyl substituents per anhydroglucopyranose unit) were highly iridescent and birefringent. The solutions had high optical rotations indicating the mesophase has a superhelicoidal structure and is cholesteric in nature (1,2).

Since this initial observation the field has expanded rapidly and there are numerous reports of cellulose derivatives that form lyotropic liquid crystals. Some of them form both lyotropic and thermotropic liquid crystals. Gray (3) has tabulated various cellulose derivatives reported to form liquid crystals prior to early 1982.

There have been several reviews of the field published (2-8) but in view of the continuing high level of activity an update appears desirable.

Investigators of cellulosic liquid crystals have two main motivations: to study mesophase formation primarily from a scientific viewpoint or a technological viewpoint. The main focus of the latter has been on the potential of preparing high strength/high modulus regenerated cellulose fibers. Another potential use of cellulosic liquid crystal derivatives is as chiroptical filters (9,10).

0097–6156/90/0433–0259$06.00/0

Sandwich chiroptical filters from aqueous hydroxylpropyl cellulose mesophases were prepared between parallel glass plates (10). The optical properties depended on the mesophase thickness. Thin filters selectively reflected up to 36% of normal incident light, and an increase in reflected intensity was observed at higher sample thickness but with an accompanying loss of selectivity. There is an early patent (11) which describes a temperature indicating device consisting of liquid crystals of hydroxypropyl cellulose and a variety of other cellulose derivatives such as β-hydroxybutyl cellulose, α-methyl-β-hydroxypropyl cellulose. Ogata et al. (12) describe liquid crystal compositions which are useful as biomaterials. They are formed from mixtures of hydroxypropyl cellulose and organic compounds containing quaternary ammonium groups and hydrophobic groups containing rigid chain segments. However, to date, the greatest number of publications in the technological area are those dealing with fiber formation (13-20).

This review will attempt to cover both aspects of the field.

Liquid Crystals

The term "liquid crystals" represents the state of matter intermediate between the long-range and high degree of orientational order of solid crystals and the statistical long-range disorder of ordinary liquids. Liquid crystals have also been described as the fourth state of matter (17). The phenomenon was first observed by the Austrian botanist, F. Reinitzer, in 1888, when, on heating cholesteryl benzoate, he noticed the solid crystals melt to a turbid fluid, and, at some degrees higher to a clear liquid. In 1889, Lehman coined the term "Flussige Kristalle" (liquid crystal) to describe the physical state where a substance flows like a liquid but is optically anisotropic. Later the terms "crystalline liquids," "mesophases," and "mesomorphic phases" were adopted – all are used interchangeably today (21). There are two classes of liquid crystals. Those which occur on heating or cooling a substance are termed thermotropic. The substance will pass from a solid state into the mesophase at a characteristic temperature and will pass into the isotropic state at some higher temperature. The process is reversible.

Lyotropic liquid crystals are those which occur on the addition of a solvent to a substance, or on increasing the substance concentration in the solvent. There are examples of cellulose derivatives that are both thermotropic and lyotropic. However, cellulose and most cellulose derivatives form lyotropic mesophases. They usually have a characteristic "critical concentration" or "A point" where the molecules first begin to orient into the anisotropic phase which coexists with the isotropic phase. The anisotropic or ordered phase increases relative to the isotropic phase as the solution concentration is increased in a concentration range termed the "biphasic region." At the "B point" concentration the solution is wholly anisotropic. These A and B points are usually determined optically.

In both the thermotropic and lyotropic classes, liquid crystals may be of three general types, depending on their specific molecular arrangement. Nematic liquid crystals result from the alignment of the individual molecules with their long axes essentially parallel but with their centers of gravity randomly arranged (21). Smetic liquid crystals have stratified structures, with the long axes of the molecules parallel to each other in the layers and their centers of gravity in equidistant planes. As many as eight distinguishable smetic phases have been identified, differing by the specific orientation of the molecules with the layers (22). Cholesteric liquid crystals have been described as skewed nematics. Layers of nematic planar structure are arranged in a superhelix with each layer rotated by an angle ϕ from the previous one. The long axes of the molecules in the layers are perpendicular to the twist axis of the helix.

Substances that form cholesteric mesophases have a chiral center(s). Presumably the chiral perturbation minimizes the free energy of the stacked array, relative to other arrangements, in which the molecules in each layer are twisted with respect to those above and below. Two such superhelicoidal structures are possible, one right-handed and one left-handed, but one is of lower energy than the other (22). For a given temperature and pressure, and for lyotropic mesophases, a given solvent concentration, the cholesteric helix will have a characteristic pitch (23). More recently it has been discovered a change in solvent will result in a change in the cholesteric twist sense for some cellulose derivatives (24). The same is also true for cellulose (17). Polymers which form cholesteric liquid crystals in solution have been observed to form rigid α helices stabilized by hydrogen bonding (2,25,26).

A given compound may exhibit one type of liquid crystalline behavior, or several types, each at characteristic temperature and solvent concentration or solvent type, or it may, of course, only form an isotropic phase.

High molecular weight mesophases were first studied during the late 1930's using suspensions of tobacco mosaic virus (TMV). Bawden and Pirie (26) reported a solution of TMV separated into two phases as the concentration was increased, one of which was birefringent. Bernal and Fankuchen (27) observed that phase separation occurred in suspensions containing as low as 1.8% of the needle-like TMV particles. Onsager (28) in 1949 first presented a theoretical explanation for these observations. Elliot and Ambrose (29) reported in 1950 that a chloroform solution of poly γ-benzyl-L-glutamate spontaneously formed a birefringent phase as the solution concentration was increased. Stimulated by Elliot and Ambrose's observations Isihara (30) and Flory (31,32) developed theoretical explanations for phase separation in polymer solutions. Flory was the first to suggest that cellulosic polymers could form liquid crystalline solutions. Flory's lattice model treatment has proved to be the most generally useful for rigid and semirigid polymers. Later, Flory and Ronca (33) extended the lattice model to a wide variety of systems. There are two molecular features necessary for liquid crystallinity (34). These are (1) asymmetry of molecular shape and (2) anisotropy of intermolecular forces. Asymmetry of molecular shape is the dominant feature, especially for polymers. The latter feature is more prominent in low molecular liquid crystalline compounds and in polymers having highly anisotropic polarizabilities. It is also usually responsible for thermotropic behavior.

The lattice theory deals with rod-like particles which do not have interactions with their neighbors except, of course, repulsions occur when the particles overlap. Above a certain critical concentration (V_2^C) that depends on the axial ratio x the theory predicts the system will adopt a state of partial order (biphasic region). Below V_2^C the system is isotropic. For rigid polymers of axial ratio x phase separation is predicted (32,35) to occur at

$$V_2^C = \frac{8}{x}\left(1 - \frac{2}{x}\right) \tag{1}$$

As the polymer concentration is increased a highly anisotropic solution is formed.

Most polymers that form mesophases are not completely rigid but have some chain flexibility. Using the lattice theory Flory (36) treated the case of polymers having rigidities intermediate between the rigid rod and the random coil and showed equation 1 holds if the semi-flexible chains are considered to consist of rigid rods connected by completely flexible joints. This is the model originally introduced by Kuhn (38). The Kuhn segment length is twice the persistence length of the real chain. However, in Flory's model (36) this is not a requirement and the segment length and the number of them are chosen to duplicate the persistence length of the real chain and not the chain length. The persistence length is defined (39) as the distance a molecule extends in the direction of the first link, i.e. as a rod, taken anywhere along a chain of indeterminate length. Critical concentrations of cellulose esters (40,41) and cellulose ethers (2,42,43) agree quite well with this model.

Werbowyj and Gray (2) suggest that for long chains obeying random flight statistics that the Kuhn segment length be used to estimate a value for the rod length. However, in the case of the chain conformation of a semi-flexible polymer it may be better modelled by the worm-like chain (44). The rigidity of a worm-like chain may be estimated from a persistence length q, or by the equivalent Kuhn statistical segment length, $k_w = 2$ q. Persistence length, of course, is a measure of chain stiffness.

Conio et al. (43) conclude that for polymers having a rigidity between the rigid rod and the random coil, equation 1 still controls mesophase formation if the axial ratio refers to the length of the statistical "rigid" segment, i.e. the persistence length of the real chain. They reached this conclusion using data obtained for hydroxypropyl cellulose (HPC) in H_2O by Werbowyj and Gray (1,2) and for HPC in DMAC. Gray (3) suggests the obvious choice for cellulosics is to set $L = k_w$ or $x = \frac{2q}{d}$. He compared the predicted values for the phase separation of cellulose derivatives, based on current theories, with experimental observations (45). For many cellulosic mesophases V_2^C values range from 0.3 to 0.5 for high molecular weight samples at ambient temperatures, which are generally in agreement with values predicted by the freely jointed chain using Kuhn segment lengths obtained from dilute solution measurements.

However, as discussed below critical concentrations for cellulose, in a variety of solvents, and based on optical observations under crossed polars are much lower than predicted using equation 1 and $k_w = 2$ q. Conio et al. (43) point out one has to consider the possibility that the lattice model does not accurately predict the values of V_2^C and that V_2^C values using the Onsager (28) and Isihara (30) theories are about half that predicted by equation 1.

The rigidity of cellulose and its derivatives are affected by the solvent and by the type and degree of substitution (46) and thus V_2^C is also dependent on the nature of the solvent and the presence or absence of substituents. In general, the more acidic the solvent the lower is the intrinsic viscosity and the lower is the critical concentration (42,47). In addition, some cellulose derivatives, viz, cellulose triacetate, tricarbanilate, trinitrate, will form mesophases in some solvents but not in others. For example, cellulose triacetate forms ordered solutions in trifluoroacetic acid (TFA)-chlorinated alkane mixtures (17,48) but not in LiCl-DMAC mixtures (49).

Liquid Crystal Solutions of Cellulose

Chanzy and Peguy (13) were the first to report that cellulose forms a lyotropic mesophase. They used a mixture of N-methyl-morpholine-N-oxide (MMNO) and water as the solvent. Solution birefringence occurred at concentrations greater than 20% (w/w) cellulose. The concentration at which an ordered phase formed increased as the cellulose D.P. decreased. The persistence length of cellulose in MMNO-H_2O is not known but presumably it has an extended chain configuration in this solvent. Again the question arises as to what is the relevant axial ratio to be used for cellulose. This will be discussed further below.

Simple shearing of an anisotropic solution produced a highly oriented polymer film which after washing and drying was shown to have the cellulose II morphology. Long fibers could be pulled from the anisotropic solutions. They also had the cellulose II morphology. Navard and Haudin (18) studied the rheological behavior of cellulose/MMNO-H_2O solutions at various temperatures and concentrations. This is the first reported study of the spinning of mesomorphic cellulose solutions, but no fiber properties were given. Quenin et al. (14) used a dry-jet, wet spinning system to spin cellulose/MMNO-H_2O solutions. Fiber properties equivalent to the best viscose rayon fiber were obtained.

Patel and Gilbert (50) showed that mixtures of TFA and chlorinated alkanes (1,2-dichloroethane, CH_2 Cl_2) are excellent solvents for cellulose. Lyotropic mesophases were obtained in 20% (w/w) solutions of cellulose as shown by optical microscopy under crossed polars. High and positive optical rotary values show the lyotropic mesophase is cholesteric, as expected due to the chirality of cellulose, and the superhelicoidal structure is right-handed. More recently, Hawkinson (17,51) and Kohout (52) showed little or no trifluoracetylation of cellulose occurs in the dissolution of cellulose in TFA-CH_2Cl_2 mixtures. This is also shown by Myasoedova et al. (53). Degradation of cellulose occurs in TFA-CH_2Cl_2, presumably by attack of the TFA at the glycosidic linkages. The rate of degradation decreases as the TFA/CH_2Cl_2 ratio is decreased (17,51). A number of Russian workers (53–58) have studied the cellulose-TFA and cellulose-TFA-$ClCH_2CH_2Cl$ systems. Papkov et al. (54) reviewed the effect of the structural characteristics of the TFA-$ClCH_2CH_2Cl$ system on mesophase formation of cotton cellulose in this solvent system. Myasoedova et al. (53,55,56) prepared cellulose solutions up to 20% concentration in 70/30 TFA-$ClCH_2CH_2Cl$. Dissolution occurred without cellulose modification or oxidative degradation. They studied the phase transitions for wood and cotton cellulose dissolved in TFA and admixtures with CH_2Cl_2, $CHCl_3$, and $ClCH_2CH_2Cl$, and attribute mesophase formation to enhanced rigidity of the cellulose by solvent. Krestov et al. (57) claim lyotropic solutions of cellulose in TFA-chlorinated alkanes are nematic within certain temperature and concentration ranges depending on the solvent ratio. Yang et al. (15) report nematic mesophases of cellulose in the NH_3/NH_4SCN solvent system depending on the solvent composition. This is of significance for fiber spinning to obtain high strength/high modulus fibers. Either an original nematic phase or untwisting of the cholesteric phase by, for example, shear would be necessary. However, for film formation the cholesteric mesophase could possibly result in biaxial orientation in the film.

Cemeris et al. (58) confirm dissolution in TFA is accompanied by ionic adduct formation and trifluoroacetylation is low.

Solvent viscosity *vs.* concentration plots for cellulose dissolved in TFA-CH_2Cl_2 (70/30, v/v) do not exhibit a maximum (17,51) in contrast to the typical behavior of polymer liquid crystal solutions. This same behavior is exhibited by other cellulose-solvent systems (59,60). Conio et al. (59) suggest that due to the close proximity of the cholesteric mesophase to its solubility limit, it is only observed in a metastable condition.

Cholesteric lyotropic mesophases of cellulose in LiCl-DMAC solutions at 10–15% (w/w) concentration have been observed by Ciferri and coworkers (19,59,61,62) and McCormick et al. (63). LiCl/DMAC ratios between 3/97 and 11/89 (w/w) were used. LiCl-DMAC does not degrade cellulose and does not react with the polymer (59). It does form a complex with the OH groups on cellulose which is believed to result in dissolution (62). Optical rotary dispersions are negative, indicating the superhelicoidal structure has a left-handed twist.

Bianchi et al. (19) spun fibers from isotropic and anisotropic solutions of cellulose (D.P. 290) in LiCl (7.8%)-DMAC solutions. The fiber mechanical properties increased through the isotropic-anisotropic transition with elastic moduli as high as 22 GPa (161 g/d) being obtained.

Solutions of cellulose in NH_3/NH_4SCN (27:73 w/w) are liquid crystalline at concentrations from 10–16% (w/w) depending on the cellulose molecular weight (64). Optical rotations of the solutions indicate the mesophase is cholesteric with a left-handed twist. The solvent does not react with cellulose. Recently, Yang (60) found that cellulose (D.P. 210) formed a mesophase at 3.5% (w/w) concentration at a NH_3/NH_4SCN of 30:70 (w/w).

The concentrations at which ordering of cellulose in the differing solvents occurs (15,17,54,59,60,63) are much lower than that observed for rigid polymers such as poly(*p*-phenylene terephthalamide) whose V_2^C value is about 9% in 100% H_2SO_4. They are also lower than V_2^C (0.3–0.5 volume fraction) values for cellulose derivatives (45). As noted above, Flory's lattice theory is probably not applicable to cellulose, and even using the Kuhn segment length to calculate V_2^C does not give the very low values reported for cellulose. Conio et al. (59) calculate axial ratios for cellulose as high as 135 when it is dissolved in LiCl-DMAC. This would mean the persistence length is of the order of 0.1 μm, much higher than previously reported (4). Obviously, the cellulose molecule is stiffened by complexation with the solvent (62,65) but presently there is a lack of careful studies to permit a theoretical treatment of the behavior of cellulose in ordered phases.

Yang et al. (15) spun fibers from a partially (biphasic) nematic solution of cellulose in NH_3/NH_4SCN. Tenacities of ca. 3 g/d and moduli of ca. 155 g/d were obtained.

Liquid Crystalline Cellulose Derivatives

Presently, there are a large number of reports of cellulose derivatives that form lyotropic mesophases (1-9,16,17,19,20,24,40-43,45,47-49,53,65-119).

Probably the most widely studied cellulose derivative is hydroxypropyl cellulose (HPC) as the HPC-H_2O system is very tractable. Werbowyj and Gray's report on HPC mesophases (1) stimulated numerous fundamental investigations by Gray and coworkers (2,10,65,69,75,78-80,90,99,112-115,117, 118,121-124), White and coworkers (42,68,77,81), Sixou and Navard and coworkers (76,82-88).

Aspler and Gray (65,69) used gas chromatography and static methods at 25°C to measure the activity of water vapor over concentrated solutions of HPC. Their results indicated that the entropy of mixing in dilute solutions is given by the Flory-Huggins theory and by Flory's lattice theory for rod-like molecules at very high concentrations.

Werbowyj and Gray (2) showed HPC forms ordered solutions in polar organic solvents, such as CH_3OH, C_2H_5OH as well as water. The concentration to form the ordered phase depended on the solvent (from 42–47 wt %) but was relatively insensitive to the molar mass of HPC. White and coworkers (42,68,81) also showed that V_2^C for HPC depends on the solvent, varying from 0.21 gm/mL for Cl_2CH COOH, 0.30 for CH_3COOH, 0.38 for dimethylacetamide, 0.42 for water, and 0.43 for C_2H_5OH and V_2^C is temperature dependent. The development of high levels of birefringence and orientation during flow of HPC solutions were studied by Onogi et al. (68) and by Asada (70). Tsutsui and Tanaka (72) showed HPC formed cholesteric mesophases of HPC in pyridine, C_2H_5OH, 2-methoxyethanol and 1,4-dioxane.

Phase diagrams of HPC in water at 20°C and in DMAC from 20° to 130°C, up to ~80% (w/w) concentration, were determined by Conio et al. (43). They found V_2^C is not markedly influenced by the HPC molecular weight in DMAC, as shown by Werbowyj and Gray (1,2) for HPC in water. It is ~0.35 at 25°C for HPC in DMAC and increases with temperature.

Werbowyj and Gray (79) examined the relationships between the cholesteric pitch and optical properties of HPC in water, CH_3COOH and CH_3OH. The reciprocal pitch varied as the third power of the HPC concentration. Optical rotatory dispersion results show HPC has a right-handed superhelicoidal structure regardless of structure. As will be discussed below, a change in solvent can reverse the handedness of other cellulose derivatives.

Navard and Haudin studied the thermal behavior of HPC mesophases (87,88) as did Werbowyj and Gray (2), Seurin et al. (85) and, as noted above, Conio et al. (43). In summary, HPC in H_2O exhibits a unique phase behavior characterized by reversible transitions at constant temperatures above 40°C and at constant compositions when the HPC concentration is above ca. 40%. A definitive paper has been recently published by Fortin and Charlet (89) who studied the phase-separation temperatures for aqueous solutions of HPC using carefully fractionated HPC samples. They showed the polymer-solvent interaction differs in the cholesteric phase (ordered molecular arrangement) from that in the isotropic phase (random molecular arrangement).

The texture of HPC mesophases in water has been studied by Marsano et al. (9), Fried and Sixou (91), Shimamura (92), and Marruci et al. (93).

Formation of band textures in HPC liquid crystalline solutions was studied by Navard (86) and Takaheshi et al. (94). Upon shearing, some polymeric liquid crystals develop a particular texture which is called "band texture" consisting of fine equidistant lines when viewed under crossed polars. In HPC-H_2O mesomorphic solutions these bands are exhibited when the solution is allowed to freely relax after shearing.

Dayan et al. (82) used NMR spectroscopy to study mesophases of HPC.

Navard and Haudin (100) examined the rheology of HPC-acetic acid solutions. The anisotropic solutions were strongly viscoelastic.

Suto et al. (102) studied the effects of salts on the turbidity and viscometric behavior of HPC mesophases in water and the rheology of liquid crystalline solutions of HPC in *m*-cresol (103). Suto (104) found that crosslinking HPC mesophases in water destroyed their order.

Patel and Gilbert (48) showed cellulose triacetate (CTA) forms a mesophase in mixtures of TFA and chlorinated alkanes. More extensive studies of this system are reported by Kohout (52) and Hong et al. (17). In TFA-CH_2Cl_2 (60/40 v/v) a biphasic solution is obtained at 20% (w/w). As in the case of cellulose, CTA is slowly degraded in TFA-CH_2Cl_2 (52). Myasoedova et al. (55) and Krestov and coworkers (57) have also studied this system. Patel and Gilbert (95,96), using PMR spectroscopy, showed that cholesteric to nematic transitions of CTA in TFA-CH_2Cl_2 or TFA-$ClCH_2CH_2Cl$ are induced by a magnetic field. Solution flow times and cholesteric pitches indicate the solvent power is TFA-CH_2Cl_2, TFA-$ClCH_2CH_2Cl$, TFA-$CHCl_2$. This is in the order of decreasing solvent acidity and confirms the observations of Aharoni (40) that mesophase formation is influenced by the solvent acidity in the case of secondary cellulose acetate.

Bheda et al. (42) showed that cellulose triacetate forms a mesophase in dichloroacetic acid. Navard and Haudin (18) examined the thermal behavior of liquid crystalline solutions of CTA in TFA. Navard et al. (73) studied the isotropic to anisotropic transitions of solutions of cellulose triacetate in TFA using differential scanning calorimetry. Navard and Haudin (87) studied the mesophases of cellulose and cellulose triacetate calorimetrically. Navard et al. (83) report similar studies. Meeten and Navard (97) showed the twist of the cholesteric helicoidal structure of CTA and secondary cellulose in TFA is left-handed.

Sixou et al. (101) showed the circular dichroism of cholesteric CTA solutions in TFA depends on the CTA molecular weight. The intensity of the circular dichroic peak increases with molecular weight. Meeten and Navard (97) studied gel formation and liquid crystallinity in TFA-H_2O solutions of CTA. When water was added to a liquid crystalline solution of CTA in TFA a gel phase formed; presumably by the formation of crosslinks due to hydrogen bonding. They interpreted their results that liquid crystalline ordering involves both inter- and intramolecular forces.

Bheda et al. (20) wet spun CTA fibers from liquid crystalline solutions in TFA. Surprisingly, they found the fibers have the CTA-I morphology. O'Brien (16) spun CTA fibers from TFA-CH_2Cl_2 and TFA-H_2O solutions as did Hong et al. (17) from TFA-CH_2Cl_2 solutions. In each case the fibers had the CTA-I morphology. Bheda et al. (20) also spun fibers from anisotropic solutions of cellulose acetate butyrate in dimethylacetamide.

Zugenmaier and coworkers (24,105-110) have published a series of basic studies on the mesophases of various cellulose derivatives in a variety of solvents.

Vogt and Zugenmaier (105) determined the pitch of cellulose tricarbanilate (CTC, D.P. = 100) in 2-pentanone and methyl ethyl ketone (MEK) and ethyl cellulose (EC) in glacial acetic acid as a function of temperature, concentration, solvent, and degree of polymerization. The pitch of the helicoidal structure of CTC/MEK and CTC/2-pentanone is right-handed but EC in glacial acetic acid is left-handed. This is the first report that the substituent will influence the sense of the cholesteric superhelicoidal structure.

Zugenmaier and Haurand (108) showed the twist of the cholesteric structure varies from left- to right-handed as the amount of dichloroacetic acid is increased in the ethyl cellulose/acetic acid/dichloroacetic acid system. The twist angle becomes zero, resulting in a quasi-nematic at 60% volume percent of $CHCl_2COOH$. The quasi-nematic phase showed shear thinning at high shear rates in accordance with the model proposed by Onogi and Asada (111). Again, this is the first recognition that the solvent will affect the cholesteric twist.

As noted above, Yang et al. (15) found the CELLOH/NH_3/NH_4SCN system changed from cholesteric to nematic as the NH_3/NH_4SCN ratio was varied.

The influence of molar mass on the cholesteric phases' properties for the system CTC/diethylene glycol monoethyl ether was investigated by Siekmeyer and Zugenmaier (106). The pitch of the cholesteric phase rapidly changes at small molar masses, remains constant at higher molar masses, but eventually at very high molar mass a gel is obtained. The increase in pitch with molar mass for CTC is opposite to results for acetoxypropyl cellulose reported by Laivins and Gray (112).

Steinmeier and Zugenmaier (107) showed a strong influence of the solvent on the pitch of (3-chlorophenyl) cellulose urethane (3-CPCU). The behavior was similar to that of CTC (105) but the 3-CPCU forms a right-handed cholesteric mesophase in triethyleneglycol monoethyl ether while CTC cholesteric mesophase in the same solvent has a left-handed twist. However, the (4-chlorophenyl) cellulose urethane did not form mesophases in any of the solvents studied, viz., 2-pentanone, MEK, and diethyleneglycol monoethyl ether (DEME).

Quasi-nematic or compensated cholesteric phases were formed by CTC dissolved in mixtures of methylpropyl ketone (MPK) and DEME. CTC/MPK has a right-handed twist but CTC/DEME a left-handed one (109). Siekmeyer et al. (110) studied the phase behavior of the ternary lyotropic system CTC/3-chlorophenylurethane/triethyleneglycol monoethyl ether.

Ritchey et al. (113) showed the introduction of trifluoroacetate groups at the unsubstituted hydroxyls of cellulose acetate causes a reversal in handedness of the cholesteric structure. Likewise the introduction of an aceto group in acetoxypropyl cellulose changes the twist (116).

Guo and Gray (114) found that acetylation of the unsubstituted groups in ethyl cellulose changes the sense of the helicoidal cholesteric twist from left-handed to right-handed in either $CHCl_3$ or *m*-cresol, aqueous phenol and acetic acid. In $CHCl_2COOH$ both (ethyl) cellulose and (acetyl)(ethyl) cellulose (AEC) are right-handed. AEC in $CHCl_3$ changes from a left-handed to right-handed cholesteric structure with an increase in the acetyl content (115).

As Guo and Gray (114) point out, "the relationship between the handedness of the supramolecular helicoidal structure and the molecular structure of cellulose derivatives and solvents is not at all understood" at the present time.

Giasson et al. (117) give direct electron microscopy evidence for the helicoidal structure of films of the cellulose acetates and of cellulose regenerated from the CTA films by aqueous ammonium hydroxide. The films were cast from anisotropic solutions of either secondary cellulose acetate or cellulose triacetate in TFA and they showed a lamellar fingerprint texture. Cellulose films formed by slow precipitation from LiCl/DMAC solutions also showed the same structure, but less clearly. Similar data are given by Ritchey et al. (118). It is noteworthy that the helicoidal structure of the cellulose triacetate is unchanged during deacetylation. That is, the molecular order in the cellulose triacetate is preserved during deacetylation.

Thermotropic Cellulose Derivatives

There are now numerous examples of cellulose derivatives that form both lyotropic and thermotropic mesophases. Of course, cellulose itself is unlikely to form a thermotropic liquid crystalline phase because it decomposes prior to melting.

Hydroxypropyl cellulose was shown by Shimamura et al. (77,119) to form a thermotropic mesophase. Tseng et al. (75) report that acetoxypropyl cellulose behaves as a thermotropic liquid crystal below 164°C. Pawlowski et al. (116) demonstrated acetoacetoxypropyl cellulose forms a thermotropic mesophase using DSC and hot stage microscopy. Aharoni (120) showed trifluoro-acetoxypropyl cellulose is also thermotropic using the same techniques. Bhadani and Gray (121) and Bhadani et al. (122) report on the thermotropic mesophase of the benzoate ester of hydroxypropyl cellulose, Ritchey and Gray (123) on that of (2-ethoxypropyl) cellulose, and Tseng et al. (124) on the thermotropicity of the propanoate ester of (2-hydroxypropyl) cellulose.

Navard and Zachariades (125) examined the optical properties of shear deformed trifluoroacetoxypropyl cellulose and observed band phenomena identical to that for thermotropic nematic copolyesters. Steinmeier and Zugenmaier (107) demonstrated that the phenylacetate and 3-phenylpropionate of hydroxypropyl cellulose and the (3-chlorophenyl) urethane of cellulose all form thermotropic liquid crystalline phases. Ritchey et al. (118) showed (ethoxypropyl) cellulose is both lyotropic and thermotropic.

These thermotropic cellulose derivatives are of course of interest from the viewpoint of their structure and properties and might be considered for such applications as chiroptical filters. However, they are unlikely to be considered for fiber formation and certainly not for regenerated fibers, as essentially they are ethers of cellulose and desubstitution would be difficult. Pawlowski et al. (126) prepared a series of cellulose derivatives, namely phenylacetoxy, 4-methoxyphenylacetoxy-, and *p*-tolylacetoxy cellulose and trimethylsilyl cellulose that are thermotropic (and lyotropic) and showed the hydroxyl substituents may be readily removed under mild conditions to regenerate cellulose.

Literature Cited

1. Werbowyj, R. S.; Gray, D. G. Mol. Cryst. Liq. Cryst. 1976, 34, 97.
2. Werbowyj, R. S.; Gray, D. G. Macromolecules 1980, 13(1), 69.
3. Gray, D. G. J. Appl. Polym. Sci., Appl. Polym. Symp. 1983, 37, 179.
4. Gilbert, R. D.; Patton, P. A. Prog. in Polym. Sci. 1983, 9(2/3), 115.
5. Kulichikhin, V. G.; Golova, L. K. Khim. Drer. 1985, (3), 9. Chem. Abstr. 103, 23906.
6. Sixou, P.; Dayan, S.; Gilli, G. M.; Fried, F.; Maissa, P.; Vellutini, M. J.; Ten Bosch, A. Carbohydr. Polym. 1982, 2(4), 2381.
7. Gray, D. G. Polym. Sci. Technol. (Plenum) 1985, 28 (Polym. Liq. Cryst.); p 369.
8. Gray, D. G. Faraday Discus. Chem. Soc. 1985, 79, 257.
9. Marsano, E.; Carpaneto, L.; Ciferri, A. Mol. Cryst. Liq. Cryst. 1988, 158(B), 267.
10. Charlet, G.; Gray, D. G. Macromolecules 1987, 20, 33.
11. Maeno, J. U.S. Patent 4,132,464, 1979 (to Ishii Hideki).
12. Ogata,T.; Yanagi, H.; Horimoto, H. Jpn. Kokai Tokyo Koho JP 6160, 734. Chem. Abstr. 105, 135796.
13. Chanzy, H.; Peguy, A. J. Polym. Sci., Polym. Phys. Ed. 1980, 18, 1137.
14. Quenin, I.; Chanzy, H.; Paillet, M.; Peguy, A. In Integration of Fundamental Polym. Sci. and Tech., Kleinties, L. A.; Lemstra, P. J., Eds.; Elsevier Appl. Science Publ.: 1986; p 57.
15. Yang, K-S.; Theil, M. H.; Cuculo, J. A. In Polymer Association Structures; American Chemical Society Symposium Series 384, El-Nokaly, M. A., Ed.; 1989; p 156.
16. O'Brien, J. P. U.S. Patent 4,464,323; 4,501,886, 1984 (Assigned to duPont).
17. Hong, Y. K.; Hawkinson, D. E.; Kohout, E.; Garrard, A.; Fornes, R. E.; Gilbert, R. D. In Polymer Association Structures; American Chemical Society Symposium Series 384, El-Nokaly, M. A., Ed.; 1989; p 184.
18. Navard, P.; Haudin, J. Br. Polym. J. 1980, 12(4), 174.
19. Bianchi, E.; Ciferri, A.; Conio, G.; Teoldi, A. J. Appl. Polym. Sci. 1989, 27, 1477.
20. Bheda, J.; Fellers, J. F.; White, J. L. J. Appl. Polym. Sci. 1981, 26, 3955.
21. Saeva, F. S. In Liquid Crystals, The Fourth State of Matter; Saeva, F. S., Ed.; Marcel Dekker: New York, 1979; Chapter 2.
22. DeVries, A., *ibid.*, Chapter 1; Petrie, S. E. B., *ibid.*, Chapter 4.
23. Gibson, H. W., *ibid.*, Chapter 3.
24. Zugenmaier, P.; Haurand, P. Carbohydr. Res. 1987, 160, 369.
25. Morawetz, H. Macromolecules in Solution, 2nd Ed.; Wiley: New York, 1975; p 67.
26. Bawden, F. C.; Pirie, N. W. Proc. R. Soc., London, Ser. B 1937, 123, 274.
27. Bernal, J. D.; Fankuchen, I. J. Gen. Physiol. 1941, 25, 111.
28. Onsager, L. Ann. N.Y. Acad. Sci. 1949, 51, 627.
29. Elliot, A.; Ambrose, E. J. Discuss. Faraday Soc. 1950, 9, 246.

30. Isihara, A. J. Chem. Phys. 1950, 18, 1446.
31. Flory, P. J. Proc. R. Soc., London, Ser. A 1956, 234, 60.
32. Flory, P. J. Proc. R. Soc., London, Ser. A 1956, 234, 73.
33. Flory, P. J.; Ronca, G. Mol. Cryst. Liq. Cryst. 1979, 54, 311.
34. Flory, P. J. In Recent Advances in Liquid Crystalline Polymers; Chapoy, L. L., Ed.; Elsevier, 1985, p 99.
35. Flory, P. J. Advan. Polym. Sci. 1984, 59, 1.
36. Flory, P. J. Macromolecules 1978, 11, 1141.
37. Matheson, R. R.; Flory, P. J. Macromolecules 1981, 14, 954.
38. Kuhn, W. Kolloid-Z. 1936, 76, 258; 1939, 87, 3.
39. Peterlin, A. Polym. Prepr. Am. Chem. Soc., Div. of Polym. Chem. 1968, 9, 323.
40. Aharoni, S. M. Mol. Cryst. Liq. Cryst. Lett. 1980, 56, 237.
41. Dayan, S.; Maissa, P.; Vellutini, M. J.; Sixou, P. J. Polym. Sci., Polym. Lett. Ed. 1982, 20, 33.
42. Bheda, J.; Fellers, J. F.; White, J. L. Coll. Polym. Sci. 1980, 258, 1335.
43. Conio, G.; Bianchi, E.; Ciferri, A.; Teoldi, A.; Aden, M. A. Macromolecules 1983, 16, 1264.
44. Kratky, O.; Porod, G. Rec. Trav. Chim. Pay-Bas 1949, 68, 1106.
45. Gray, D. G. In Polymeric Liquid Crystals; Blumstein, A., Ed.; Plenum: New York, 1983; p 371.
46. Ciferri, A. In Polymer Liquid Crystals; Ciferri, A.; Krigbaum, W. R.; Meyer, R. B., Eds.; Academic Press: 1982; p 90.
47. Aharoni, S. M. J. Macromol. Sci.-Phys. 1982, B21, 105.
48. Patel, D. L.; Gilbert, R. D. J. Polym. Sci., Polym. Phys. Ed. 1981, 19, 1449.
49. Marsano, E.; Bianchi, E.; Ciferri, A.; Ramis, E.; Tealdi, R. Macromolecules 1986, 19(3), 62.
50. Patel, D. L.; Gilbert, R. D. J. Polym. Sci., Polym. Phys. Ed. 1981, 19, 1231.
51. Hawkinson, D. M.S. Thesis, North Carolina State University, 1987.
52. Kohout, E. M.S. Thesis, North Carolina State University, 1987.
53. Myasoedova, V. V.; Adamova, O. A.;. Krestov, G. A. Vysokomol. Soedin., Ser. B 1984, 26(3), 215.
54. Papkov, S. P.; Belousov; Yu Ya; Kulichikhin, V. E. Khim. Volokna 1983, 3, 8.
55. Myasoedova, V. V.; Alekseeva, O. V.; Krestov, G. A. Zh. Prikl. Khim. (Leningrad) 1987, 60(10), 2526.
56. Myasoedova, V. V.; Belov, S. Yu; Krestov, G. A. Vysokomol. Soedin, Ser. A. 1987, 29(6), 1149.
57. Krestov, G. A.; Myasoedova, V. V.; Alekseeva, O. V.; Belov, S. Yu Dokl. Akad. Nauk SSSR, 1987, 293(1), 174.
58. Cemeris, M.; Musiko, N.P.; Cemeris, N., Kim. Drev., 1986, (2), 29.
59. Conio, G.; Corazzo, P.; Bianchi, E.; Teoldi A.; Ciferri, A. J. Polym. Sci., Polym. Lett. Ed. 1984, 22, 273.
60. Yang, K-S. Ph.D. Thesis, North Carolina State University, 1988.
61. Bianchi, E.; Ciferri, A.; Conio, G.; Coseni, A.; Terbojevick, M. Macromolecules 1985, 18, 646.
62. Terbojerick, M.; Coseni, A.; Conio, G.; Ciferri, A.; Bianchi, E. Macromolecules 1985, 18, 640.
63. McCormick, C. L.; CaVair, P. A.; Hutchinson, B. H. Macromolecules 1985, 18, 2394.

64. Chen, Y. S.; Cuculo, J. A. J. Polym. Sci., Polym. Chem. Ed. 1986, 24, 2075.
65. Aspler, J. S.; Gray, D. G. Macromolecules 1979, 12(4), 563.
66. Patel, D. L.; Gilbert, R. D. J. Polym. Sci., Polym. Phys. Ed. 1982, 20, 1019.
67. Luise, P. R.; Morgan, P. W.; Panar, M.; Willcox, O. B. Abstract of paper presented at 53rd Colloid and Surface Sci. Symp., Univ. of Missouri, Rolla, June, 1979.
68. Onogi, Y.; White, J. L.; Fellers, J. F. J. Non-Newt. Fluid. Mech. 1980, 7, 121.
69. Aspler, J. S.; Gray, D. G. Macromolecules 1981, 14, 1546.
70. Asada, T. Polym. Prepr. Am. Chem. Soc., Div. Polym. Chem. 1979, 20(1), 70.
71. Patel, D. L.; Gilbert, R. D. J. Polym. Sci., Polym. Phys. Ed. 1982, 20, 1019.
72. Tsutsui, T.; Tanaka, R. Polym. J. 1981, 12, 473.
73. Navard, P.; Haudin, J. M.; Dayan, S.; Sixou, P. J. Polym. Sci., Polym. Lett. Ed. 1981, 19(8), 379.
74. Patton, P.; Gilbert, R. D. J. Polym. Sci., Polym. Phys. Ed. 1983, 21(4), 515.
75. Tseng, S-L.; Valente, A.; Gray, D. G. Macromolecules 1981, 14, 715.
76. Dayan, S.; Maissa, P.; Vellcatini, M. J.; Sixou, P. J. Polym. Sci., Polym. Lett. Ed. 1982, 20(1), 33.
77. Shimamura, K.; White, J. L.; Feller, J. F. J. Appl. Polym. Sci. 1981, 26, 2165.
78. Werbowyj, R. S.; Gray, D. G. Polym. Prepr. Am. Chem. Soc., Div. of Polym. Chem., 1979, 20(1), 102.
79. Werbowyj, R. S.; Gray, D. G. Macromolecules 1984, 17(8), 1512.
80. Tseng, S-L.; Valente, A.; Gray, D. G. Macromolecules 1981, 14(3), 715.
81. Onogi, Y.; White, J. L.; Fellers, J. F. J. Polym. Sci., Polym. Phys. Ed. 1980, 18(4), 663.
82. Dayan, S.; Fried, F.; Gilli, J. M.; Sixou, P. J. Appl. Polym. Sci., Appl. Polym. Symp. 1983, 37, 193.
83. Navard, P.; Haudin, J. M.; Dayan, S.; Sixou, P. J. Appl. Polym. Sci., Appl. Polym. Symp. 1983, 37, 24.
84. Seurin, M. J.; Ten Bosch, A.; Sixou, P. Polym. Bull. (Berlin) 1983, 9, 450.
85. Seurin, M. J.; Gilli, J. M.; Fried, F.; Ten Bosch, A.; Sixou, P. In Polymer Liquid Crystals; Blumstein, A., Ed.; Plenum, New York, 1983; p 377.
86. Navard, P. J. Polym. Sci., Polym. Phys. Ed. 1986, 24(6), 435.
87. Navard, P.; Haudin, J. M. Calorim. Anal. Therm. 1983, 14, 207.
88. Navard, P.; Haudin, J. M. In Polymer Liquid Crystals; Blumstein, A., Ed.; Plenum: New York, 1983; p 389.
89. Fortin, S.; Charlet, G. Macromolecules 1989, 22(5), 2286.
90. Charlet, G.; Gray, D. G. J. Appl. Polym. Sci. In Press.
91. Fried, F.; Sixou, P. J. Polym. Sci., Polym. Chem. Ed. 1984, 22, 239.
92. Shimamura, K. Makromol. Chem., Rapid Comm. 1983, 4, 107.
93. Marruci, G.; Grizzuti, M.; Buonaurio, A. Mol. Cryst. Liq. Cryst. 1987, 153, 26.

94. Takaheshi, J.; Shibata, K.; Nomura, S.; Kurokawa, M. Seni. Gakkaishi 1982, 38, 375.
95. Patel, D. L.; Gilbert, R. D. J. Polym. Sci., Polym. Phys. Ed. 1982, 20, 1019.
96. Patel, D. L.; Gilbert, R. D. J. Polym. Sci., Polym. Phys. Ed. 1983, 21, 1079.
97. Meeten, G. H.; Navard, P. Polymer 1982, 23, 1727.
98. Meeten, G. H.; Navard, P. Polymer 1983, 24, 815.
99. Charlet, G.; Gray, D. J. Appl. Polym. Sci. 1989, 37(9), 2517.
100. Navard, P.; Haudin, J. M. J. Polym. Sci., Polym. Phys. Ed. 1986, 24(1), 189.
101. Sixou, P.; Lematre, J.; Ten Bosch, A.; Gilli, J. M.; Dayan, S. Mol. Cryst. Liq. Cryst. 1983, 91(3-4), 277.
102. Suto, S.; Nishibor, W.; Kudo, K.; Karasawa, M. J. Appl. Polym. Sci. 1989, 37(3), 737.
103. Suto, S.; Gotoh, H.; Nishibori, W.; Karasawa, M. J. Appl. Polym. Sci. 1989, 37(4), 1147.
104. Suto, S. J. Appl. Polym. Sci. 1989, 37(9), 2781.
105. Vogt, V.; Zugenmaier, P. Ber. Bunsenges. Phys. Chem. 1985, 89, 1217.
106. Siekmeyer, M.; Zugenmaier, P. Makromol. Chem., Rapid Commun. 1987, 8, 511.
107. Steinmeier, H.; Zugenmaier, P. Carbohyd. Res. 1988, 173, 75.
108. Zugenmaier, P.; Haurand, P. Carbohyd. Res. 1987, 160, 369.
109. Siekmeyer, M.; Zugenmaier, P. In Press.
110. Siekmeyer, M.; Steinmeier, H.; Zugenmaier, P. In Press.
111. Onogi, S.; Asada, T., Rheology, 1980, 1, 127.
112. Laivins, G. V.; Gray, D. G. Polymer 1985, 26, 1435.
113. Ritchey, A. M.; Holme, K. R.; Gray, D. G. Macromolecules 1988, 21(5), 2194.
114. Guo, J-X.; Gray, D. G. Macromolecules 1989, 22, 2082.
115. Guo, J-X.; Gray, D. G. Macromolecules 1989, 22, 2086.
116. Pawlowski, W. P.; Gilbert, R. D.; Fornes, R. E.; Purrington, S. T. J. Polym. Sci., Polym. Phys. Ed. 1987, 25, 2293.
117. Giasson, J.; Revol, J-F.; Ritchey, A. M.; Gray, D. G. Biopolymers 1988, 27, 1999.
118. Ritchey, A. M.; Giasson, J.; Revol, J-F.; Gray, D. G., J. Appl. Polym. Sci., Polym. Symp. Ed. In Press.
119. Shimamura, K.; White, J. L.; Feller, J. F. J. Polym. Sci., Polym. Lett. Ed. 1982, 20(1), 33.
120. Aharoni, S. M. J. Polym. Sci., Polym. Lett. Ed. 1981, 19(10), 495.
121. Bhadani, S. N.; Gray, D. G. Makromol. Chem., Rapid Commun. 1982, 3(6), 449.
122. Bhadani, S. N.; Tseng, S-L.; Gray, D. G. Makromol. Chem. 1983, 184(8), 1727.
123. Ritchey, A. M.; Gray, D. G. Macromolecules 1988, 21(5), 1251.
124. Tseng, S-L.; Laivins, G. V.; Gunar, O.; Gray, D. G. Macromolecules 1982, 15(5), 1262.
125. Navard, P.; Zachariades, A. E. J. Polym. Sci., Polym. Phys. Ed. 1987, 25(5), 1089.
126. Pawlowski, W. P.; Gilbert, R. D.; Fornes, R. E.; Purrington, S. T. J. Polym. Sci., Polym. Phys. Ed. 1988, 26, 1101.

RECEIVED January 18, 1990

AGRICULTURAL POLYMER UTILIZATION
Corn-Based Feed Stocks

Chapter 23

Specialty Starches

Use in the Paper Industry

Kenneth W. Kirby

Penford Products Company, 1001 First Street, SW, Cedar Rapids, IA 52404

The paper industry continues to be the largest single market for starch. Recent growth in coated and specialty papers has lead the way toward total consumption of starch in the industry during 1987 of about 3.5 billion pounds. Specialty uses include wet end cationic starches and neutral starch ethers for size press and coating starches. Increasing demands of higher speed machines require thinner viscosity specialty starches which resist retrogradation. Specialty starches for increased gloss and water resistance will soon allow natural polymer based products to replace many synthetic petrochemical compounds.

The paper industry is a huge business and is in the strongest growth period in recent history. Paper production in the world in 1988 is headed for the fifth consecutive year of growth. Paper and paperboard production will approach 77 million tons in the USA, who is the leading producer of pulp, paper and paperboard and the leading per capita consumer. Demand continues to outpace production. The USA continues to dominate by producing 31% of the world's output. Outstanding among all countries were the increases seen in the Republic of Korea and the Peoples Republic of China. However, neither of these countries appears on the list of the top 20 in per capita consumption. When this does occur, pulp production will take a significant increase.

0097–6156/90/0433–0274$06.00/0

A review of the data for 1986[1] shows the top 10 producers of paper and paperboard, pulp and the 10 top per capita consumers. (Tables I, II, III)

Although paper and paperboard production is growing, certain segments of the industry are growing faster than the total. This is particularly true of coated paper. Coated paper ranges from heavily coated stock used in coated board, annual reports and slick advertising grades to very light weight coated stock seen as newspaper supplements and varied advertising forms. The main reason for growth is the demand for four color advertising copy mainly filled by the lower coating grades such as No. 5. Growth in coated paper in 1988 shipments from 1987 was more than 12%.

The paper industry constitutes the largest single industrial market for starch in the world. If the world production of paper is about 235 MM tons, starch production for paper may be estimated at about 3 MM tons. This recognizes that large volume items such as newsprint and tissue are essentially non-starch users.

The growth in coated and specialty papers has meant growth in the demand for specialty starches to accomplish the quality increase that has been demanded by the papermaker. Some of the quality demands in the USA have come from attempting to meet the quality of imported coated papers. Other quality demands have come from changes in machine speeds, newer methods of applying films and the need for improved rheology of very thin starch pastes. The latter reasons in particular are helping to move the use of specialty starches to a more important position in starch usage.

A specialty starch is defined as one where some form of chemical modification takes place to give properties to the starch that are lacking in the native product. The remainder of this paper will discuss the chemistry of modification and the chemical and physical reasons for the use of specialty starches in treating paper.

This paper will also assume that the reader has basic starch and cellulose knowledge and that it is not necessary to review the structure of the molecules. It is, however, important to know that starch from native, non-genetically selected sources, is a mixture of two molecules and not simply one compound. Amylose is an essentially linear molecule and differs from amylopectin, which has about 4-6% α-(1$\rightarrow$6) branches, even though both molecules are mainly α-(1$\rightarrow$4) linked D-glucose. The differences in these two molecules and their chemical modifications are the basis of application technology and the reason for the growing importance of specialty starches.

Table I. Paper and Paperboard Production
(000 Metric Tons)

	1986	% Change 86/85
1. U.S.A.	67,544	+ 5.0
2. Japan	22,537	+ 7.0
3. Canada	16,046	+ 5.1
4. People's Rep. China	11,411	+ 14.3
5. U.S.S.R.	11,036	+ 6.2
6. Fed. Rep. Germany	9,924	+ 5.5
7. Finland	8,012	+ 6.1
8. Sweden	7,811	+ 6.1
9. France	5,832	+ 4.3
10. Italy	4,987	+ 6.9
14. Rep. of Korea	3,163	+ 14.1

Source: Pulp & Paper August 1988

Table II: Pulp Production
(000 Metric Tons)

	1986	% Change 86/85
1. U.S.A.	54,036	+ 4.5
2. Canada	23,021	+ 6.1
3. U.S.S.R.	11,610	+ 5.9
4. Sweden	9,973	+ 6.2
5. Japan	9,733	+ 5.3
6. Finland	8,467	+ 6.8
7. People's Rep. China	7,204	+ 9.4
8. Brazil	4,031	+ 0.9
9. Fed. Rep. Germany	2,259	+ 1.8
10. France	2,092	+ 3.5

Source: Pulp & Paper August 1988

Table III: Paper - Per Capita Consumption
(lb./year)

	1986	% Change 86/85
1. U.S.A.	640	+ 0.4
2. Sweden	524	+ 3.9
3. Netherlands	488	+ 24.9
4. Canada	469	+ 1.2
5. Finland	444	- 0.3
6. Switzerland	416	- 0.7
7. Fed. Rep. Germany	407	- 0.1
8. Denmark	403	- 0.2
9. Japan	381	- 0.2
10. Belgium	374	0.0

Source: Pulp & Paper August 1988

Starches used for paper applications range in viscosity from thick-boiling unmodified viscosity to very highly converted, low molecular weight products used in high solids coatings. The unmodified thick-boiling starches are used in the wet end of a paper machine and in corrugating pastes. Size press, calendar stack, wire applications and coatings all require viscosity modified starches. Several methods are used to lower the molecular weight of starch. The principal ones used are alpha-amylase digestion and acid catalyzed hydrolysis. Both are practiced at the mill site with the latter using a pressure cooker to lower the molecular weight with the acid catalyst present. The physical nature of the products produced by these processes is the principal reason for the growing use of specialty starches.

Aqueous Dispersion Properties of Starch

Products from enzyme and acid hydrolysis are generally regarded to form gels and develop strong retrogradation tendencies. They also form less continuous films, suffer from poor water holding properties and give poor viscosity control. These conditions are all related to the dynamic changes occurring in a starch paste once it is prepared.

After the chains are totally hydrated in a hot water colloidal dispersion, the movement of the chains causes them to bump into each other and as this happens they will begin to associate. Chain re-association is commonly called retrogradation. During cooling the association accelerates and if the solids are high enough, a gel will form. Generally, in paper applications this is a condition to be avoided and starch pastes are best used relatively quickly. Shorter chains associate much more quickly than do longer chains and this is amply demonstrated by the use of an acid modified thin boiling starch for forming gum candies.

The same effect occurs in varying degrees with any acid modified starch regardless of the acid modification system used. Prevention of chain association is done by a number of processes including derivatization, oxidation, cross-linking, addition of complexing agents or alkalis. Retrogradation can also be retarded by constant stirring which physically keeps the chains moving and avoids some association. As the molecules get shorter, however, when a paste cools, retrogradation will take place. As chain association or retrogradation, also known as set-back, progresses, viscosity effects are noted and solids relationships change. The papermaker is usually quite concerned about these changes and means of avoiding them are of utmost importance.

Uses of Starch in Paper

Paper production benefits from starch application in a variety of ways. Pulping procedures range from chemical digestion for removal of lignins, hemicellulose and other soluble and semi-soluble components of wood, to ground wood. The latter involves pressing a barked log against a pulpstone and reducing the wood to a mass of relatively short fibers.

Cellulose fibers prepared by chemical pulping or as ground wood do not hold together very well. If they are beaten in water, the hydroxyl groups are hydrated and hydrogen bonds are formed between fibers. Beating requires energy and to save a portion of this expense, paper mills will add starch as an adhesive to add bonding of fibers and to increase the strength of the formed sheet. Although unmodified starch can be used it is extremely inefficient. About one-half of that added is lost to the mill effluent, an additional cost for subsequent BOD treatment.

Cellulose fibers can be caused to develop significant strength increases by addition of cationic starch to the wet end of a paper machine. The positive charge on the starch will be bonded to anionic groups on the fiber, usually carboxyl and sulfonate groups, and a firm coulombic bond is formed. Cationically charged starch added to a pulp mixture will also attract filler substances such as clay, titanium dioxide and other anionically charged particles. In this manner the ash of a sheet can be increased, sometimes with an attendant loss in strength.

Cationic starch in a paper mill furnish can have additional benefits beyond ash retention and strength. Properly added cationic starch can improve formation in a sheet. With an even distribution of fibers, the natural attraction of water for ionized anionic groups can be counteracted by the addition of cationic counter ions in the form of cationic starch. The flocculation effect that occurs produces much improved drainage on the paper machine. The result is increased speed on the machine yielding greater production rates and overall efficiency. To a paper mill, increased production means increased profitability.

Surface sizing of paper with starch is probably the most common method of applying starch to paper. Surface application is done to improve the appearance of the paper and more importantly, to improve the printing characteristics. Surface sizing will coat over and lay fibers, it will flow into the sheet and fill voids and it will add strength to the sheet by bonding fibers and by forming a film. Generally, the size applied is intended to remain on the surface of the sheet for printing purposes. However, some penetration is necessary to effect bonding. The degree of penetration is usually controlled by the viscosity of the starch applied. The chemical and physical nature of the film applied is critical to determining paper quality and is directly related to the kind of starch applied. The factors of retrogradation, water holding properties, film continuity, reduced initial swelling temperature of the starch during cooking and viscosity control differentiate starch products chosen for surface sizing. Surface sizes are also applied to paper sheets to be used for coated paper to prevent penetration of the coating into the sheet. Surface sizes are sometimes reacted with aldehyde or resin type insolubilizing agents to prevent loss of starch in the printing operation when the sheet is wetted with water. The ease and speed of reaction with these reagents is somewhat dependent on the structure of the starch. Specialty starches generally react faster and more completely.

Surface sizing with cationic starches offers the added property of substantivity to the cellulose fiber. Most mills will use broke in their furnish. Repulping of broke that previously had a starch treatment can mean loss of the starch to the mill effluent and subsequent BOD treatment. Cationic surface sizes are substantive to the fibers and are not lost to the effluent. Thus pollution problems are minimized.

Coated paper is comprised of a base sheet, a pigment such as clay or calcium carbonate or titanium dioxide and a binder such as starch. Coatings are added to paper to provide surfaces for printing that are superior to a sheet with only a starch size. Pigment adds opacity, whiteness, gloss and brightness. Pigment also provides a smoother surface which is required by some forms of printing such as letterpress and gravure. Coated paper will also use synthetic latexes in the binder system. Starches are found to be very compatible with latexes and in particular with styrene-butadiene latex, the dominant kind used. Coated papers designed for offset printing will require that an insolubilizer be added in the same manner as described above for surface sized sheets.

Current developments in coating technology are placing extreme demands on the binder and the viscosity of the binder. In some low pigmented systems, machine speeds are at 1500 meters/minute and faster. The thixotropy-dilatancy response of binders is putting high stress on non-derivative starches. This leads to the need for maintaining a coating in the nip and not allowing the system to run dry.

Another factor is the need for thinner viscosity binders. This puts a heavy demand on the water holding property of the starch. A retrograded starch will result from a non-derivative made to a low viscosity. Specialty starches are providing the properties needed for these operations. These conditions relate directly to the retrogradation discussion in previous paragraphs.

Starch Cooking

Good sizing demands good cooking of the starch. The most frequent complaint fielded by starch suppliers is to correct the method of cooking. Good cooking requires sufficient water to hydrate the granules, sufficient temperature to go beyond the gelatinization point, sufficient agitation to stir the mixture, and sufficient time to thoroughly cook out the granules. A papermaker is not getting full value from his starch as an adhesive and as a film former if the product is not thoroughly cooked and dispersed. Batch cooking is most likely to produce variable results because of the many human variables introduced. Jet cooking tends to minimize the chances for poorly hydrating a starch. However, experience shows that a poor choice of conditions can result in a poorly dispersed starch.

Atmospheric Cooking

Atmospheric or kettle cooking is still practiced in many mills. Steam is added by direct injection or to a jacketed kettle. For best results, starch should be added from bags or a bulk system to room temperature water with good agitation. Mixers should have a bottom blade as close to the bottom of the tank as possible. A second upper blade is also recommended. Stirring must continue during addition of the starch to break up all of the lumps and give a smooth uniform mixture. Steam may then be added to raise the temperature to 190-195°F for at least 20 minutes. Longer cooking may not be necessary but will not ordinarily be detrimental unless significant evaporation occurs and the water-solids ratio is changed. This will result in a different viscosity and size pick-up will be affected. Direct addition of steam can also add dilution to a starch paste and must be accounted for in the cooking procedure. Starch sizes prepared by batch cooking are frequently diluted to obtain a running viscosity. Dilution must always be with warm water to prevent shock and potential lumping due to localized retrogradive effects. Prepared size may be stored for machine application. Storage temperatures should be 140°F or more to prevent microorganisms from degrading the starch.

Thermomechanical/Thermochemical Cooking

Jet cooking or thermomechanical cooking may be practiced with[2] or without[3] added catalysts. When acid forming catalysts or peroxides are used in jet cooking it is termed thermochemical conversion. The principal difference between batch cooking and jet cooking is the degree of hydration given to the starch. Jet cooking with "excess" steam in a continuous cooker will yield significantly lower viscosities at equal solids because the starch chains are separated by water molecules instead of having one starch chain touch and drag on another. The mechanical shear developed during flashing the steam to the atmosphere may cause some lowering of molecular weight although no evidence has been found to substantiate this possibility.

A principal advantage in thermal conversion is the potential for using unmodified starch at a cost saving over pre-converted, viscosity controlled starches. Other advantages may include decreased manpower demand, automation and uniform viscosities. If, however, acid modified starch or acid catalysts are used, a strong potential for retrogradation and set-back results and users may find sludges and precipitates forming if the starch is not stored properly and used promptly. Use of specialty starch will alleviate this problem. Operation of the equipment requires constant attention because of the potential for malfunction and allowing uncooked starch to go forward into the cooked starch without being gelatinized. Storage tank temperatures are generally high enough to partially swell the passed starch. Viscosity will increase dramatically and once in the size press bath the lumps formed will be rejected and accumulate giving severe running problems.

Thermochemical conversion is generally practiced using ammonium persulfate as the catalyst[2]. A holding chamber attached allows molecular weight reduction so that 37% solids may be reduced to workable viscosities. The variables of time, temperature, and amount of catalyst allow considerable flexibility. The principal action is hydrolysis and molecular weight reduction yielding a high population of short chain molecules. These conditions may lead to serious color problems, retrogradation and set-back with precipitation of sludges and what some term as "amylose", a misnomer for precipitated starch.

Starches for Wet End Application

Commercially available cationic starches for wet end application are quaternary and tertiary products. These products have been available since about the mid 1950's and no new basic chemistry has been developed since that time. The development in the late 1940's and early 1950's of starch ethers and esters made in the original granule form led to a torrent of starch derivatives for industrial use. Very few of these became commercial. This is possibly because the functions that were required by the industrial and food markets were far more limited than the imagination of the chemist. Cationic starches, however, have filled a basic need in the paper industry and will continue to for many years to come. All wet end cationics are used as native unmodified viscosity starches.

The first commercially successful cationic starch was prepared by reacting 2-diethylaminoethyl chloride with starch in a highly alkaline aqueous suspension[4]. The reaction proceeded through nucleophilic displacement of chloride and a starch ether was formed. The ether derivative was quaternized and made cationic by addition of a mineral acid. Teriary amino starches so made are used predominantly at acidic pH where the charge remains maximum. As neutral pH is approached, the charge is diminished until at weakly alkaline pH the charge is lost.

Other teriary amino starch derivatives can be made using glycidyl reagents[5]. These must also be quaternized with a mineral acid. It is indicated that the PK values are about 7.5[6].

Quaternary ammonium starch ethers are prepared by reacting the chlorohydrin form of 2,3-epoxypropyltrimethylammonium chloride with starch under conditions very similar to the reaction of starch with tertiary reagents. The quaternary ammonium ethers have the advantage of having a stable cationic charge over the pH range of that used in paper making. The starches are not dependent on pH adjustment to maintain the cationic charge since the reagent itself is cationic. No data are available for comparison. However, it is very likely that in world wide use, the quaternary derivative is dominant.

Starches containing both cationic and anionic groups in the same molecule perform as amphoteric starches[7]. The cationic charge may be tertiary or quaternary. The anionic charge may be a phosphate, carboxylate or sulfonate ion. The proper balance of cationic and anionic charges is recommended for specific situations and isoelectric points can be adjusted. Amphoteric starches are most commonly used in paper systems containing high amounts of alum and in systems which are described as "dirty". "Dirty" systems frequently have measurable to high amounts of black liquor in the furnish and have a significantly high demand for cationic or amphoteric starch. Potato starch, when made cationic, is an amphoteric starch because of naturally occurring phosphate groups in the potato starch.

Emulsification

Cationic starches will have an increasing importance in papermaking with the revived interest in alkaline sizing of paper. Systems using the alkaline sizing agents, alkenyl succinic anhydride (ASA) or alkyl ketene dimer (AKD) will need a cationic polymer to emulsify the oleophilic sizing agent and cationic starch is the polymer of choice. Alkaline sizing has many benefits, particularly if calcium carbonate, precipitated, is used as the filler. The sheet has increased brightness and fiber substitution becomes a reality. Thus quality and economics are positive results of the changeover from alum-rosin sizing. Additional cationic starch may be added at the wet end to give increased sheet strength.

For alkaline sizing both corn and potato cationic starches are suitable. In practical usage, however, potato cationics with about 0.3% nitrogen added appear to be the product of choice. These are used at the wet end and for emulsification.

Base starches used for cationics are corn, potato and waxy. Corn remains dominant world wide although potato is a close second. Waxy starches are mainly used in the USA. The amount of waxy starch available is dependent on demands for waxy starch in the food industry.

Starches for Size Press Application

Before discussing specialty starches for size press applications, it should be noted that the largest volume of starch applied at the size press is either enzyme converted or ammonium persulfate hydrolyzed native corn starch. These are economic considerations and the properties resulting from these treatments do not compare to the properties provided by specialty starches. It is also fair to state that all grades of paper do not require a specialty starch and it is only through the need for improved paper performance that specialty starches are required.

The dominant specialty starch used in size press applications is the hydroxyalkyl derivative commonly made with ethylene oxide to yield a hydroxyethyl starch[8]. Low substituted hydroxypropyl starch has similar properties. These starches were introduced to the industry in 1950 and they have grown in importance each year. Recent expansions in production by major producers has indicated their continuing dominance of the specialty starch surface sizing and coating field.

Hydroxyethyl starch is manufactured by addition of ethylene oxide to a strongly alkaline aqueous slurry of starch and reacting for 17-24 hours, at 100-110 ^{o}F[9]. Sodium chloride is added as a swelling inhibitor and the starch is recovered in the original granule form after washing and drying. Because specialty starches are used in higher solids conditions than would be allowed by unmodified starch, the prepared starch derivative is acid-modified in the same reaction tank to controlled viscosity levels. In this manner, the manufacturer can supply a product with the by-products of reaction, the soluble starch generated by acid treatment and the salts used in reaction washed out. Specialty starch prepared in this manner is in contrast to those products prepared at the mill site by enzyme hydrolysis or ammonium persulfate catalyzed depolymerization wherein the entire preparation must be used.

Significant volumes of low oxidized starch are used at the size press. These starches are made by treatment in alkaline suspension with sodium hypochlorite so that from 1 to 2% active chlorine acts on the starch. The reaction is simple to perform. However, the reaction products are complex. Chain scission occurs at the same time that carboxyl and carbonyl groups are formed in the starch. It is most desirable to prepare the highest ratio of carboxyl to carbonyl as possible and this reaction is a function of the pH in the slurry. Carbonyl groups contribute to cross-linking reactions where hemiacetals are formed. These are the greatest influence on viscosity build-up when low oxidized starches are used at neutral to acidic pH. When highly oxidized starches are used, the influence of the carbonyl group is much less because the molecular weight has been lowered and solubility is increased. The complexity of the reaction is shown in Figure 1.

Surface sizing with cationic starches is increasing in popularity because of the substantivity of the starch to the fiber. Most pulp formulas include a significant percentage of broke. Paper surface sized with cationic starch does not lose the starch to the effluent on repulping and significant environmental improvement is obtained through having a lower BOD in the plant effluent. As environmental concerns grow, so will the use of cationic surface sizes. Finland is an example of a country that has shown this concern and where cationic surface sizes are in general use.

Starches for Coating Applications

Considerable volumes of enzyme converted starch and thermochemically converted starch are consumed as binders for coating colors. Certain qualities of paper can tolerate the physical characteristics given to the coating by these binders wherein one would experience retrogradation and set-back, varying viscosities, loss of water holding properties, lowered film clarity and smoothness and poor storage properties, particularly for viscosity stability. All of these detrimental qualities can be overcome with the use of a specialty derivatized starch containing hydroxyethyl substitution. These premier products will allow the coater to upgrade quality and enter into the specialty coated market. Hydroxyethyl starches are neutral molecules that show extreme resistance to retrogradation and exhibit all of

Principal Products

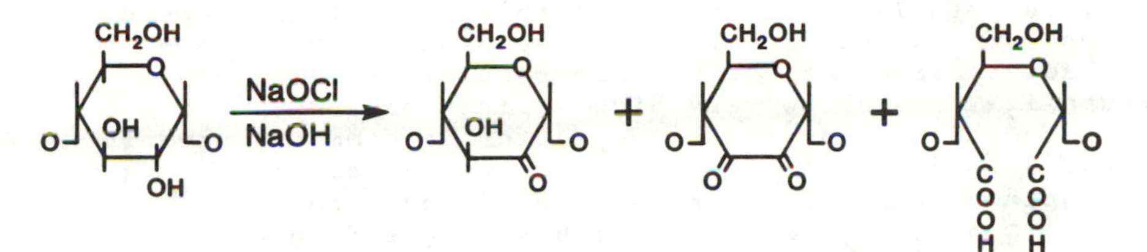

Other Species

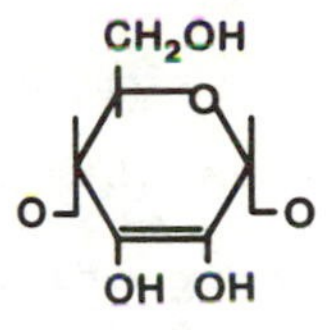

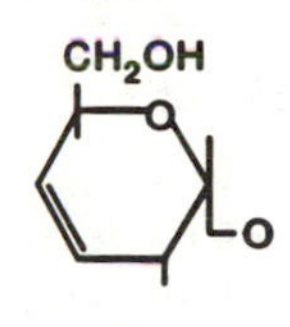

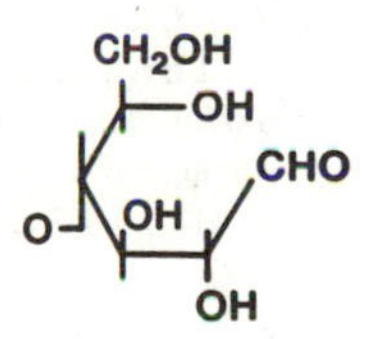

At 0.05 Mole Cl_2 per AGU

Carboxyls	1 per 36–97 AGU
Carbonyls	1 per 110–200 AGU

Figure 1. Oxidation of Starch

the positive features in a coating color as contrasted to the properties of enzyme converted and thermochemically converted starches mentioned above.

Many coatings will use latex as a binder either singly or in combination with starch. Addition of specialty starch allows cost control and depending on the quality demanded, the level of starch can be varied. Many formulas will use 16 to 18 parts of binder based on the clay and frequently ratios of 5 starch and 12 latex are used.

Coating speeds are increasing and with this comes increasing demands on the binder to be thixotropic. Hydroxyethyl starch excels in this property and speeds approaching 5000 ft/min have been successfully used. Blade coating has become more popular and Bill Blade applications allow very high solids coatings putting less demand on drying. The latter is commonly used for No. 5 coated grades usually seen as newspaper supplements and for varied advertising forms.

Generally coated sheets have higher gloss and will print better if the surface fibers have been laid and do not come out on the offset blanket.

Starch Research Frontiers

For all the excellent properties that starch exhibits there are still continuing areas where starch does not match the properties of synthetic molecules. Starch is very hydrophilic and generally this is used to its advantage. However, water resistance in a starch film has long been a property that the starch user has sought. Insolubilizers such as glyoxal, melamine-formaldehyde resins, and other formaldehyde based resins have been an attempt to render the film water insensitive. Restrictions on formaldehyde based resins have almost eliminated this form of insolubilizer. Thus, research is needed to take advantage of the hydrophilicity of starch and at the same time render it water resistant.

Starch has found a small but growing market in biodegradable bags and other degradable packages including some diaper products. Starch as a food for organisms to cause gradual disappearance of the starch based container seems like a logical use.

Starch can be changed in character by grafting molecules onto it by free radical reactions[10]. Most of these grafted starches are extremely hydrophilic. The product named Super Slurper was developed at U.S.D.A. laboratories in Peoria and the patents resulting from this work have been widely licensed[11]. Other interesting approaches to grafting are demonstrated by Kightlinger[12] using a starch derivative as the starting product. More grafting work will undoubtedly produce new and different properties while still making use of starch as the base material.

Starch is one of the most abundantly produced carbon sources to be renewable on an annual basis. The chemistry of glucose and the possibility of having starch be the raw material of choice for synthetics has attracted researchers for many years. Fermentation reactions as well as chemical modification continue to be attractive. Economics has been the principal reason for lack of progress in using starch as a carbon source. However, as supplies of natural gas and petroleum decrease, new interest in starch will be found.

Literature Cited

1. P. Sutton, J. Pearson and H. O'Brien, Paper Production Growth Continues for Fifth Record Year in a Row, Pulp and Paper, 48-57, August 1988.
2. Lauterbach, G.E., U.S. Patent 3,211,564 (1965). Brogly, D.A., TAPPI 61, No. 4, pp 43-45 (1978).
3. Winfrey, V.L., Black, W.C., U.S. Patent 3,133,836 (1964).
4. Caldwell, C.G., Wurzburg, O.B., U.S. Patent 2,813,093 (1957).
5. Y. Merle, Compt. Rend., 246, 1425. Harrier, G.C., Leonard, R.A., U.S. Patent 3,070,594 (1962).
6. Wood, J.W., Mora, P.T., J. Org. Chem., 27, pg. 2115 (1962). Antal, M., Toman, R., Die Starke, 36, 143 (1984).
7. Caldwell, C.G., Jarowenko, W., Hodgkin, I.D., U.S. Patent 3,459,632 (1969).
8. Kesler, C.C. and Hjermstad, E.T., U.S. Patent 2,516,633 (1950).
9. Kesler, C.C., Hjermstade, E.T., Methods in Carbohydrate Chemistry, Vol. IV, 304-306 (1964), Academic Press, NY, NY.
10. Fanta, George F., Doane, W.M., Grafted Starches in Modified Starches: Properties and Uses, Chapter 10, CRC Press, Boca Raton, Florida, 1986. O.B. Wurzburg, Ed.
11. Weaver, M.O., Bagley, E.B., Fanta, G.F., Doane, W.M., U.S. Patents 3,935,099, 3,981,100, 3,985,616, 3,997,484 (1976).
12. Kightlinger, A.P., U.S. Patent 4,301,017 (1981).

RECEIVED January 24, 1990

Chapter 24

Saponified Starch-*g*-poly(acrylonitrile-*co*-2-acrylamido-2-methylpropanesulfonic acid)

Influence of Reaction Variables on Absorbency and Wicking

George F. Fanta and William M. Doane

Northern Regional Research Center, Agricultural Research Service, U.S. Department of Agriculture, Peoria, IL 61604

Mixtures of acrylonitrile and 2-acrylamido-2-methylpropanesulfonic acid ($AASO_3H$) were graft polymerized onto gelatinized starch, and the resulting graft copolymers were saponified with sodium hydroxide to yield absorbent polymers. To simulate a commercial process, saponifications were carried out by adding sodium hydroxide to reaction mixtures immediately after graft polymerization, as opposed to carrying out saponifications on isolated and dried graft copolymers. A number of variables in the graft polymerization, saponification, and methanol-precipitation procedures were investigated. Addition of aluminum chloride to the saponificate was studied as a method to reduce gel blocking and to improve wicking. Fast-wicking absorbents were obtained only if an equivalent amount of sodium hydroxide was added to the saponificate along with aluminum chloride and if methanol-precipitated polymers were neutralized with acid before drying. Based on these results, a mechanism for interaction between aluminum chloride and absorbent polymer was proposed. A series of graft polymerizations and saponifications was carried out with monomer mixtures containing 0-5 mole % $AASO_3H$. Conversion of monomers to polymer increased with decreasing percentages of $AASO_3H$; however, incorporating even small amounts of $AASO_3H$ in monomer mixtures increased the absorbency of final products.

The reaction of starch with ceric salts leads to formation of macroradicals which can initiate the polymerization of a wide variety of monomers to yield starch graft copolymers. Graft copolymers of acrylonitrile and starch are easily prepared by this technique (Equation 1), and saponification of starch-g-polyacrylonitrile (Equation 2) yields absorbent polymers that can take up hundreds of times their weight in water as they swell to form a gel-like mass (1). Absorbent polymers have a number of practical uses, such as absorption of body fluids (e.g. in disposable diapers), coating of seeds to enhance germination, coating of

$$\text{Starch} + CH_2{=}CHCN \xrightarrow{Ce^{+4}} \text{Starch-}(CH_2\text{-}\underset{\underset{CN}{|}}{CH})_x\text{—} \tag{1}$$

$$\text{Starch-}(CH_2\text{-}\underset{\underset{CN}{|}}{CH})_x\text{—} \xrightarrow{NaOH} \text{Starch-}(CH_2\text{-}\underset{\underset{CO_2Na}{|}}{CH})_y\text{—}(CH_2\text{-}\underset{\underset{CONH_2}{|}}{CH})_z\text{—} + NH_3 \tag{2}$$

Absorbent Polymer

plant roots to reduce or eliminate transplant shock, addition to sandy soils to enhance their water-holding capability, and incorporation into fuel filters to remove aqueous contaminants.

If a mixed-monomer system containing 90-99 mole % acrylonitrile and 1-10 mole % 2-acrylamido-2-methylpropanesulfonic acid ($AASO_3H$) is substituted for acrylonitrile, the final product has a higher fluid absorbency than one prepared from acrylonitrile alone (2). $AASO_3$ also shortens the reaction time needed for alkaline saponification, possibly by interrupting the highly conjugated intermediate that first forms, thereby increasing its accessibility to alkali in the second step of the saponification.

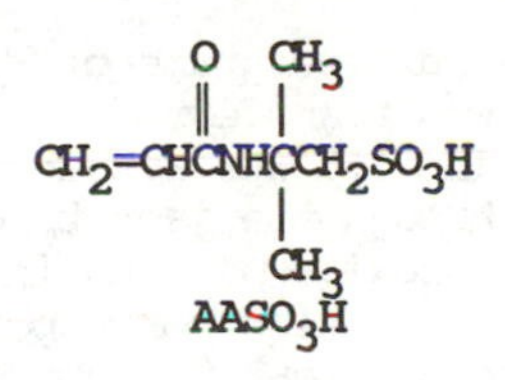

In our first experiments with this monomer system (2), graft polymerizations were run on a small scale, and intermediate graft copolymers were isolated and vacuum dried prior to saponification. Although these laboratory experiments established the value of incorporating $AASO_3H$ as a comonomer in the reaction mixture, potential scale-up to a practical commercial process required that some modifications be made. In a 1987 publication (3), we therefore examined some variables that must be considered if the $AASO_3H$-containing absorbent polymer is to be produced economically in larger than laboratory quantities. The present research extends this 1987 work and also includes a study of aluminum chloride addition (4) as a practical method to promote rapid wicking of water into absorbent polymer particles.

Experimental

Materials. Cornstarch was Globe 3005 from CPC International, Inc. Acrylonitrile (99+%) was from Aldrich Chemical Company, Inc., and $AASO_3H$ was Lubrizol 2404 from the Lubrizol Corporation. Both monomers were used as received. Ceric ammonium nitrate was Fisher Certified A.C.S. Grade.

Graft Polymerization. In the first series of polymerizations, a 2 L resin flask equipped with paddle stirrer, condenser, and nitrogen inlet was charged with 40.0 g (dry basis) of starch and 668 mL of deionized water. In later polymerizations, 48.0 g of starch was used to obtain a product containing a higher percentage of the less expensive starch component. A slow stream of nitrogen was passed through the slurry, and starch was gelatinized by heating the stirred dispersion for 1 h at 95°C. The starch slurry was cooled to 25°C, and 11.72 g of $AASO_3H$ (5 mole %) and 57.0 g of acrylonitrile (95 mole %) were added followed after 5 min by a freshly prepared solution of 1.352 g of ceric ammonium nitrate in 12 mL of 1 N nitric acid. The mixture was stirred for 1 h at 25°C (exotherm controlled with ice bath), a second 1.352 g portion of ceric ammonium nitrate in 1 N nitric acid was added, and the mixture was stirred for another hour at 25°C. Sodium hydroxide solution (6 N) was added to adjust the pH to 7-8, and 600 g of the mass was then transferred to a stirred sigma reactor for saponification. The remaining material was diluted with excess ethanol; and the precipitated graft copolymer was isolated by filtration, washed with ethanol, and allowed to air-dry.

Saponification. Saponifications were carried out in a Readco steam-jacketed sigma double-arm mixer with 1 quart working capacity (Teledyne Readco, York, PA). There were holes in the lid to accommodate a reflux condenser and to permit addition of reagents.

Procedure A. Six hundred grams of reaction mass from the graft polymerization was transferred to the mixer and washed in with 5 mL of water. The lid was set in place, steam at atmospheric pressure was passed through the jacket for 30 min, and a solution of 22.8 g of sodium hydroxide pellets in 81 mL of water was added. This amount of sodium hydroxide corresponds to 0.73 mole per mole of acrylonitrile charged. Heating was continued for 90 min, and cooling water was then passed through the jacket. After 15 min, 25 mL of water was added, and cooling was continued for another 15 min to yield a viscous saponificate with a final temperature of 20-25°C.

The absorbent polymer was isolated by stirring 100 g of saponificate with 400 mL of methanol in a Waring blender for 1 min at high speed (initial precipitation). Supernatant was decanted from settled solid, 200 mL of fresh methanol was added, and the mixture was again blended for 1 min. The mixture was transferred to a beaker, stirred for 30 min, and the solid was collected on a filter. The methanol-wet solid was weighed and divided into four equal portions, which were then stirred for 30 min with 25 mL of either pure methanol or solutions prepared by dissolving concentrated HCl (usually 0.5 g or 1.0 g) in 100 mL of methanol.

Polymers were isolated by filtration, dried under vacuum at 40°C, and ground to pass 40 mesh. The pH was determined after dispersing 0.5 g of polymer in 50 mL of water.

<u>Procedure B</u>. Nine grams (dry basis) of ethanol-precipitated and air-dried graft copolymer was mixed with 81 mL of 0.7<u>N</u> sodium hydroxide in a 250-mL Erlenmeyer flask. The flask was heated on a steam bath for 10 min, loosely stoppered, and placed in a 100°C oven for either 1.5 or 3 h. Only about 1 g of volatile material was lost during heating. The cooled reaction mass was methanol-precipitated as described in Procedure A.

<u>Procedure C</u>. Six hundred grams of the reaction mixture from graft polymerization was added to the sigma mixer and was washed in with 48 mL of water. A solution of 21.3 g of sodium hydroxide pellets in 200 mL of water was added, and atmospheric pressure steam was passed through the jacket of the stirred mixer for 1 h and 40 min. Cooling water was then run through the jacket for 15 min, and 90 g of saponificate (temperature about 30°C) was removed for precipitation. A solution of either 10, 12.5 or 15 g of $AlCl_3.6H_2O$ in 25 mL of water was then added. The saponificate was stirred for 30 min, and another 90 g portion was removed for precipitation. An aqueous solution of sodium hydroxide equivalent to that needed to convert aluminum chloride to aluminum hydroxide was then added (for 12.5 g $AlCl_3.6H_2O$, a solution of 5.53 g NaOH pellets in 20 mL of water was used). The saponificate was again stirred for 30 min, and a final 90 g portion was removed for precipitation. The three portions of saponificate were methanol-precipitated as described in Procedure A.

<u>Tests for Absorbency and Wicking</u>. To measure water absorbency, 1-2 mg of polymer (accurately weighed) was allowed to swell for 30 min in 50 mL of distilled water and the weight of water-swollen gel was then determined after separation of unabsorbed water by screening (<u>5</u>) (30 min drain time). For 0.9% NaCl absorbencies, 20-25 mg of polymer was used.

Wicking was measured by adding 50±0.5 mg of absorbent polymer to a folded (fluted) filter paper cone prepared from an accurately tared circle of 9 cm Whatman 54 paper. The cone was lightly tapped to settle the polymer into the tip, and the tip of the cone was then held for 60 sec in a 9 cm diameter Petrie dish containing 25 mL of water. Water wicked up the entire length of the paper in 60 sec. Excess water was then allowed to drain from the paper by contacting the tip for 60 sec with a circle of dry filter paper on a square of absorbent towel. The weight of wet paper plus water-swollen polymer was determined, and the absorbency of the polymer in g/g was then calculated after correcting for the weight of dry paper and the amount of water absorbed under identical conditions by the paper alone in the absence of polymer. Each test was repeated five times and the results were averaged.

<u>Acid Hydrolysis</u>. Ethanol-precipitated and air-dried graft copolymers were acid-hydrolyzed to determine both weight % synthetic polymer in the graft copolymer (% add-on) and the $AASO_3H$ content

of grafted branches. Two grams of graft copolymer (accurately weighed) were heated under reflux for 1.5 h with 150 mL of 0.5N hydrochloric acid. The mixture was centrifuged in a tared bottle and the solid was washed four times by suspending in water and centrifuging. The solid was freeze-dried in the centrifuge bottle and the weight of dry solid was determined. Percent add-on was calculated from weight loss due to acid hydrolysis of starch. $AASO_3H$ content of isolated polymer grafts was determined from sulfur analysis (6).

Results and Discussion

Initial Experiments to Define Variables. This series of graft polymerizations was carried out with acrylonitrile-$AASO_3H$ mixtures containing 5 mole % (17 weight %) $AASO_3H$, and conditions were similar to those reported by us earlier (3). Previous research showed that grafting is the predominant reaction in these polymerizations and that only minor amounts of homopolymer are formed. Dimethylformamide (DMF) extraction of products prepared from gelatinized starch and monomer mixtures containing 1-5 mole % $AASO_3H$ removed less than 10% of the total polymer as DMF solubles (2). Moreover, these soluble fractions were not all homopolymer, as evidenced by carbohydrate absorptions in infrared spectra. Intrinsic viscosity determinations carried out previously indicate that graft molecular weights are high. In the absence of $AASO_3H$, we have observed $[\eta]$ values of about 10, corresponding to molecular weights on the order of 7-8 x 10^5, as calculated from the Mark-Houwink equation (7). A monomer system composed of 25% $AASO_3H$ and 75% acrylonitrile also produced high molecular weight polymer, since $[\eta]$ values of 5 and 10 were observed for polymer isolated from DMF-soluble and DMF-insoluble fractions, respectively (8). The question of crosslinking in these absorbent polymers during graft polymerization and alkaline saponification has been addressed in a previous publication (9).

To more closely simulate a process that might be used commercially, graft copolymers were not isolated before saponification. Instead, reaction mixtures after graft polymerization were transferred directly to a sigma mixer and were allowed to react with sodium hydroxide without prior removal of unreacted monomers. Residual monomers would be converted to sodium acrylate under the alkaline saponification conditions and would then be removed from the final product by precipitation and washing with methanol. Alternatively, unpolymerized acrylonitrile could be easily removed by azeotropic distillation prior to saponification, and we have successfully used this technique in earlier preparations of starch-g-polyacrylonitrile carried out in the absence of $AASO_3H$ (10). Although residual cerium salts cannot be easily removed from the final product, these salts are present in only low concentrations. Moreover, cerium salts have a low order of toxocity (11).

The influence of methanol precipitation conditions on water absorbencies of final products is shown in Table I, and results of this study may be summarized as follows: 1) If the saponificate is stirred for 1 h open to the atmosphere to allow some of the ammonia to escape before methanol precipitation, the polymer has about the

Table I. Influence of Precipitation Conditions and Aging of Saponificate[a]

Treatment of Saponificate	Aging Time Before Precip.	Initial Precip., mL MeOH per 100 g Saponificate	Final wash, g HCl per 100 mL MeOH	pH of Dry Product	H_2O Absorb., g/g
Precip. Immediately	0	300	0	10.3	1330
			1.0	6.8	1410
Stir 1 hr, open to atmosphere	0	300	0	10.5	1220
			1.0	7.0	1480
	0	400	0	10.7	1390
			1.0	7.0	1650
	0	500	0	10.8	1420
			1.0	6.9	1440
	6 days	400	0	10.6	1510
			1.0	7.0	1725

[a] 40 g of starch used for graft polymerization. Add-on: 58%; Conversion of monomers: 80%. Procedure A (see Experimental) used for saponification.

same properties as one isolated immediately after saponification. 2) Allowing the saponificate to age in a covered beaker for 6 days at room temperature before methanol precipitation does not adversely affect and may even improve absorbency. 3) Increasing the amount of methanol used in the initial precipitation from 300 to 500 mL per 100 g of saponificate does not greatly change the absorbency of the final product.

Methanol-precipitated polymers are alkaline (pH 10.3-10.8) when dispersed in water; however, products having a neutral pH are easily obtained by stirring precipitated polymers with a solution of concentrated HCl in methanol before vacuum drying. In most cases, neutralized products have higher absorbencies than those isolated without HCl treatment. After considering the results of Table I and also the convenience of various work-up procedures, we precipitated the absorbent polymers in subsequent reactions immediately after reaction with alkali without allowing the saponificate to either lose ammonia or to age at room temperature. Also, 400 mL of methanol per 100 g of saponificate was used for the initial precipitation.

We next studied the influence on water absorbency of some selected variables in both the graft polymerization reaction and the alkaline saponification (Table II). Results presented in the table lead to the following conclusions: 1) Vacuum drying the absorbent polymer at 60°C rather than 40°C reduces absorbency slightly. 2) Subjecting the saponificate to reduced mechanical shear by stirring intermittently rather than continuously during alkaline saponification has no influence on water absorbency. 3) Gelatinizing starch under milder conditions (30 min at 85°C rather than 60 min at 95°C) has little effect on water absorbency. 4) A lower nitric acid concentration in the graft polymerization recipe (0.1N rather than 1N HNO_3 used to dissolve ceric ammonium nitrate) does not alter absorbency properties. 5) Initiation with only one portion of ceric ammonium nitrate (rather than with two portions added one hour apart) reduces both the conversion of monomer to polymer and the water absorbency of the final product. Based on these results, starch in future reactions was gelatinized by heating for 60 min at 95°C, to more closely simulate conditions of continuous steam jet cooking (12), which would be used commercially. Also, polymerizations were initiated with two portions of ceric ammonium nitrate dissolved in 1N HNO_3, saponification mixtures were stirred continuously rather than intermittently (as a matter of convenience), and absorbent polymers were dried at 40°C.

We next reinvestigated some saponification conditions reported in our first publication (2) on the starch-acrylonitrile-$AASO_3H$ system, since these conditions led to water absorbencies in excess of 2000 g/g. In this early work, graft copolymers were first isolated from reaction mixtures and dried. Saponifications were then carried out in small Erlenmeyer flasks without stirring, after first dispersing 1 g of dry graft copolymer in 9 mL of 0.7N sodium hydroxide solution. For the current study (Table III), ethanol-precipitated and air-dried graft copolymer fractions were again saponified in unstirred flasks; and conditions were the same as those of our early publication (2), except reactions were scaled up

Table II. Influence of Selected Reaction Variables[a]

Change in Reaction Procedure	Add-on, %	Monomer Conversion, %	H_2O Absorb, g/g
None (control)[b]	58	81	1480
Dry absorbent at 60°C	58	81	1350
Stir intermittently during saponification[c]	58	80	1450
Gelatinize starch for 30 min at 85°C	58	80	1440
Use 0.1N HNO_3 to dissolve ceric ammonium nitrate	58	81	1500
Use only one portion of ceric ammonium nitrate	54	67	1040

[a]40 g of starch used for graft polymerization. Procedure A (see Experimental) used for saponification. Polymers treated with a solution of 1.0 g conc. HCl in 100 mL MeOH; pH of final products: 6.6-7.0.

[b]Conditions are those given in the Experimental Section.

[c]Stirred for a 10 sec. period every 10 min.

by a factor of 9. Methanol precipitation of these saponificates, treatment with methanolic HCl, and drying were carried out by the same procedure used for reactions run in the stirred sigma reactor. As shown in the first two entries of Table III, absorbencies in the 2200-2300 g/g range were obtained, in agreement with our earlier work. It is interesting that lower absorbency values are obtained if the NaOH normality is increased to 0.95 (the calculated normality of the saponification mixture used for Procedure A, Experimental) or if an amount of 0.7N NaOH equivalent to 81 mL of 0.95N alkali is used.

Table III. Saponification of Isolated and Dried Graft Copolymer[a]

Volume and Normality of NaOH[b]	Saponification Time, hr	H_2O Absorb., g/g
81 mL of 0.7N	3	2180
81 mL of 0.7N	1.5	2340
81 mL of 0.95N	1.5	1290
110 mL of 0.7N	1.5	1540

[a]40 g of starch used for graft polymerization. Polymer isolated by ethanol precipitation and air dried. Procedure B (see Experimental) used for saponification. Polymers treated with solutions of either 0.5 g or 1.0 g conc.HCl in 100 mL methanol; pH of final products: 6.5-6.9.

[b]Used to saponify 9 g of graft copolymer.

To better understand how absorbency varies with saponification method, (i.e. stirred sigma reactor vs. unstirred flask), another graft polymerization was carried out; and the reaction mixture after polymerization was divided into two portions, which were treated as follows (Table IV): 1) The first portion (300 g) was mixed with a solution of 10.65 g of sodium hydroxide pellets in 124 mL of water (quantities calculated to give a final concentration equal to 9 g of graft copolymer in 81 mL of 0.7N NaOH, assuming complete conversion of monomers to polymer). Ninety grams of this alkaline mixture was saponified by heating for 1.5 h in an unstirred Erlenmeyer flask at 100°C, while the rest of the material was saponified in the stirred sigma reactor. 2) Graft copolymer was precipitated from the remaining portion of graft polymerization mixture by addition of ethanol. Nine grams of air-dried graft copolymer were then saponified in an unstirred flask with 81 mL of 0.7N NaOH, using the procedure of Table III; while 45 g of the dry copolymer was similarly saponified in 405 mL of 0.7N NaOH in the stirred sigma reactor.

Table IV. Influence of Saponification Method [a]

Graft Copolymer used for Saponification	Method used for Saponification	H_2O Absorb., g/g
Saponified without Isolation	Unstirred Flask	1580
	Stirred Reactor	1470
Ethanol-Precipitated and Dried	Unstirred Flask	2130
	Stirred Reactor	1580

[a]48 g of starch used for graft polymerization. Polymers treated with a solution of 0.5 g conc. HCl in 100 mL methanol. pH of final products: 6.8-7.0.

Results of these four saponifications (Table IV) show that an absorbency of over 2000 g/g is obtained only if isolated and dried graft copolymer is saponified in an unstirred flask. Saponifications in the stirred sigma reactor with both isolated and unisolated graft copolymer and saponification in an unstirred flask with unisolated graft copolymer all gave final products with lower water absorbencies (1470-1580 g/g). Reasons for these absorbency differences are still obscure. It is possible that precipitating and drying the graft copolymer before saponification increases hydrogen bonding between starch segments and thus reduces the amount of soluble polymer produced in the saponification reaction. Since solubles contribute nothing to the fluid absorbency of a given weight of polymer, absorbency values could increase with a decrease in the magnitude of the soluble fraction. The advantage of a glass reaction vessel over our stainless steel reactor might be due to minute amounts of dissolved iron and other metals produced through reaction of hot alkali with reactor components. It is well-known that multivalent ions sharply decrease the fluid absorbency of these products via ionic crosslinking of carboxyl constituents. Final answers to these questions must await further research.

Since isolation and drying of graft copolymers prior to saponification would not be a viable commercial procedure, future saponifications were carred out as before in the stirred sigma reactor with unisolated graft copolymer. However, the amount and concentration of sodium hydroxide used were the same as for Table IV.

<u>Addition of Aluminum Chloride to Improve Wicking</u>. Wicking is a term used to describe the diffusion of water or other aqueous fluids into a mass of absorbent polymer. Wicking is inhibited if

polymer near the surface of the sample absorbs water and swells rapidly, since further diffusion is blocked by the resulting layer of aqueous gel. Fast wicking is a desirable property, since a significant reduction in overall absorbency may result if gel blocking inhibits diffusion. We have attempted to quantitate the measurement of water wicking into absorbent polymers by a simple test described in the Experimental Section. In this test, the tip of a fluted paper filter containing 50 mg of absorbent polymer is placed in water, and the amount of water absorbed by the polymer in 60 sec is recorded.

Addition of aluminum chloride to saponified starch-g-polyacrylonitrile has been described in the patent literature (4) as a method for reducing gel blocking and improving wicking. The success of this method has been attributed to ionic crosslinking by formation of aluminum salts of carboxylate groups due to ion exchange between sodium and trivalent aluminum. Since many commercial uses for absorbent polymers require rapid wicking of aqueous fluids, we have examined the addition of aluminum chloride to saponified $AASO_3H$-containing graft copolymers prepared from a monomer mixture containing 5 mole % $AASO_3H$. The object of this research was not only to prepare a fast wicking, high absorbency polymer but also to clarify the mechanism of interaction between aluminum chloride and polymer.

Table V shows absorbencies and wicking values of a series of products prepared with different amounts of aluminum chloride. After alkaline saponification, each reaction mixture was divided into three portions which were treated as follows: 1) Absorbent polymer was methanol-precipitated as is with no aluminum chloride treatment. These products served as controls and provided absorbency values that could be compared with absorbencies presented in earlier tables. 2) Absorbent polymer was methanol-precipitated after addition of either 15, 12.5, or 10 g of $AlCl_3.6H_2O$ to the saponificate. 3) Polymer was methanol precipitated after adding aluminum chloride followed by an amount of sodium hydroxide solution equivalent to that needed to convert aluminum chloride to aluminum hydroxide. Wicking as well as absorbency was then determined both as is and after treatment with HCl in methanol to neutralize excess alkali.

A number of facts are apparent from Table V: 1) Wicking is poor in the absence of aluminum chloride. 2) Addition of aluminum chloride (but no NaOH) improves wicking but not to a large extent. There is an accompanying reduction in water absorbency to about 55% of its original value, irrespective of the amount of aluminum chloride added. 3) Addition of aluminum chloride plus NaOH yields absorbent polymers having widely different properties, depending on whether or not polymers are treated with methanolic HCl to reduce the pH before drying. Without HCl treatment, pH's are in the 9.4-10.0 range, wicking is poor, and water absorbencies are higher than those observed after treatment with aluminum chloride only. In fact, with 12.5 and 10 g of $AlCl_3.6H_2O$, water absorbencies are about the same as the control values with no aluminum chloride treatment. When products are neutralized with methanolic HCl before drying, absorbency is again reduced, and wicking increases dramatically. In summary, fast wicking absorbents are obtained

Table V. Absorbency and Wicking with Aluminum Chloride[a]

$AlCl_3 \cdot 6H_2O$, g	Treatment of Saponificate	Final Wash, g HCl per 100 mL MeOH	pH of Dry Product	Absorb., g/g		Wicking, g/g
				H_2O	0.9% NaCl	
15.0	None	0	10.2	1230		
		0.5	7.0	1390	74	8
	$AlCl_3 \cdot 6H_2O$	0	6.8	770		25
		0.5	6.5	770		31
	$AlCl_3 \cdot 6H_2O$;NaOH	0	9.4	980		9
		1.0	6.8	330	44	109
12.5	None	0	10.0	1390		
		0.5	6.9	1420	77	
	$AlCl_3 \cdot 6H_2O$	0	6.8	780	56	30
		0.5	6.4	790		36
	$AlCl_3 \cdot 6H_2O$;NaOH	0	9.4	1350		9
		0.5	7.6	620	52	92
		1.0	6.6	450	46	101
10.0	None	0	10.0	1220		
		0.5	6.9	1310	74	
	$AlCl_3 \cdot 6H_2O$	0	7.2	690		37
		0.5	6.6	690		50
	$AlCl_3 \cdot 6H_2O$;NaOH	0	10.0	1270		8
		0.5	7.7	790		41
		1.0	6.7	640	52	38

[a] 48 g of starch used for graft polymerization. Procedure C (see Experimental) used for saponification.

only if an equivalent amount of sodium hydroxide is used in conjunction with aluminum chloride and if the absorbent polymer is neutralized with acid before drying.

Although we have not rigorously established the mechanism for interaction between aluminum chloride and polymer, the data in Table V suggest a plausible hypothesis. In the absence of an equivalent amount of NaOH, addition of aluminum chloride apparently leads to some ionic crosslinking, as evidenced by reduced absorbency. This crosslinking, however, provides only minor improvement in wicking properties and is not overly dependent on the amount of aluminum chloride added. Since $AlCl_3.6H_2O$ has some solubility in methanol, it is probably largely removed from the polymer during methanol precipitation.

Since aluminum is amphoteric, water soluble salts, such as $Na^+Al(OH)_4^-$, are formed under the high pH conditions caused by addition of an equivalent amount of NaOH. The reported insolubility of sodium meta-aluminate in alcohol (13) suggests that these salts should be insoluble in methanol and should thus be largely incorporated into the final product when it is methanol-precipitated. However, if the polymer is not neutralized with methanolic HCl before drying, it remains sufficiently alkaline to allow these aluminum salts to be removed again from the polymer by solution in the excess water used to perform the absorbency test. Also, any ionic crosslinks previously formed by reaction with aluminum can be destroyed by ion exchange with sodium. When the absorbent polymer is neutralized with methanolic HCl, aluminum salts are no longer water-soluble to a significant extent and are thus locked into the final product. These water-insoluble aluminum salt domains should greatly facilitate the diffusion of water and are probably responsible for most of the desirable wicking properties imparted to the polymer.

Influence of $AASO_3H$ Content of the Graft Copolymer. Since all previous experiments were run with monomer mixtures containing 5 mole % $AASO_3H$, a series of graft polymerizations and saponifications were carried out with reduced percentages of $AASO_3H$ to determine the influence of monomer composition on % add-on, % conversion of monomers to polymer, and absorbency properties of the saponified polymer. With ceric ammonium nitrate initiation, $AASO_3H$ does not graft polymerize onto starch as readily as acrylonitrile, and Table VI shows that both % add-on and % conversion increase with decreasing percentages of $AASO_3H$ in the monomer mixture. Conversion of monomers to polymer is over 90% with monomer mixtures containing about 1 mole % $AASO_3H$ or less. It is also significant that the % $AASO_3H$ found in the grafted polymer more closely resembles that in the starting monomer mixture when the $AASO_3H$ content of the monomer mixture is low.

Alkaline saponifications of the four graft copolymers of Table VI were next carried out to determine the influence of mole % $AASO_3H$ on absorbent properties of final products. Saponifications were carried out under the conditions described in Table V; and in addition to a control product isolated with no aluminum treatment,

Table VI. Dependency of Add-on and Conversion on Percent $AASO_3H$ in the Monomer Mixture

$AASO_3H$ in Monomer Mixt.				
Mole %	Weight %	Add-on,[b] %	Conversion,[c] %	% $AASO_3H$ in Grafted Polymer[d]
5	17.05	54.3	82.9	9.2
2.33	8.53	55.9	88.6	5.9
1.13	4.26	57.0	92.4	4.1
0	0	58.2	97.3	--

[a]48 g of starch used for graft polymerization.

[b]Calculated from weight loss on acid hydrolysis.

[c]Calculated from % add-on.

[d]Calculated from sulfur analysis of grafted polymer isolated after removal of starch by acid hydrolysis.

a fast wicking product was also obtained from each saponificate by adding 12.5 g of $AlCl_3.6H_2O$ along with an equivalent amount of NaOH. Differences in color between the four saponificates were immediately apparent, and these color differences were also present in isolated and dried products. Incorporation of $AASO_3H$ into graft copolymers produced light yellow saponificates, and lighter colors were obtained with higher mole percentages of $AASO_3H$. Absorbencies and wicking properties of products isolated from the four saponificates are presented in Table VII. In both the presence and absence of aluminum chloride, the data clearly show the beneficial effects of incorporating even small amounts of $AASO_3H$ into monomer mixtures. Cost-benefit calculations will determine the optimum amount of $AASO_3H$ to be used in a commercial process.

Table VII. Absorbency and Wicking with Aluminum Chloride. Influence of $AASO_3H$ Content [a]

Mole % $AASO_3H$	Treatment of Saponificate	Final Wash, g HCl per 100 ml MeOH	pH of Dry Product	Absorb., g/g		Wicking, g/g
				H_2O	0.9% NaCl	
5	None	0.5	6.9	1420	77	
	$AlCl_3.6H_2O$;NaOH	0.5	7.6	620	52	92
		1.0	6.6	450	46	101
2.33	None	0.5	6.9	1310	71	
	$AlCl_3.6H_2O$;NaOH	0.5	7.8	500	51	97
		1.0	6.3	530	48	76
1.13	None	0.5	6.8	1000	60	
	$AlCl_3.6H_2O$;NaOH	0.5	7.2	530	46	65
		1.0	6.2	470	44	62
0	None	0.5	6.7	670	57	
	$AlCl_3.6H_2O$;NaOH	0.5	7.6	350	41	94
		1.0	6.1	330	39	77

[a] 48 g of starch used for graft polymerization. Procedure C (see Experimental) used for saponification with 12.5 g of $AlCl_3.6H_2O$ plus an equivalent amount of NaOH.

Acknowledgments

We are indebted to M. I. Schulte for sulfur analyses.

The mention of firm names or trade products does not imply that they are endorsed or recommended by the U.S. Department of Agriculture over other firms or similar products not mentioned.

Literature Cited

1. Fanta, G. F.; Doane, W. M. In Modified Starches: Properties and Uses; Wurzburg, O. B., Ed.; CRC Press: Boca Raton, FL, 1986, p 149.
2. Fanta, G. F.; Burr, R. C.; Doane, W. M.; Russell, C. R. Stärke. 1978, 30, 237-42.
3. Fanta, G. F.; Burr, R. C.; Doane, W. M. Stärke. 1987, 39, 322-5.
4. Elmquist, L. F. U.S. Patent 4 302 369, 1981.
5. Fanta, G. F.; Bagley, E. B.; Burr, R. C.; Doane, W. M. Stärke. 1982, 34, 95-102.
6. White, D. C. Mikrochim. Acta. 1962, 807-12.
7. Taylor, N. W.; Fanta, G. F.; Doane, W. M.; Russell, C. R. J. Appl. Polym. Sci. 1978, 22, 1343-57.
8. Burr, R. C.; Fanta, G. F.; Doane, W. M. J. Appl. Polym. Sci. 1979, 24, 1387-90.
9. Fanta, G. F.; Burr, R. C.; Doane, W. M. In Graft Copolymerizations of Lignocellulosic Fibers; Hon, D. N. S., Ed.; ACS Symposium Series No. 187; American Chemical Society: Washington, DC, 1982; pp 195-215.
10. Weaver, M. O.; Montgomery, R. R.; Miller, L. D.; Sohns, V. E.; Fanta, G. F.; Doane, W. M. Stärke, 1977, 29, 413-22.
11. Sax, N. I. Dangerous Properties of Industrial Materials; Van Nostrand Reinhold: New York, 1984; p 659.
12. Winfrey, V. L.; Black, W. C. U.S. Patent 3 133 836, 1964.
13. Handbook of Chemistry and Physics; Weast, R. C., Ed.; 56th Edition; CRC Press: Cleveland, 1975; p B-139.

RECEIVED January 22, 1990

Chapter 25

Use of Hemicelluloses and Cellulose and Degradation of Lignin by *Pleurotus sajor–caju* Grown on Corn Stalks

D. S. Chahal[1] and J. M. Hachey[2]

[1]University of Quebec, Institut Armand-Frappier, Laval, Québec, Canada H7V 1B7
[2]University of Quebec at Chicoutimi, Chicoutimi, Québec, Canada G7H 2B1

Pleurotus sajor-caju is capable of utilizing polysaccharides (cellulose + hemicelluloses) from corn stalks pretreated with 1.5% sodium hydroxide at 121°C for 1 h. The final product, mycelial biomass, contained about 40% crude protein which can be used as a food or feed. During fermentation of polysaccharides, lignin (oligolignols) of corn stalks was depolymerized into oligolignols of progressively lower molecular weight (MW). However, there is some evidence that repolymerization of oligolignols of low MW into oligolignols of high MW is also occurring.

Agricultural polymers, especially the polysaccharides (cellulose and hemicelluloses), form a major portion of plant cell wall and are potential feedstocks for the production of energy, chemicals and microbial biomass; the latter to be used as food or feed (1,2). Lignin another agricultural polymer bonded to hemicelluloses (3) in plant cell walls, covers cellulose (4). It is necessary, therefore, to remove, solubilize or degrade lignin into compounds of smaller molecular weight to expose the polysaccharides. By doing so various enzymes (cellulases and hemicellulases) can convert the exposed polysaccharides to monomeric sugars for their fermentation into useful products.

Pleurotus sajor-caju utilizes polysaccharides (cellulose and hemicelluloses) from various agricultural residues to produce mushrooms for human consumption throughout the world (5). It is also known to degrade lignin (6,7).

Pleurotus sajor-caju was selected in the present study for the utilization of polysaccharides from corn stalks for production of protein-rich mycelial biomass in submerged fermentation rather than for production of mushrooms (fruiting bodies). Production of mycelial biomass of various mushrooms including morels in submerged cultural conditions is another way to use them as food, food additives or mushroom flavor agents (8,9).

0097–6156/90/0433–0304$06.00/0

Materials and Methods

Organism. The culture of Pleurotus sajor-caju (Fr.) Singer (a fungus - an edible mushroom) was maintained on yeast malt agar slant (Difco) at 30°C. When fully grown after a week or so, the cultures were stored at 4°C.

Substrate and Pretreatment. Sweet corn (hybrid Lingodor) of W.H. Perron Laval, Quebec was grown in well prepared soil in a plot of 3 x 2 meters. Corn stalks were ground to 20 mesh to be used as a substrate. It was pretreated with 1.5% sodium hydroxide (NaOH) wt/vol with substrate:water ratio of 1:10 at 121°C for 60 minutes. The substrate was not washed after the pretreatment, and all the solubilized polymers (hemicelluloses and lignin) were retained along with the insoluble polymer (cellulose) in the fermentation medium. The composition of corn stalk is presented in Table 1.

Table 1. Composition of Corn Stalks

Polysaccharides		
Holocellulose (hemicelluloses+cellulose)	=	43.30%
Lignin (Klason)	=	29.30%
Protein (N x 6.25)	=	7.00%
Ash	=	4.97%
Other materials	=	15.43%

- Holocellulose was measured by the method of Myhre and Smith (10).
- Lignin was estimated by dissolving the corn stalks in 72% H_2SO_4.
- Protein was estimated by Micro-Kjeldahl method of A.O.A.C. (11).
- Ash was estimated by incinerating corn stalks at 550°C for overnight.

Fermentation Medium. Mandels and Weber (12) medium containing (g/l): pretreated substrate, 10 g (dry wt. basis); KH_2PO_4, 2.0; $(NH_4)_2SO_4$, 1.4; urea, 0.3; $MgSO_4 . 7H_2O$, 0.3; $CaCl_2$, 0.3; and (mg/l) $FeSO_4 . 7H_2O$, 5; $MnSO_4 . 7H_2O$, 1.56; $ZnSO_4 . 7H_2O$, 1.4; $CoCl_2$, 2.0. Proteose peptone in the above medium was replaced with 0.5 g of yeast extract/l (Difco). The pH of the medium was adjusted to 6.0. The 250 ml Erlenmeyer flasks containing 100 ml of the above medium were sterilized at 121°C for 20 minutes.

Inoculum. Pleurotus sajor-caju was grown at 30°C on the above medium containing 1% glucose as a carbon source in 250 ml Erlynmeyer flasks on rotary shaker at 200 rpm for 60 h. The mycelial biomass thus produced was blended in a Waring blender for 1 min. under aseptic conditions in order to obtain an homogenous inoculum of well dispersed mycelial bits. The inoculum was used at the rate of 10% vol/vol. The inoculated experimental Erlynmeyer flasks were incubated at 30°C on a rotary shaker at 200 rpm for various intervals of time.

Dry Weight of Final Product. The dry weight of the final product (containing mycelial biomass and unutilized insoluble substrates was determined by filtering 100 ml of sample through a pre-weighed filter paper Whatman No. 4. The residual biomass was washed with distilled water, dried overnight at 80°C, then weighed.

Degradation of Lignin. Degradation of lignin was evaluated by determining the distribution of fractions of lignin (oligolignols) having different molecular weights with gel permeation chromatography (GPC). The gel permeation chromatography was performed on a glass column of 17 x 1.7 cm (ID) using Sephadex G-10. Demineralized water was used as eluant at the rate of 0.45 ml/min. The column was pretreated at pH 2 with HCl. A 500 ul volume was injected and detector wavelength was fixed at 280 nm.

Determination of Molecular Weight. The molecular weights of various oligolignols produced during the fermentation of corn stalks were determined by using carbonic anhydrase, MW 29,000; cytochrome C, MW 12,400; aprotinine, MW 6,500; and riboflavine, MW 376 (Sigma), as molecular weight markers. Dextran Blue 2000 MW 2×10^6 was used for the determination of the exclusion volume. The use of these markers is in good agreement with previous studies (13,14). Hemicelluloses were not removed prior to GPC analysis. Polystyrene as used by Chum et al. (15) was not included in our list of standards for MW determination, because it is not soluble in water whereas lignin in our sample was in soluble form. Moreover, polystyrene calibration was found unsuitable in the analysis of lignin derivatives by other workers (14).

Results and Discussion

Utilization of Hemicelluloses and Cellulose (Polysaccharides). The pretreated corn stalk medium (contained solubilized hemicelluloses and lignin and insoluble cellulose) was fermented with *P. sajor-caju* for 28 days. The data presented in Table 2 indicated that the dry weight of final product continued to decrease due to the utilization of polysaccharides (cellulose and hemicelluloses). The accumulation of mycelial biomass was indicated by the steady increase in the protein synthesized by the organism into its mycelium. Optimal protein production was obtained between 2.5 and 3 days of fermentation. At this point the final product (mycelial biomass + unutilized substrates) contained 39.82% protein and only 9.6% cellulose. There was only 0.5 g residual cellulose/l at this time. Residual hemicelluloses could not be estimated because it is difficult to estimate hemicelluloses when mixed with mycelial biomass of fungi. It is assumed that most of the hemicelluloses should have also been utilized according to our previous studies (5). The final product, protein-rich mycelial biomass, thus obtained can be used as a food for human consumption or as a feed for animals. After 3 days of fermentation the dry weight of mycelial biomass synthesized remained almost constant up to 28 days of fermentation. The fermentation was carried on beyond 3rd day of fermentation to see the effect of growth of *P. sajor-caju* on the degradation of lignin present in the medium.

Analysis of Lignin of Corn Stalks Before Fermentation. During the pretreatment of corn stalks with NaOH, the lignin and hemicelluloses were solubilized. The solubles obtained after filtering through

Table 2. Utilization of hemicelluloses and cellulose from corn stalks by Pleurotus sajor-caju

Time of fermentation (days)	Dry weight of final product* (g/l)	Protein content of final product (%)	Total crude protein produced (g/l)	Residual cellulose (g/l)
0 (not inoculated)	10.85	1.73	0.188	-
0 (inoculated)	9.54	2.98	0.274	5.7
1	7.72	8.62	0.666	-
2	6.12	23.16	1.415	2.6
2.5	5.21	36.75	1.915	-
3	4.98	39.82	1.983	-
4	5.26	38.18	2.008	0.5
7	5.27	28.67	1.510	0.3
10	4.81	29.88	1.437	0.2
14	5.14	29.21	1.504	-
17	4.78	29.90	1.432	-
21	4.95	25.49	1.262	0.4
24	4.51	25.62	1.155	-
28	4.58	25.01	1.146	0.2

* Dry weight of final product contains the mycelial biomass synthesized and the unutilized cellulose and hemicelluloses.

Whatman filter paper #4 were examined for the distribution of lignin fractions (oligolignols), produced during the pretreatment. The results given in Table 4 at 0 time (uninoculated) indicated that the solubilized lignin just after pretreatment and prior to its fermentation with P. sajor-caju contained oligolignols of different MW.

Degradation of Lignin of Corn Stalks During Fermentation. Degradation of lignin during the fermentation of polysaccharides was measured as breakdown of oligolignols of high MW to oligolignols of progressively lower MW. The percentage of various oligolignols obtained after different time intervals of fermentations are presented in Table 3 and their MW are presented in Table 4.

Immediately after inoculation at 0 time, the oligolignols of high MW were degraded into oligolignols of lower and lower MW. This degradation could have happened during storage at 4°C because the samples were examined after 4 weeks of storage. The oligolignols of lowest MW (V_6) disappeared within first two days of fermentation, indicating that some of the oligolignols of lowest MW had disappeared by further degradation to CO_2 and some might have been repolymerized into oligolignols of high MW. Disappearance of oligolignols of low MW as CO_2 indicated that lignin was simultaneously being degraded when the organism was utilizing polysaccharides for synthesis of protein in its mycelial biomass and reappearances of oligolignols of high MW (V_1) on the 4th day of fermentation indicated that repolymerization of oligolignols of low MW was also going on. Recently repolymerization of oligolignols had been reported by Cyr et al.

Table 3. Distribution of oligolignols during the fermentation of lignin with P. sajor-caju (Gel Permeation Chromatography on Sephadex G-10)

Time of Fermentation (days)	V_1(ml) (S_1(%))	V_2(ml) (S_2(%))	V_3(ml) (S_3(%))	V_4(ml) (S_4(%))	V_5(ml) (S_5(%))	V_6(ml) (S_6(%))
0 (uninoculated)	7.9 (1)	13.7 (8.3)	17.3 (7.1)	18.5 (17.5)	23.2 (66.2)	- (-)
0 (inoculated)	-	13.7 (1.5)	17.3 (1)	18.5 (2.4)	23.2 (29.1)	25.7 (66)
2	-	13.7 (10.1)	16.2 (9.2)	18.5 (34.6)	23.2 (46.1)	- -
4	7.9 (1.4)	13.7 (4.1)	-	19.1 (42.7)	21.4 (51.8)	-
5	- -	15.1 (7.5)	- -	20.7 (34.4)	24.3 (58.1)	-
10	- -	13.1 (1)	-	19.6 (4.5)	24.1 (94.5)	- -
14	- -	13.1 (4.8)	- -	19.4 (10.3)	24.1 (84.9)	- -
28	- -	13.1 (1.5)	15.1 (9)	19.6 (11.9)	24.9 (77.6)	-

V_1: Elution volume at the first maximum of absorption.
V_2 : Elution volume at the second maximum of absorption and so on.
S_1: Area of the first peak relative to the area of all peaks (%).
S_2: Area of the second peak relative to the area of all peaks (%) and so on.

(16) and Trojanowski et al. (17). The data presented in Tables 3 and 4 indicated that the residual lignin became homogenous after 5 days of fermentation with majority of population (77-94%) of oligolignols of medium MW (about 1,400).

The studies on quantitative analysis of residual lignin by ^{13}C NMR, FAB/MS are underway. Studies on disappearing of lignin as CO_2 are being carried on by using ^{14}C-labelled lignin, and will be reported in the future.

Table 4. Distribution of molecular weight (MW) of various oligolignols produced during fermentation of corn stalks lignin with P. sajor- caju

Time of Fermentation (days)	V_1	V_2	V_3	V_4	V_5	V_6
0 (uninoculated)	81,625	21,160	8,603	3,914	1,983	-
0 (inoculated)	-	21,164	8,850	5,643	1,781	986
2	-	15,552	7,476	5,186	1,591	-
4	83,963	16,900	-	4,141	2,567	-
5	-	13,880	-	3,701	1,383	-
10	-	22,388	-	4,381	1,383	-
14	-	24,354	-	4,767	1,344	-
28	-	23,027	6,871	4,894	1,781	-
Range of MW	81,625 to 83,963	13,880 to 24,354	6,871 to 8,850	3,701 to 5,643	1,344 to 2,567	986
Average MW	82,794	19,607	7,732	4,673	1,666	986

V_1,V_2... indicate the elution volume as explained under Table 3.

Acknowledgments

The authors are very grateful to the University of Quebec for the grant to carry out this study. They are also thankful to Miss Johanne Lemay and Mr. Remy Larouche for their technical help.

Literature Cited

1. Chahal, D.S. Med. Fac. Landbouww. Rijksuniv. Gent. 1987, 52(4), 1565-1624.
2. Tsao, G.T., Ladish, M., Ladish, C., Hsu, T.A., Dale, B. and Chow, T. Ann. Report. Fermentation Processes. 1978, 2, 1-22.
3. Fengel, D. and Wegener, G. Wood, Chemistry, Ultrastructure, Reactions. Walter de Gruyter, New york, 1984, p. 613.
4. Higuchi, T. Adv. Enzyymol. 1971, 34, 207-277.
5. Chahal, D.S. J. Ferm. Bioeng. 1989, 68, 334-338.
6. Bisaria, R. and Madan, M. Current Sci. 1984, 53, 322-323.
7. Bourbonnais, R. and Paice, M.G. Biochem J. 1988, 255, 445-450.
8. Humfeld, H. and Sugihara, T.F. Food Technol. 1949, 3, 355-356.

9. Litchfield, J.H. In: Microbiol Technology. Peppler, H.J. (ed.), Reinhold, New York. 1967 p. 107-143.
10. Myhre, D.V. and Smith, J.J. J. Agric. Food Chem. 8, 359-364, 1960.
11. A.O.A.C.: Official Methods of Analysis, 12th edition, Washington, D.C. 1975.
12. Mandels, M. and Weber, J. Adv. Chem. Ser. 1969, 95, 391-414.
13. Kern, H.W. Holzforschung. 1983, 37, 109-115.
14. Pellinen, J. and Salkinoja-Salonen, M. J. Chromatogr. 1985, 328, 299-308.
15. Chum, H.L., Johnson, D.K., Tucker, M.P. and Himmel, H.E. Holzforschung. 1987, 41, 97-108.
16. Cyr, N., Elofron, R.M., and Ripmeester, J.A. J. Agric. Food Chem. 1988, 36, 1197-1201.
17. Trojanowski, A., Milstein, O., Majcherczyk, A., and Haars, A. Colloq. INRA 40 (Lignin Enzymic Microb. Degrad.), 1987, 223-229.

RECEIVED February 5, 1990

INDEXES

Author Index

Albertsson, Ann-Christine, 60
Ando, Tadanao, 136
Angle, Jay S., 149
Bailey, William J., 149
Benicewicz, Brian C., 161
Bezwada, R. S., 167
Blackburn, James W., 13
Chahal, D. S., 304
Chubin, David E., 220
Clarke, M. A., 210
Clemow, Alastair J. T., 161
Cole, Michael A., 76
Dexter, L. B., 65
Dirlikov, Stoil K., 176
Doane, William M., 288
Dowd, Patrick F., 33
Fanta, George F., 288
Gandini, Alessandro, 195
Gilbert, Richard D., 259
Glass, J. Edward, 52
Gordon, S. H., 65
Gould, J. Michael, 65
Hachey, J. M., 304
Han, Y. W., 210
Jones, Frank N., 220
Kaczmarski, James P., 220
Kaleem, Kareem, 230
Karlsson, Sigbritt, 60
Kawai, Fusako, 110
Kirby, Kenneth W., 274
Kravetz, L., 96
Kuruganti, Vijaya K., 149
Ma, Zeying, 220
Matsumura, Shuichi, 124
Newman, H. D., Jr., 167
Oser Zale, 161
Rowell, Roger M., 242
Sayler, Gary S., 13
Shalaby, S. W., 161,167
Shen, Samuel K., 33
Suzuki, Tomoo, 136
Swanson, C. L., 65
Swift, Graham, 2
Takeda, Kiyoshi, 136
Thames, Shelby F., 230
Thayer, Ann M., 38
Tokiwa, Yutaka, 136
Wang, Daozhang, 220
Yoshikawa, Sadao, 124

Affiliation Index

Chemical and Engineering News, 38
Eastern Michigan University, 176
Ecole Francaise de Papeterie (INPG), 195
Ethicon, Inc., 161, 167
Fermentation Research Institute, 136
Johnson and Johnson Orthopedics, 161
Keio University, 124
Kobe University of Commerce, 110
North Carolina State University, 259
North Dakota State University, 52,220
Penford Products Company, 274
Rohm and Haas Company, 2
Shell Development Company, 96
Sugar Processing Research, Inc., 210
The Royal Institute of Technology, 60
U.S. Department of Agriculture, 33,65,210,242,288
University of Illinois, 76
University of Maryland– College Park, 149
University of Quebec–Chicoutimi, 304
University of Quebec–Laval, 304
University of Southern Mississippi, 230
University of Tennessee, 13

Subject Index

A

Absorbent polymers
formation, 288–289
practical applications, 289
2-(Acrylamido)-2-methylpropanesulfonic acid
formation of absorbent polymers, 289
structure, 289
Addition polymers
biodegradability, 150
biodegradation, 5–8
microbial growth in hydrocarbons, 5*t*
Aerobic biodegradation, description, 96–97
Agricultural polymers, applications, 304
Agricultural product utilization in plastics
historical perspective, 54
need for new applications, 52–53
nonnutrition application areas, 55
Agricultural resources, source of furan-containing polymers and oligomers, 195
Alkaline sizing, advantages, 283
2-Alkenylfurans, cationic polymerization, 200–201
1-Alkylfurans, chemical modification of polymers, 207
Alkylphenol ethoxylates
hydrophobe preparation, 97
structural features, 97,99*f*
ultimate biodegradation by CO_2 evolution, 102,103f
Alternative Crops Program, function, 220
Aluminum chloride, effect on wicking, 297–298,299*t*,300
Amphoteric starches, applications, 282
Amylopectin, structure, 275
Amylose, structure, 275
Anionic polymerization, furans, 201–202
Aqueous dispersion properties, starch, 217
Atmospheric cooking of starch, 280

B

Bacteria that use PEG, 111–112
Bacterial colonization on starch–plastic blends, 82
Bacterial populations in subsurface soils, estimation of genetic potential, 21,23*t*
Bioabsorbable fibers of *p*-dioxanone copolymers
breaking strength studies, 168
characterization methods, 168
comparative properties of *p*-dioxanone and copolymer fibers, 169,172t
Bioabsorbable fibers of *p*-dioxanone copolymers–*Continued*
conversion of copolymers to fibers, 168
copolymer characterization, 168–171
copolymerization conditions, 168,169*t*
copolymer properties, 168,169*t*
effect of initiator type, 168,171*t*
experimental materials, 167
in vivo absorption studies, 168
polymerization scheme, 168
preparation of copolymers, 168
tensile properties of fibers, 168–170
Bioanalytical methods to assess specific activity of microorganisms, 24–26
Biodegradable, definition, 66
Biodegradable additives, 61
Biodegradable poly(carboxylic acid), 124–134
Biodegradable polymers
generalizations, 9–10
produced by free-radical ring-opening polymerization
biodegradability test procedure, 158
biodegradation in soil, 156*t*,157
CO_2 liberation, 152,153t,f
copolymerization of styrene and cyclic ketene acetal I, 154t
copolymerization with styrene, 152*t*
degradation apparatus, 155*f*,156
experimental material preparation, 157–158
soil biodegradation test procedure, 158
synthetic schemes, 150–151
synthetic polymers, history of research, 3
Biodegradation
addition polymers, 5–8
description, 2
medical polymers, 9
nonionic ethoxylates, *See* Nonionic surfactant(s)
nonionic surfactants, *See* Nonionic surfactant(s)
poly(carboxylic acid), 124–134
polyethers, *See* Polyether(s)
polyethylene glycol, 111–115
poly(L-lactide)-braided multifilament yarn, *See* Poly(L-lactide)-braided multifilament yarn, degradation
polymer blends, 8–9
polypropylene glycol, *See* Polypropylene glycol
polysaccharide–plastic blends, 76–94
poly(sodium carboxylate) containing glycopyranosyl group, 133–134
polytetramethylene glycol, 120–122
polyvinyl-type poly-(sodium carboxylate), *See* Polyvinyl-type poly(sodium carboxylate), biodegradation

Biodegradation–*Continued*
starch-containing plastics, *See* Starch-containing plastics, biodegradation
step-growth polymers, 8
synthetic polymers containing ester bonds, *See* Synthetic polymers containing ester bond
See also Degradation
Biodegradative processes for control of environmental contaminants
analytical monitoring techniques, 16,17*t*,18
factors influencing application, 16
molecular tools, 16–26
rate limitations, 14
system analysis, 26,27*f*,28
system identification, 28,29*f*,30
Biodegradative system, prediction of genetic potential, 18
Biodegradative testing, *See* Testing protocols for biodegradability
Biological waste treatment, biodegradative processes, 14–30
Bioluminescent report strains
development, 24
example of bioluminescent light emitted by bacteria, 24,25f
measurement of light emission, 24,25*f*,26
Biotransformation
naphthalene, system analysis, 26,28,29*f*
soil, analysis, 28
Bottle bills, effectiveness, 40
Branched primary alcohol ethoxylates
hydrophobe preparation, 97
structural features, 97,99*f*
ultimate biodegradation by CO_2 evolution, 102,103f
Brassylic acid
economic viability, 222
formation via ozonolysis of erucic acid, 221–222
melting points, 228*t*
polymer-related uses, 222
use in coatings, 222
Brassylic acid containing coatings
coating test procedures, 225
experimental materials, 223–224
nylons, physical properties, 222*t*,223
prepared from cyclohexanedimethanol, 226
prepared from hexakismethoxymethyl-melamine, 226
prepared from polyester resins, 225
properties of dodecanedioic acid coatings, 226,228*t*
properties of hexakismethoxymethylmelamine coatings, 226,227*t*
Brassylic acid containing coatings–*Continued*
properties of polyester resin coatings, 226*t*
synthesis of polyester resins, 224
unpigmented coatings, 224–225

C

Carbohydrate–synthetic polymer blends
evaluation of commercialization viability, 53–54
market restrictions, 53
technical impositions, 53–54
Cationic polymerization, furans, 200–201
Cellobiose
applications, 191
characterization, 190–191
description, 189–190
polymerization, 190
synthesis, 190
Cellulose
as liquid crystal(s)
discovery, 259
formation, 263–264
preparation, 260
solvent viscosity vs. concentration plots, 264
derivatives as liquid crystals
cellulose triacetate, 266–267
cellulose tricarbanilate, 267–268
(hydroxypropyl)cellulose, 265–266
factors influencing rigidity, 261
lyotropic mesophase formation, 263–264
utilization by *Pleurotus sajor-caju*, 305
Cellulose triacetate, formation of lyotropic mesophases, 266–267
Cellulose tricarbanilate, formation of lyotropic mesophases, 267–268
Center for Plastics Recycling Research
collection procedure, 46–47
establishment, 41–42
manual for setup and operation of recycling plant, 42–43
resin recovery system, 44
Chain polymerizations, furans, 196,199–202
Characterization, chlorinated rubber, *See* Chlorinated rubber from low-molecular-weight guayule rubber
Chemical degradation, description, 3
Chemical modification of lignocellulosic fibers to produce high-performance composites
acetylation rate, 243,244*f*
biological resistance, 252–255
oven-dry thickness, 248,250,251*f*
chemical composition of materials 246,247*t*
deflection–time curves for flakeboards, 253,254*f*

Chemical modification of lignocellulosic fibers to produce high-performance composites–*Continued*
dimensional stability, 248,249–251*f*
equilibrium moisture content, 243,246*t*
equilibrium moisture content reduction vs. bonded acetyl content, 243,245*f*,246
fire properties, 256
fungal cellar tests on flakeboards, 252,253*t*
moisture sorption, 243,245–248
rate of swelling in liquid water of flakeboard, 248,249*f*
reaction chemistry, 243
UV resistance, 255*t*
Chlorinated rubber from low-molecular-weight guayule rubber
chlorination of guayule rubber, 231–232
^{13}C-NMR spectrum, 236t,237f
DSC, 236,240*f*
effect of azobisisobutyronitrile on chlorine content, 233,234*f*,236,239*f*
effect of molecular weight on physical and mechanical properties, 232
empirical formula, 231
experimental conditions for synthesis, 233*t*
experimental materials and methods, 231
FTIR spectra, 233,235*f*,236
GPC, 236,238–239*f*
^{1}H-NMR spectroscopy, 232–233,234f
properties vs. applications, 236,240
4-Chlorobiphenyl catabolic gene abundance, detection by blot hybridization, 21–24
Cholesteric liquid crystals, 261
Chromatography, use in studying polyethylene degradation, 60–61
Cigarette beetle, detoxification by symbionts, 34–35
Coated paper
composition, 279
growth, 275
use of starches, 279
Coatings
applications, starches, 284,286
based on brassylic acid, *See* Brassylic acid containing coatings
Collections of plastics for reuse, 3
Commodity plastics
definition, 2
degradability, overview, 2–12
stability as environmental problem, 52
Compatibilizer, description, 53
Composites, high-performance, produced by chemical modification of lignocellulosic fibers, 242–255
Consumer packaging, resins used, 51
Copolymers of acrylonitrile and starch, preparation, 288–289
Corn stalks, to grow *Pleurotus sajor-caju*, 304
Council on Plastics & Packaging in the Environment, function, 47–48
Council on Solid Waste Solutions, function, 49
Crambe oil, source of erucic acid, 221

D

Degradability
definition, 2
overview, commodity plastics and specialty polymers 2–12
See also Biodegradation
Degradable plastics
effect of noncost factor on commercial success, 56
market restrictions, 53
realization of commercial process, 55–56
technical impositions, 53–54
Degradation
lignin by *Pleurotus sajor-caju*, 307
mechanisms, types, 2–3
poly(L-lactide) braided multifilament yarns, 161–166
polyethylene, See Polyethylene degradation
See also Biodegradation
Design of biodegradable poly(carboxylic acid)
biodegradation of poly(sodium carboxylate) containing a glycopyranosyl group, 133,134*f*
biodegradation of polyvinyl-type poly(sodium carboxylate), 128–132
detergency building performance, 133
detergency test, 126
five-day biochemical oxygen demand, 126,127*t*
GPC profiles, 128,129*f*
material preparation, 125
measurement procedures, 126
polymer structures, 125
polymer-degrading bacteria, 126
Detoxification by insect symbionts
cigarette beetle, 34–35
occurrence, 34
potential for use, 35–36
Dextrans, production of polymeric compounds for industrial use, 210
1,2:5,6-Dianhydro-3,4-*O*-isopropylidene-D-mannitol
applications, 189
characterization, 189
sources, 189
synthesis, 187–189

1,6-Dichloro-1,6-dideoxy-3,4-*O*-isopropylidene-D-mannitol, experimental procedure, 192–193
1,4:3,6-Dilactone of mannosaccharic acid
applications, 186
characterization, 186
synthesis, 185–187
Dimethyl isosorbide
applications, 180–181
characterization, 180
structure, 180
synthesis, 180
p-Dioxanone copolymers, bioabsorbable fibers, 167–172
Disaccharides, as basis for monomers and polymers, 176–194
DNA probes, use in detection of degradative genes, 18t,20–26
Double-radiolabeled nonionic surfactants
biodegradation of labeled hydrophobe groups to ^3H_2O, 102,106–107,108f
fate in biotreatment, 106*t*
radiochemical product distribution during biodegradation, 105,106*t*
ratio of ethoxylate to hydrophobe during biodegradation, 105*t*
structures, 102,103*f*
ultimate biodegradation of alkyl portion, 102,104*f*,105
Dow and plastics recycling, 49

E

Emulsification, starches, 282–283
Environmental biotechnology for hazardous wastes, components for development, 14,15*t*
Environmental erosion, description, 3
Enzymes, detoxifying, insect symbionts as source, 33–37
EPA's Office of Solid Waste, goal of solid waste treatment methods, 49
Erucamide, market, 221
Erucic acid derivative, *See* Brassylic acid
Ester bonds, *See* Synthetic polymers containing ester bond

F

Fate analysis, importance, 4
Fibers, bioabsorbable, *See* Bioabsorbable fibers of *p*-dioxanone copolymers
Forestry resources, source of furan-containing polymers and oligomers, 195
Free-radical ring-opening polymerization, schemes for production of biodegradable polymers, 150–151
Fructan(s)
production, 213–214
classification, 210
Fructan-producing bacterium, isolation scheme, 211,212f
Fructan sucrase, biosynthesis, 213
Fungus, *Pleurotus sajor-caju*, use of hemicelluloses and cellulose, 305–309
Furan(s) in polymers and oligomers
anionic polymerizations, 201–202
cationic polymerization, 200–201
chain polymerizations, 196,199–202
chemical modification of polymers, 207
radical polymerization, 196,199–200
step polymerizations, 203–206
2-Furancarboxyaldehyde
source, 195
transformation into furanic monomers, 196–197
Furanic polymers, synthesis via step polymerization
polyamides, 204
polycarbonates, 205
polyesters, 203–204
polyethers, 204–205
polyurethanes, 205–206
2-Furfurylidene acetone, anionic polymerization, 202
2-Furfuryl methacrylate, radical polymerization, 199
2-Furfuryl vinyl ether, cationic polymerization, 200
2-Furyl isocyanate, synthesis via step polymerization, 206
2-Furyl oxirane, anionic polymerization, 202

G

Genpak, plastics recycling, 48
D-Glucose methacrylate
applications, 191
characterization, 191
polymerization, 192
synthesis, 191
Graft copolymers of acrylonitrile and starch, preparation, 288–289
Group-transfer polymerization, lack of commercial success, 56
Guayule rubber
applications of coproducts, 230–231
isolation, 230
use in synthesis of chlorinated rubber, 231–24

H

Hazardous wastes, biotechnological control, 14
Hemicelluloses, utilization by *Pleurotus sajor-caju*, 305
High-density polyethylene, reclamation technologies, 44
High-erucic oils, potential end uses, 221
High-molecular-weight mesophases, discovery, 261
High-performance composites produced by chemical modification of lignocellulosic fibers, 242–255
Hydrophobe structure, influence on biodegradation pathways of nonionic ethyoxylates, 96–108
Hydroxyethyl starch, preparation, 283–284
2-(Hydroxymethyl)-2-furancarboxyaldehyde, transformation into furanic monomers, 196,198
(Hydroxypropyl)cellulose
  formation of lyotropic mesophases, 265–266
  formation of thermotropic mesophase, 268

I

Industrial polymer science, laws, 53
Insect(s)
  detoxification of toxins, 33–34
  development of resistance to insecticides, 34
  species, 33,
Insect symbionts
  detoxification abilities, 34
  detoxification of cigarette beetle, 34–35
  potential for use in detoxification, 35–36
Isomannide
  applications, 180
  characterization, 179–180
  synthesis, 179
Isosorbide
  applications, 178–179
  characterization of polymers, 178–179
  polymerization, 178
  structure, 177
  synthesis, 178

J

Johnson Controls, plastics recycling, 49

K

Ketene acetals, use in synthesis of biodegradable addition polymers, 7–8
Kuhn segment length, description, 262

L

1,4-Lactone of 3,6-anhydrogluconic acid
  applications, 185
  characterization, 183
  experimental procedures, 192
  properties of linear soluble polyurethanes, 183–185
  synthesis, 182–183
  synthesis of linear soluble polyurethanes, 183
Levans, *See* Microbial fructans
Lignin
  analysis in corn stalks before fermentation, 306–307,308t
  composition of corn stalks, 305*t*
  degradation in corn stalks, 307,308–309*t*
  determination of dry weight of final product, 305–306
  determination of molecular weight, 306
  distribution of molecular weight, 307,309*t*
  evaluation, 306
  fermentation medium preparation, 305
  inoculum preparation, 305
  oligolignol distribution, 307,308*t*
  organism preparation, 305
  substrate preparation, 305
Lignocellulosic materials
  chemical modification to produce high-performance composites, 243–256
  description, 242
  undesirable properties, 242
Linear primary alcohol ethoxylates
  hydrophobe preparation, 97
  structural features, 97,99*f*
  ultimate biodegradation by CO_2 evolution, 102,103f
Linear secondary alcohol ethoxylates
  hydrophobe preparation, 97
  structural features, 97,99*f*
  ultimate biodegradation by CO_2 evolution, 102,103f
Liquid crystal(s)
  classes, 260
  description, 260
  discovery, 260
  lattice theory, 261
  solutions of cellulose, 263–264
  solutions of cellulose derivatives, 265–268
Low-density polyethylene, volumetric content of degradable component and degradability, 77,78*f*,80*f*
Low-molecular-weight guayule rubber, *See* Chlorinated rubber from low-molecular-weight guayule rubber
Lyotropic liquid crystals, definition, 260

M

Material recovery facilities, development, 47
Medical polymers
biodegradability, 3
biodegradation, 9
2-Methylfuran, chemical modification of polymers, 207
Microbial degradation of natural polymers, processes, 149
Microbial fructans
alditol acetate peaks, 216,217*f*
^{13}C-NMR spectra, 214,215f,216
composition, 214
GLC, 216,217*f*
identification, 211
methylation analysis, 216*t*
production, 213–214
properties, 214
structure, 214,215*f*,216*t*,217*f*
Microbial process, system analysis and identification, 26
Mixed plastics, 44–45
Mobil Chemical, plastics recycling, 48
Molecular analytical monitoring techniques
advantages, 16
assessment of specific activity of microorganism, 24–26
examples, 16,17*t*,18
gene- or DNA-specific methods, 16,18–24
Molecular weight, effect on polymer degradation, 61–62
Monomers based on mono- and disaccharides, 176–194
Monosaccharides, as basis for monomers and polymers, 176–194
Multifilament yarn, *See* Poly(L-lactide) braided multifilament yarn, degradation
Mushrooms, produced by *Pleurotus sajor-caju*, 304–309

N

Naphthalene
biotransformation, 26,28,29f
degradative bacterial colonies, autoradiographic detection, 21,22*f*
1-Naphthyl acetate esterase, 35
National Association for Plastic Container Recovery, function, 47
Natural rubber, development of domestic sources, 230
Nematic liquid crystals, description, 260
Noninvasive microbial growth on starch–plastic blends
limitations to bacterial colonization, 82
oxygen availability in degrading films, 81–82
postulated mechanism of microbial decay, 79,80*f*
Nonionic ethoxylates, *See* Nonionic surfactants
Nonionic surfactant(s)
environmental concerns, 96
primary biodegradation, 98,99–100*f*
structural features, 97,99*f*
ultimate biodegradation, 98,100*f*
biodegradation
aquatic toxicity of effluents, 107,108*f*
pathway for hydrophobe, 98,101*f*
pathway determination by radiotracer techniques, 102–108
points of initial attack, 98,100*f*
primary, 98,99–100*f*
ultimate, 98,100*f*
ultimate biodegradation by CO_2 evolution, 102,103f
Nucleic acid hybridization
applications, 21,22*f*,23–24
use for molecular monitoring in biodegradation, 18,20*f*,21

O

Olefin–polycaprolactone blends, 9
Oligomers containing furan rings, 195–207

P

Paper industry
coated paper growth, 275
growth, 274
paper and paperboard production, 275,276*t*
per capita consumption of paper, 275,276*t*
pulp production, 275,276*t*
use of starch, 275,278–280
Persistence length, description, 262
Petrochemical-based plastic materials, persistence problems, 65
Photodegradation, description, 3
Plasmid fingerprinting, 18,19*f*
Plastic(s)
additives, environmental impact, 90–93
advantages, 39
banning of polystyrene packages, 40
degradability, 52–56
market restrictions, 53
realization of commercial process, 55–56
technical impositions, 53–54

Plastic(s)–*Continued*
legislation for waste management, 39–40
Plastic Bottle Institute, function, 47
Plastic Pollution Control Act, 39
Plastic recycling
collection problem, 46–47
high-density plastics, 44
industrial recycling, 48–49
markets, 44–46
Plastics Recycling Action Plan, 40
polyethylene terephthalate beverage bottles, 43–44
programs, 40
research programs, 41–42
Plastics Recycling Foundation, 41
Pleurotus sajor-caju utilization of hemicelluloses and cellulose, 305–309
composition of corn stalks, 305t
determination of dry weight of final product, 305–306
determination of molecular weight, 306
evaluation of lignin degradation, 306
experimental data, 306,307*t*
fermentation medium preparation, 305
inoculum preparation, 305
organism preparation, 305
substrate preparation and pretreatment, 305
Polyacetals, biodegradation, 7
Polyacids, biodegradation, 7
Polyamides, biodegradation, 7
Poly(carboxylic acids)
biodegradation, 6–7
biodegradation of poly(sodium carboxylate) containing a glycopyranosyl group, 133,134*f*
biodegradation of polyvinyl-type poly(sodium carboxylate), 128–132
detergency building performance, 133
detergency test, 126
five-day biochemical oxygen demand, 126,127*t*
GPC profiles, 128,129*f*
material preparation, 125
measurement procedures, 126
polymer structures, 125
polymer-degrading bacteria, 126
Polyester resins
chemical reclamation, 51
description, 225–226
Polyether(s)
applications, 110–111
biodegradation, 8
polyethylene glycol, 111–117
polypropylene glycol, 116,118*t*,119,121*f*
polytetramethylene glycol, 119,120*t*,121*f*,122
structural formula, 110
Polyether(s)–*Continued*
synthesis, 110
Polyethylene
biodegradation, 5–6
degradation
analytical techniques, 60–61
effect of biodegradable additives, 61
effect of evolution of low molecular weight degradation products, 60
effect of molecular weight, 61–62
effect of rate, 60
mechanisms, 63
products, 62–63
use in consumer packaging, 51
Polyethylene glycol
analysis, 111,112*f*
applications, 110
bacteria that use, 111,113,115*f*
biochemical routes, 113–114,115*f*
biodegradation, 111–116
symbiotic degradation, 114,116,117*f*
Polyethylene terephthalate
beverage bottles, recycling, 43–45
biodegradation, 8
production of new unsaturated polyester resins, 50
reuse requirements, 50
use in methanolysis reaction, 50
use in consumer packaging, 51
Poly(2-hydroxyacrylic acid)
biodegradation in soil, 156*t*,157
synthesis, 154–155
Poly(isosorbide–hexamethylene diisocyanate), 193
Poly(L-lactide), potential use for bioabsorbable surgical devices, 161
Poly(L-lactide)-braided multifilament yarn, degradation
comparison of breaking strength retention with that of commercial sutures, 165f
determination of breaking strength, 162
experimental procedures, 161–162
in vitro absorption, 163,164*t*
in vitro breaking strength retention, 163*t*
in vivo breaking strength retention, 164*f*,165*t*
properties of yarns 162,163*t*
Polymer(s)
based on mono- and disaccharides, 176–194
blends, biodegradation, 8–9
containing furan rings, 195–207
synthetic, *See* Synthetic polymers containing ester bond
Polymerization, free-radical ring-opening, *See* Biodegradable polymers produced by free-radical ring-opening polymerization
Polypropylene, use in consumer packaging, 51

Polypropylene glycol
applications, 110–111
biodegradation
by immobilized cells, 116,119
GC–MS analysis, 119,121*f*
metabolic route, 119
utilizing bacteria, 116,118*t*
structure, 110
Polysaccharide(s), production of polymeric compounds for industrial use, 210
Polysaccharide–plastic blend(s)
additive release upon degradation, 90,92,93*t*
CO_2 evolution, 77,80f
critical film properties, 85–88
degradability, 77
developmental incentives, 76–77
effect of plastic additives, 89–90
effect of pore dimensions, 85–86,87*f*
enzyme exclusion, 88–89,91*f*
examples of degradable, 76
inhibitory properties of photodegradable polyethylene, 90,91*f*
kinetics, 79
mechanism for microbial decay, 79,80*f*
noninvasive microbial growth, 79,81–82
pore size distribution, 86,87*f*
scanning electron micrograph, 77,78*f*
starch degradation mechanism, 83,84*f*,85
surface porosity vs. rate, 86,88*t*
volumetric content of degradable component, 77
Poly(sodium carboxylate) containing a glycopyranosyl group
biodegradation curves, 133,134*f*
GPC profiles, 133,134*f*
Polystyrene
banning of packages, 40
use in consumer packaging, 51
Polytetramethylene glycol
application, 111,119
biodegradation, 120–122
Polyurethanes, biodegradation, 8
Poly(vinyl alcohol), biodegradation mechanism, 6
Poly(vinyl chloride), use in consumer packaging, 51
Polyvinyl-type poly(sodium carboxylate), biodegradation
curves, 130,131–132*f*
GPC profiles, 130,132*f*
mechanism, 128,130
substrate specificity for growth of degrading strain, 130*t*
Primary biodegradation
definition, 98
intermediates, 98,100*f*
Primary biodegradation-*Continued*
schematic representation, 98,99*f*
Procter & Gamble, plastics recycling, 49

Q

Q-resin technology, 56
Quaternary ammonium starch ethers, synthesis, 282

R

Radical polymerization, furans, 196,199–200
Recyclable Materials Science & Technology Development Act, 39–40
Recycling plastic
collection problem, 46–47
high-density plastics, 44
industrial recycling, 48–49
markets, 44–46
Plastics Recycling Action Plan, 40
polyethylene terephthalate beverage bottles, 43–44
programs, 40
research programs, 41–42
Resins in common use in consumer packaging, 51
Ring-opening polymerization, *See* Biodegradable polymers produced by free-radical ring-opening polymerization
Rubber, chlorinated, synthesis from low-molecular-weight guayule rubber, 230–240

S

Saccharides, as basis for monomers and polymers, 176–194
Saponified starch-*g*-poly(acrylonitrile-*co*-2-acrylamido-2-methylpropane-sulfonic acid)
acid hydrolysis procedure, 291–292
effect of 2-(acrylamido)-2-methylpropane-sulfonic acid content on water absorbency, 300,301–302t
effect of 2-(acrylamido)-2-methylpropane-sulfonic acid content on wicking, 301,302*t*
effect of aluminum chloride addition on wicking, 297–298,299*t*,300
effect of methanol precipitation conditions on water absorbencies, 292,293*t*,294
effect of saponification conditions on water absorbencies, 294,296*t*
effect of saponification method on water absorbency, 296,297*t*

Saponified starch-*g*-poly(acrylonitrile-*co*-2-acrylamido-2-methylpropane-sulfonic acid)–*Continued*
effect of variables on water absorbency, 294,295*t*
experimental materials, 290
experiments to define variables, 292–297
graft polymerization procedure, 290
saponification procedures, 290–291
test procedures for absorbency and wicking, 291
Size press application, starches, 283–284,285*f*
Smetic liquid crystals, 260–261
Soil biotransformations, analysis, 28
Solid waste management
crisis situation, 38
future developments, 49–50
landfilling problems, 38–39
local, state, and federal legislation, 39
recycling, 38–39
Sorbitol, 177
Specialty polymers, degradability, overview, 2–12
Specialty starches
advantages, 279–280
definition, 275
in paper industry, 275
atmospheric cooking, 280
coating applications, 284,286
emulsification, 282–283
future research, 286–287
properties, 287
size press application, 283–284,285*f*
starch cooking, 280
thermomechanical–thermochemical cooking, 280–281
wet end applications, 281–282
Specific activity of microorganisms, assessment via biological methods, 24
Starch
aqueous dispersion properties, 277
atmospheric cooking, 280
cooking method, 280
degradation mechanism
cellulose-containing blends, 85
diffusion-based limitations, 83,85
moist or submerged films, 83,84*f*
scheme, 83
future research, 286–287
oxidation, 284,285*f*
saponified graft copolymers, *See* Saponified starch-g-poly(acrylonitrile-co-2-acrylamido-2-methylpropane-sulfonic acid)
specialty, See Specialty starches
thermomechanical/thermochemical cooking, 280–281
Starch–*Continued*
use in enhancement of biodegradability, 9
use in paper, 277–280
wet end applications, 281–282
Starch-containing plastics, biodegradation
analytical methods, 69
assay development, 74
effect of amount of starch on rate, 74
effect of amylolytic bacteria on FTIR spectrum, 70,73*f*
effect of starch-degrading bacteria on tensile strength of films, 70,71*f*
effect of starch-degrading bacteria on weight of films, 70,71*f*
FTIR spectra of components used in film production, 70,72*f*
influencing factors, 74
possible routes, 66,67*f*
preparation of microorganisms, 69
preparation of plastics, 69
rate of starch disappearance from films incubated with bacteria, 70,73*f*
Starch–plastic blend(s), oxygen availability in degrading films, 81–82
Starch–plastic blend degradation, *See* Polysaccharide–plastic blend degradation
Starch–plastic composites
biodegradability, 66,69–75
composition, 69–70
degradation routes, 66,67*f*
production method using gelatinized starch, 66,68*f*
production method using granular starch, 66,67*f*
Starch-g-polyacrylonitrile, saponification, 288–289
Step-growth polymers, biodegradation, 8
Step polymerizations
furans, 203–206
polyamides, 204
polycarbonates, 205
polyesters, 203–204
polyethers, 204–205
polyurethanes, 205–206
Substituted acetoxycelluloses, formation of thermotropic mesophase, 268–269
Surface sizing, use of starch in paper, 278–279
Surfactants, applications, 96
Synthesis, chlorinated rubber, *See* Chlorinated rubber from low-molecular-weight guayule rubber
Synthetic polymers containing ester bond
aliphatic polyester degradation by fungi, 138,139*f*
analysis of polyester-degrading enzyme, 138
assay of enzymatic hydrolysis of synthetic solid polymers, 137

Synthetic polymers containing ester bond–
Continued
carboxylic group terminal determination, 138
culture procedure, 137
effect of e-caprolactone and adipic acid molar ratio on hydrolysis by lipase, 141,143f
effect of chemical structure on hydrolysis by lipase, 145,146*f*
effect of molecular weight on hydrolysis by lipase, 141,145,146*f*
effect of particle size of powders on hydrolysis by lipase, 141,142*f*
effect of polyester molecular weight on hydrolysis by lipase, 141,142*f*
estimation of molecular weight distribution of polyamide blocks, 138
hydrolysis of copolyamide–esters, 145*t*,147*f*,148
hydrolysis of copolyesters containing aromatic and aliphatic ester blocks by lipase, 145,147*f*
hydrolysis of polyurethanes by lipase, 141,145,146*f*
importance of environmental stability to performance, 149
material preparation, 136–137
melting point measurement, 137
molecular weight determination, 137
polyester hydrolysis by lipase, 139,140*t*
relationship between T_m and biodegradability of polyester by lipases, 141,144*f*
System analysis of microbial processes
biotransformation of naphthalene, 26,28,29*f*
conceptual approval, 26,27*f*
potential applications, 26
System identification of microbial processes
analysis of soil biotransformations, 28
procedure, 28
reactors, 28,29*f*,30

T

Tertiary amino starch derivatives, synthesis, 282
Testing protocols for biodegradability, 4
Thermomechanical–thermochemical cooking of starch, description, 280–281
Thermotropic cellulose derivatives
(hydroxpropyl)cellulose, 268
substituted acetoxycelluloses, 268–269
(trifluoroacetoxypropyl)cellulose, 268
Thermotropic liquid crystals, 260
1,4:2,5:3,6-Trianhydromannitol
applications, 182
characterization, 182
synthesis, 181
(Trifluoroacetoxypropyl)cellulose, formation of thermotropic mesophase, 268

U

Ultimate biodegradation
definition, 98
schematic representation, 98,100*f*
Uses of starch in paper
adhesive, 278
improvement of sheet formation, 278
strengthener, 278
surface sizing, 278–279

V

2-Vinylfuran
cationic polymerization, 201
radical polymerization, 199
2-Vinyl furoate, radical polymerization, 199
2-Vinyltetrahydrofuran, cationic polymerization, 201

W

Waste, solid, management
crisis situation, 38
future developments, 49–50
landfilling problems, 38–39
recycling, 38–39
Water-insoluble polymers, testing for biodegradability, 4–5
Water-soluble polymers, testing for biodegradability, 4
Wicking
definition, 297
effect of aluminum chloride, 297–298,299*t*,300
measurement, 298

Other ACS Books

Chemical Structure Software for Personal Computers
Edited by Daniel E. Meyer, Wendy A. Warr, and Richard A. Love
ACS Professional Reference Book; 107 pp;
clothbound, ISBN 0–8412–1538–3; paperback, ISBN 0–8412–1539–1

Personal Computers for Scientists: A Byte at a Time
By Glenn I. Ouchi
276 pp; clothbound, ISBN 0–8412–1000–4; paperback, ISBN 0–8412–1001–2

Biotechnology and Materials Science: Chemistry for the Future
Edited by Mary L. Good
160 pp; clothbound, ISBN 0–8412–1472–7; paperback, ISBN 0–8412–1473–5

Polymeric Materials: Chemistry for the Future
By Joseph Alper and Gordon L. Nelson
110 pp; clothbound, ISBN 0–8412–1622–3; paperback, ISBN 0–8412–1613–4

The Language of Biotechnology: A Dictionary of Terms
By John M. Walker and Michael Cox
ACS Professional Reference Book; 256 pp;
clothbound, ISBN 0–8412–1489–1; paperback, ISBN 0–8412–1490–5

Cancer: The Outlaw Cell, Second Edition
Edited by Richard E. LaFond
274 pp; clothbound, ISBN 0–8412–1419–0; paperback, ISBN 0–8412–1420–4

Practical Statistics for the Physical Sciences
By Larry L. Havlicek
ACS Professional Reference Book; 198 pp; clothbound; ISBN 0–8412–1453–0

The Basics of Technical Communicating
By B. Edward Cain
ACS Professional Reference Book; 198 pp;
clothbound, ISBN 0–8412–1451–4; paperback, ISBN 0–8412–1452–2

The ACS Style Guide: A Manual for Authors and Editors
Edited by Janet S. Dodd
264 pp; clothbound, ISBN 0–8412–0917–0; paperback, ISBN 0–8412–0943–X

Chemistry and Crime: From Sherlock Holmes to Today's Courtroom
Edited by Samuel M. Gerber
135 pp; clothbound, ISBN 0–8412–0784–4; paperback, ISBN 0–8412–0785–2

For further information and a free catalog of ACS books, contact:
American Chemical Society
Distribution Office, Department 225
1155 16th Street, NW, Washington, DC 20036
Telephone 800–227–5558